AF389493

Study of the Influence of Abiotic and Biotic Stress Factors on Horticultural Plants

Editors

Agnieszka Hanaka
Jolanta Jaroszuk-Ściseł
Małgorzata Majewska

MDPI • Basel • Beijing • Wuhan • Barcelona • Belgrade • Manchester • Tokyo • Cluj • Tianjin

Editors

Agnieszka Hanaka	Jolanta Jaroszuk-Ściseł	Małgorzata Majewska
Maria Curie-Skłodowska University	Maria Curie-Skłodowska University	Maria Curie-Skłodowska University
Poland	Poland	Poland

Editorial Office
MDPI
St. Alban-Anlage 66
4052 Basel, Switzerland

This is a reprint of articles from the Special Issue published online in the open access journal *Horticulturae* (ISSN 2311-7524) (available at: https://www.mdpi.com/journal/horticulturae/special_issues/Abiotic_Biotic_Stress_Factors).

For citation purposes, cite each article independently as indicated on the article page online and as indicated below:

LastName, A.A.; LastName, B.B.; LastName, C.C. Article Title. *Journal Name* **Year**, *Volume Number*, Page Range.

ISBN 978-3-0365-3112-0 (Hbk)
ISBN 978-3-0365-3113-7 (PDF)

Cover image courtesy of Małgorzata Majewska and Agnieszka Hanaka

Contents

About the Editors

Agnieszka Hanaka, Associate Professor

From 2020 Agnieszka Hanaka is working as an Associate Professor in the Department of Plant Physiology and Biophysics, Faculty of Biology and Biotechnology at Maria Curie-Sklodowska University (UMCS), Poland. In 1997 she was awarded the title Master of Sciences in Biology (specialty: biochemistry). For a few years, she focused on the determination of ion channels functioning in the liverwort tonoplast and the obtained results were discussed in her doctoral dissertation defended in 2007. From 2009 she worked as a doctor in the Department of Plant Physiology, UMCS. The aim of her research was to investigate physiological response of plants to excess metals, like copper and biotic factors (soil bacterial isolate, leaf eating insects) modified by the presence of a signaling molecules - methyl jasmonate. This research cycle ended with the preparation of the post-doctoral accomplishment. In 2019, she obtained the post-doctoral degree (habilitation) in biological sciences in biology. She is focused on influence of abiotic and biotic stress factors (with application of signaling molecules like salicylic acid, and microorganisms) on plant physiology and biochemistry, both primary and secondary metabolites.

Jolanta Jaroszuk-Ściseł, Associate Professor

From 2019 Jolanta Jaroszuk-Ściseł is working as an Associate Professor in the Department of Industrial and Environmental Microbiology, Faculty of Biology and Biotechnology at Maria Curie-Sklodowska University (UMCS), Poland. In 1988 she was awarded the title Master of Sciences in Biology (specialty: microbiology) and she started her first work as an Assistant in the Department of Environmental Microbiology at the Faculty of Biology and Earth Sciences, UMCS. She focused on the rhizosphere microbiology and plant biological control methods and the obtained results were discussed in her doctoral dissertation defended in 1997. In 2013, she obtained the post-doctoral degree (habilitation) in biological sciences in biology and became the Head of the Department of Environmental Microbiology. She has developed a research workshop and implemented methods for elucidation of the mechanisms of plant biocontrol, bio-fertilization, and bioremediation of soils contaminated with heavy and petroleum metals using soil, rhizosphere, and endophytic microorganisms, quantitative and qualitative determination of cell wall-degrading enzymes and numerous microbiological metabolites. She is focused on the indirect mechanisms of biological plant protection induced by biotic and abiotic factors of plant resistance, and the mechanisms of direct inhibition of phytopathogen growth such as mycoparasitism, antibiotics, and competition.

Małgorzata Majewska, Assistant Professor

From 2018 Małgorzata Majewska is working as an Assistant Professor in the Department of Industrial and Environmental Microbiology, Faculty of Biology and Biotechnology at Maria Curie-Sklodowska University (UMCS), Poland. In 1996 she was awarded the title of Master of Sciences in Biology (specialty: microbiology) and she started her first work as a Trainee Assistant in the Department of Environmental Microbiology at the Faculty of Biology and Earth Sciences, UMCS. From 1997, as an Assistant, she focused on the immobilization and mobilization of cadmium in the soil environment and the impact of microbiological activity on these processes. The obtained results were discussed in her doctoral dissertation defended in 2003. From 2004 to 2016 she worked as a doctor in the Department of Environmental Microbiology, UMCS. The aim of her research was to identify the rate and efficiency of the mobilization of cadmium previously immobilized in the root biomass of two plant species (*Festuca ovina* and *Secale cereale*) and Cd immobilization by components of the soil. This research cycle ended with the preparation of the post-doctoral accomplishment. In 2018, she obtained the post-doctoral degree (habilitation) in biological sciences in biology. She is focused on metals bioavailability, bioremediation of the soil contaminated with heavy metals, microbial interactions in the rhizosphere, and plant growth promoting microorganisms in metal-induced stress.

Preface to "Study of the Influence of Abiotic and Biotic Stress Factors on Horticultural Plants"

In the face of rapidly changing environment and anthropogenic pressure, plants must develop mechanisms allowing them to survive in the adverse conditions. Understanding the alterations under stress and relationships among different organisms or environmental elements can help create more favorable conditions and holistic approach towards not only the survival of plants but also for their optimal growth and development. This reprint book is dedicated to the scientists especially interested in environmental sciences, involved in plant physiology and biochemistry, as well as horticultural and microbiological issues.

Global climate change is expected to be critical over the century, leading to influences on different parameters of the environment. First, biochemical, and physiological changes appear and affect plant biomass and consequently limit the yield of crops. Instead of the up-to-date knowledge, deeper approach about alleviating the stress effects is still vital in understanding the complexity of the problem. Such direction of research may be the basics to obtain environmentally friendly methods and tools to inhibit negative influence of stress agents on plants.

We have a great pleasure to present the scientists a set of studies entitled "Study of the Influence of Abiotic and Biotic Stress Factors on Horticultural Plants". The reprint book contains 12 papers about the influence of the stress factors on the plant growth and soil parameters. The ideas of the papers are gathered around five topics: (1) achieving better quality of plant material for food production by changes made in the growth conditions, metabolic and genetic modifications; (2) increasing the plant resistance to environmental stresses by application of exogenous compounds of different chemical character; (3) reducing plant stress caused by anthropogenic activity applying nonmodified and genetically modified plants (GMP); (4) mitigating drought stress by irrigation; and (5) the positive effect of plant growth-promoting microorganisms on horticulture plants performance during drought stress.

Agnieszka Hanaka, Jolanta Jaroszuk-Ściseł, Małgorzata Majewska
Editors

Editorial

Study of the Influence of Abiotic and Biotic Stress Factors on Horticultural Plants

Agnieszka Hanaka [1],*, Małgorzata Majewska [2] and Jolanta Jaroszuk-Ściseł [2]

[1] Department of Plant Physiology and Biophysics, Institute of Biological Sciences, Faculty of Biology and Biotechnology, Maria Curie-Skłodowska University, Akademicka 19, 20-031 Lublin, Poland

[2] Department of Industrial and Environmental Microbiology, Institute of Biological Sciences, Faculty of Biology and Biotechnology, Maria Curie-Skłodowska University, Akademicka 19, 20-031 Lublin, Poland; malgorzata.majewska@umcs.pl (M.M.); jolanta.jaroszuk-scisel@umcs.pl (J.J.-Ś.)

* Correspondence: agnieszka.hanaka@umcs.pl

Citation: Hanaka, A.; Majewska, M.; Jaroszuk-Ściseł, J. Study of the Influence of Abiotic and Biotic Stress Factors on Horticultural Plants. *Horticulturae* **2022**, *8*, 6. https://doi.org/10.3390/horticulturae 8010006

Received: 3 November 2021
Accepted: 30 November 2021
Published: 22 December 2021

Publisher's Note: MDPI stays neutral with regard to jurisdictional claims in published maps and institutional affiliations.

In changing environmental conditions, horticulture plants are affected by a vast range of abiotic and biotic stresses which directly and indirectly influence plant condition. Moreover, biomass production or some of the plant metabolites are expected to steadily increase. Such expectations lead to research on the influence of different stressors and their potential modifiers. It is extremely important to have a holistic approach to the processes taking place in a plant that is affected by stress factors. At the same time, the plant can be affected by very diverse factors that mutually affect and shape the plant growth environment, such as the state of water supply—extremely low or, conversely, too high—the impact of low or high temperatures, which in turn cause a state of drought, increased osmotic pressure and salinity.

This Special Issue (SI) was planned with a structure to consider a large range of aspects on stress factors affecting horticultural plants. It is a research summary on the influence of the stress factors on plant growth and the soil parameters. The studies were investigated at the cellular, tissue, organ and whole plant level. Authors described the impact of stress caused by both climate change and human activity resulting in disorder of the optimum temperature (low- and high-temperature stresses), water balance (water and drought stress and irrigation) and the subsequent disturbance of soil parameters. The SI gathers eleven research papers [1–11] and one review [12]. Three papers were dedicated to cold stress, two to salt stress, two to inorganic pollutants such as metals and phosphite (Phi), three to climate change (i.e., high temperature, water and drought stress) and two to irrigation. The subject of the studies were different plant species, i.e., watermelon, lettuce, kale, potato, tomato, grapevine, hops, orchid, strawberry and *Buxus megistophylla*.

Among the classical parameters used as indicators of plant condition are morphological, anatomical, physiological, biochemical and genetical ones [12]. Physiological, biochemical and anatomical changes occurring in the plant under the influence of stress factors should be especially strongly noticed and analyzed.

In the presented SI, the most frequently applied morphological features were: seed germination, plant growth, leaf and berry area, leaf number, stem diameter and plant dry weight [1–3,5,7–11]. Anatomical characteristics were based mainly on the evaluation of epidermis and mesophyll quality and number of cells and chloroplasts [8]. Physiological aspects were focused on water content in leaves, mineral elements (nutritional elements and heavy metals, e.g., N, P, K, Mg, Ca, Mn, Fe, Cu, Na, Cl, Co, Cr, Ni, Pb), photosynthetic pigments content (chlorophylls *a* and *b*, carotenoids), polyphenolic compounds content (e.g., polyphenols, phenolic acids and flavonoids), soluble sugars, acids and proline contents and plant growth regulators (e.g., abscisic acid, indole-3-acetic acid, gibberellin A3) [1–11]. Biological traits were mostly lipid peroxidation (i.e., content of malondialdehyde; MDA), level of reactive oxygen species (ROS, e.g., H_2O_2 and $O_2^{-\bullet}$), activity of antioxidative enzymes (e.g., superoxide dismutase, peroxidase and catalase) and content of non-enzymatic

1

compounds (e.g., vitamin C and lycopene) [2,4–11]. Additionally, expression of ethylene-responsive factor genes was analyzed [4]. Some of the papers studied soil properties [1,9]. At the same time, attention was drawn to the close links between hormonal balance and induction of plant immunity, in which regulatory functions are performed by signaling substances (such as salicylic acid, jasmonic acid, ethylene) produced in the pathways of phenolic compounds transformation under the influence of marker enzymes of the phenyl-propanoid pathway [12]. Stress factors affect both the uptake of mineral compounds and the process of photosynthesis, which can be monitored based on the biometric parameters and the level of photosynthetic pigments (chlorophyll and carotenoids).

All these elements have been taken into consideration in twelve publications presented in this SI, providing meaningful results revealing a series of events occurring as a consequence of the influence of various environmental factors and explaining the operation of plant defense mechanisms. In the largest number of works, the increase in polyphenols and phenolic acids content and the activity of antioxidant enzymes has been shown because of the action of the abiotic stress factors [1,2,4,9]. Accumulation of proline was observed in salinity stress after chitosan (CTS) application [2], in low temperature stress (kale, *Brassica oleracea* var. *acephala*) [3] and in the Phi supply (potato, *Solanum tuberosum*) [4].

The ideas of the research papers gathered under the titled "Study of the Influence of Abiotic and Biotic Stress Factors on Horticultural Plants" were divided into five sections. They were dedicated to the following fields: (1) achieving better quality of plant material for food production by changes made in the growth conditions, metabolic and genetic modifications; (2) increasing the plant resistance to environmental stresses by application of exogenous compounds of different chemical character; (3) reducing plant stress caused by anthropogenic activity applying non-genetically modified and genetically modified plants (GMP) and (4) mitigating drought stress by irrigation, whereas the main goal of the review paper [12] was to discuss (5) the positive effect of plant growth-promoting microorganisms (PGPM) on horticulture plant performance during drought stress.

(1)   Better quality of plant material for food production.

To achieve better plant material for food production, changes can be made in the plant growth conditions, its metabolism and genetic material [3,5–7].

Manipulations leading to an increase in the stress tolerance of a plant may not always allow us to obtain new varieties with better parameters. Such an example was the research on *Vitis vinifera* L. varieties [3] in the wine industry, it seems crucial to find the answers to two questions: (1) do the new genotypes of the pathogen-resistant grapevines (new geno-types with low sensitivity to biotic stress) keep good qualities of fruit? and (2) how do the new varieties react to global climate warming? Frioni et al. [3] proved that the production and fruit composition traits during ripening of several new cross-bred pathogen-resistant grapevine varieties (patented and admitted to cultivation) are significantly lower than two *V. vinifera* traditional varieties, Ortrugo and Sauvignon Blanc. In these studies, five white pathogen-resistant varieties (PRV) listed as UD 80-100, Soreli, UD 30-080, Sauvignon Rytos and Sauvignon Kretos were tested. All tested PRV exhibited an earlier onset of veraison and faster sugar accumulation compared to Ortrugo and Sauvignon Blanc. Such effects could suggest an earlier start of the harvest. Therefore, canopy and ripening management strategies must be significantly adjusted compared to the standard practice employed for the parental Sauvignon Blanc. Overall, PRV could perform better in cooler climates, in north-facing hillsides, or at higher altitudes, where their good resistance to mildews could match an adequate grapes' biochemical balance. Moreover, retaining adequate acidity at harvest is crucial to produce high-quality white wines [3].

Not only genetic manipulation leading to obtain the new varieties, using cold-tolerant rootstocks to efficient adaptation plantlets, or regulation of the growth temperature, but also the nearness of other plants can improve growth parameters and fruit quality of the horticultural plants in stress. Such a phenomenon was observed by Karakas et al. [5]. They showed that strawberries, as salt-sensitive plants, reacted strongly to slight or moderate salinity, reducing the crop yield and quality of fruits. Salt stress negatively affected the growth, stomatal conductance, electrolyte leakage, contents of chlorophyll, proline, H_2O_2, MDA, activity of catalase and peroxidase and content of the health-related compounds such as vitamin C and lycopene. On the other hand, when strawberry seedlings were grown in combination with *Portulaca oleracea* L. under NaCl stress condition, not only an increase in weight of the green parts of the plant and the total fruit yield of strawberry plants, but also an improvement in physiological and biochemical parameters were observed. The cultivation of strawberry plants with *P. oleracea* directly reduced the concentrations of stress metabolites and antioxidant enzyme levels, as well as indirectly contributing to an increase in vitamin C and lycopene contents. Therefore, Karakas et al. [5] suggested the use of *P. oleracea* on the areas with significant salinity as an environmentally friendly method to diminish salt stress.

The plant struggle with stress may be manifested by changes in metabolism and the accumulation of various compounds. For example, kale tolerance to low temperatures is associated with the presence of specialized metabolites such as polyphenols, carotenoids and glucosinolates, which can act not only as protective factors against environmental stress for the plant, but they can also be a source of beneficial compounds for human health [6]. Ljubej et al. [6] observed that a short (24 h) chilling period (8 °C) was beneficial for the accumulation of phytochemicals in kale. However, freezing temperatures (-8 °C) caused significant stress and decrease in pigments and phytochemical compound levels. The studies suggested that the temperature of kale cultivation should be controlled by producers to achieve production of crops with a high content of health-related compounds [6].

Lu et al. [7] proved that using cold-tolerant rootstocks may be an efficient adaptation strategy for improving stress tolerance in watermelon (*Citrullus lanatus* (Thunb.), cv. ZaoJia 8424). It was demonstrated that the improved cold tolerance was associated with gourd-grafted watermelons compared to non-grafted (control) plants. Grafted plants accumulated lower levels of ROS, consequently representing enhanced antioxidant activity. Under cold stress, higher chlorophyll and proline contents and lower MDA content were also determined [7].

(2)   Exogenous compounds in plant resistance to environmental stresses.

To mitigate stress, growth of horticultural plants can be supported with exogenous substances during agrotechnical treatments such as 24-epi-brassinolide (EBR) on tomato [4], 5-aminolevulinic acids (5-ALA) on *Buxus megistophylla* [10] and CTS on lettuce (*Lactuca sativa* L.) [11].

The exogenous EBR used by Heidari et al. [4] as analog of brassinosteroids eliminated the effects of oxidative stress induced by low temperature in cold-sensitive tomato species. 24-epi-brassinolide decreased the ROS content, simultaneously increasing antioxidant enzymes activity, auxin and gibberellin contents, then improved the growth rate of the tomato.

Yang et al. [10] detected that 5-ALA promoted the growth of *B. megistophylla*, improved plant survival, increased leaf color and enhanced the greening effect. The content of several kinds of mineral nutrient elements, such as nitrogen, phosphate, calcium, magnesium, iron, copper and boron in leaves of *B. megistophylla* was strongly increased by 5-ALA treatment. Unfortunately, a negative effect was also observed. Under this treatment, accumulation of cadmium, mercury, chromium and lead in roots increased. Luckily, these toxic elements were intercepted in roots without translocation and accumulation in leaves. The activities of antioxidative enzymes and the stress resistance of plants were enhanced. According to the results, 5-ALA, as a specific activator of biochemical pathways, can lead to both favorable and unfavorable alterations in metabolism. Therefore, Yang et al. [10] recommend application of this non-protein amino acid in urban landscapes to improve stress tolerance of ornamental plants.

It has been proven by Zhang et al. [11] that CTS, the classic, widely commercially used elicitor of plant immunity protecting plants against phytopathogens, can be effective in protecting the plant (*Lactuca sativa*) against the effects of the abiotic factor—excessive salt concentration. Most likely, its direct protective effect under salinity condition was associated with the regulation of intracellular ion concentration, controlling osmotic adjustment and increasing antioxidant enzymatic activity (e.g., peroxidase and catalase) in lettuce leaves. Moreover, results of Zhang et al. [11] showed that exogenous CTS could improve plant growth and biomass under salt stress. There is significant evidence that CTS curbed the accumulation of sodium but enhanced the accumulation of potassium in the leaves of NaCl-treated plants. This fact may be important for obtaining better-quality lettuce and supplementing the deficiencies of K in the human diet [11]. Chitosan, as natural polysaccharide, is a safe and cheap substance promoting plant growth and increasing the biotic and abiotic stress tolerance of plants.

(3) Reduction of plant stress under anthropogenic activity by application of non-genetically modified plants and GMP.

Human beings create habitats unfavorable for the development of plants. This is the result of industrial and agricultural activity. Fortunately, the plants have also learned to deal with this kind of stress. For example, Maleva et al. [8] found the *Neottia ovata* growing in the young forest community formed during the natural revegetation of the fly ash deposits (fly ash dump of Verkhnetagil'skaya Thermal Power Station). In Russia, this orchid species is included in several regional Red Data Books, and it is especially interesting to gather knowledge of the adaptive characteristics of orchids. The study of orchid adaptive responses to unfavorable factors by Maleva et al. [8] will help to run the process of the introduction of the *N. ovata* into new environments. The adaptive changes in the leaf mesostructure organization, such as an increase in epidermis thickness, the number of chloroplasts in the cell and the internal assimilating surface were found for the first time by Maleva et al. [8]. The orchid population colonizing the fly ash deposits was characterized by a relatively favorable water balance and stable assimilation indexes further contributing to its high viability.

Crop production is expensive in the areas with low phosphorus (P) availability, and for this reason Domatey et al. [2] were looking for other compounds that may serve as a useful source of assimilable P for *Solanum tuberosum* L. According to this paper, it is possible to apply Phi as fertilizer, only when plants stay resistant to this phosphorus form. Like herbicides, Phi has an inhibitory effect on plant growth. The authors try to combine these phenomena as a hypothetical advantage. Only if plant genotypes are resistant to Phi could it be used both as herbicide to weed control and the source of bioavailable P. Such a solution would be environmentally friendly. Furthermore, Domatey et al. [2] showed significant genotypic variation in tolerance indices among the five tested genotypes (Atlantic, Longshu3, Qingshu9, Longshu6 and Gannong2). Firstly, antioxidant enzyme activities and proline content increased significantly under Phi treatments compared to control without Phi. Secondly, potato genotypes with larger root systems such as Atlantic and Longshu3 were more tolerant to Phi stress than genotypes with smaller root systems (Qingshu9, Longshu6 and Gannong2) [2].

(4) Mitigation of drought stress by irrigation.

Plants require an adequate amount of water during the growing season. Nowadays, this is a rising problem because there is not enough water in many parts of the world, even in areas where such shortages did not occur in the past. Luckily, water limitations can be partially eliminated using various irrigation methods, e.g., by flooding the inter-row [1] or more precisely drip irrigation that supplies water directly to the place where the plant grows out of the soil [9].

Flooding the inter-row is still the most frequently used irrigation method for hop (*Humulus lupulus* L.) in northern Portugal. Afonso et al. [1] showed that using this type of irrigation to prevent drought stress can worsen the condition of the soil in inter-rows. The irrigation of the hop fields by flooding the inter-row for more than 20 years caused decreased porosity and increased soil bulk density in the 0–10 cm soil layer in comparison to the 10–20 cm layer. Fortunately, it did not damage the soil structure of ridges, which are the place of nutrient accumulation's gradual uptake by hop plants. Although irrigation and soil tillage have damaged the soil structures, they did not create the negative nutrient gradient along the row. Moreover, they refilled water deficiencies during plant development, and the quality of the hop cone yield was sufficient, but on the other hand, the water consumption was too high [1].

Limited water resources force more economical use of water by precise irrigation techniques. It is important to accurately follow the plants' water needs, and correctly predict the moments of deficiency supplementation. New irrigation techniques based on biological parameters and very precise calculations can serve as an adequate solution. It is important to follow exactly the first reactions of the plants to water restrictions to finally be able to use less water [9].

Ru et al. [9] devoted the article to the above topic. The authors searched a reliable method to easily quantify and monitor the grapevine water status to enable effective manipulation of the water stress of the plants. It was shown that the study on a daily stem diameter variation of grapevine planted in a greenhouse could be helpful to precise irrigation management of plants. The relative daily variation of the grapevine stem diameter from the vegetative stage to the fruit stage was related to different irrigation levels. Both signal intensity calculation of maximum daily shrinkage (SIMDS) and daily increase (SIDI) can be applied as indicators of the moisture status of grapevine and soil. Ru et al. [9] concluded that SIDI was suitable as an indicator of water status of grapevine and soil during the vegetative and flowering stages, whereas SIMDS was suitable as an indicator of the moisture status of plant and soil during the fruit expansion and mature stages. In general, SIMDS and SIDI were very good predictors of the plant water status during the growth stage and their continuous recording can offer the promising possibility of their use in programming automatic drip irrigation of the grapevine [9].

(5)   The positive effect of PGPM on plant performance during drought stress.

Trend towards temperature rise contributes to water evaporation and global warming, eventually leading to drought stress in plants. The review paper of Hanaka et al. [12] discussed the positive influence of PGPM on horticulture plant performance during drought stress. Among mechanisms of plant protection by rhizospheric or plant surface-colonizing and endophytic bacteria and fungi are the production of phytohormones, antioxidants and xeroprotectants, and the induction of plant resistance. On one hand, application of various biopreparations containing PGPM seems to be a relatively cheap, easy to apply and efficient method of alleviating drought stress in plants, with implications in productivity and food condition. On the other hand, the vital problems of using biopreparations containing PGPM include limitations in introducing the microbial inoculum to the appropriate conditions and the low repeatability of their activities. Microorganisms that promote plant growth and at the same time induce physical and chemical alterations that result in enhanced tolerance to abiotic stresses, constitute an important separate group called "induced systemic tolerance" (IST). It should be strongly emphasized that a significant enhancement of the protective effect in drought conditions is achieved by using a mixture of bacterial strains (e.g., from genus Bacillus and Serratia), which indicates the synergistic effect of IST strains. It is also very important to emphasize the special role of fungi in protecting plants against drought. Fungi are more tolerant to drought than bacteria and their abundance in soil increases in water-limiting conditions. This is due to the specific fungal growth and traits which allow an intensive soil and plant tissue exploration and colonization and taking water from resources unavailable to other microorganisms.

Studies on the protection of horticultural plants influencing drought stress indicated that the application of well-selected microorganisms can be efficient [12]. Biopreparations should be multicomponent to achieve an appropriate level of microorganism cooperation and the final desired effect. Moreover, the combination of bacterial and fungal strains into one preparation gives even better effectiveness and reliability. Crop specificity should also be taken into consideration.

Author Contributions: Conceptualization, A.H. and M.M.; writing—original draft preparation, A.H., M.M. and J.J.-Ś.; and writing—review and editing, A.H., M.M. and J.J.-Ś. All authors have read and agreed to the published version of the manuscript.

Funding: This research received no external funding.

Acknowledgments: We gratefully acknowledge all the authors that participated in this Special Issue.

Conflicts of Interest: The authors declare no conflict of interest.

References

1. Afonso, S.; Arrobas, A.; Rodrgues, M.Â. Twenty-Years of Hop Irrigation by Flooding the Inter-Row Did Not Cause a Gradient along the Row in Soil Properties, Plant Elemental Composition and Dry Matter Yield. *Horticulturae* **2021**, *7*, 194. [CrossRef]
2. Dormatey, R.; Sun, C.; Ali, K.; Qin, T.; Xu, D.; Bi, Z.; Bai, J. Influence of Phosphite Supply in the MS Medium on Root Morphological Characteristics, Fresh Biomass and Enzymatic Behavior in Five Genotypes of Potato (*Solanum tuberosum* L.). *Horticulturae* **2021**, *7*, 265. [CrossRef]
3. Frioni, T.; Squeri, C.; Del Zozzo, F.; Guadagna, P.; Gatti, M.; Vercesi, A.; Poni, S. Investigating Evolution and Balance of Grape Sugars and Organic Acids in Some New Pathogen-Resistant White Grapevine Varieties. *Horticulturae* **2021**, *7*, 229. [CrossRef]
4. Heidari, P.; Entazari, M.; Ebrahimi, A.; Ahmadizadeh, M.; Vannozzi, A.; Palumbo, F.; Barcaccia, G. Exogenous EBR Ameliorates Endogenous Hormone Contents in Tomato Species under Low-Temperature Stress. *Horticulturae* **2021**, *7*, 84. [CrossRef]
5. Karakas, S.; Bolat, I.; Dikilitas, M. The Use of Halophytic Companion Plant (*Portulaca oleracea* L.) on Some Growth, Fruit, and Biochemical Parameters of Strawberry Plants under Salt Stress. *Horticulturae* **2021**, *7*, 63. [CrossRef]
6. Ljubej, V.; Karalija, E.; Salopek-Sondi, B.; Šamec, D. Effects of Short-Term Exposure to Low Temperatures on Proline, Pigments, and Phytochemicals Level in Kale (*Brassica oleracea* var. *acephala*). *Horticulturae* **2021**, *7*, 341. [CrossRef]
7. Lu, K.; Sun, J.; Li, Q.; Li, X.; Jin, S. Effect of Cold Stress on Growth, Physiological Characteristics, and Calvin-Cycle-Related Gene Expression of Grafted Watermelon Seedlings of Different Gourd Rootstocks. *Horticulturae* **2021**, *7*, 391. [CrossRef]
8. Maleva, M.; Borisova, G.; Chikina, N.; Sinenko, O.; Filimonova, E.; Lukina, N.; Glazyrina, M. Adaptive Morphophysiological Features of *Neottia ovata* (Orchidaceae) Contributing to Its Natural Colonization on Fly Ash Deposits. *Horticulturae* **2021**, *7*, 109. [CrossRef]
9. Ru, C.; Hu, X.; Wang, W.; Ran, H.; Song, T.; Guo, Y. Signal Intensity of Stem Diameter Variation for the Diagnosis of Drip Irrigation Water Deficit in Grapevine. *Horticulturae* **2021**, *7*, 154. [CrossRef]
10. Yang, H.; Zhang, J.; Zhang, H.; Xu, Y.; An, Y.; Wang, L. Effect of 5-Aminolevulinic Acid (5-ALA) on Leaf Chlorophyll Fast Fluorescence Characteristics and Mineral Element Content of *Buxus megistophylla* Grown along Urban Roadsides. *Horticulturae* **2021**, *7*, 95. [CrossRef]
11. Zhang, G.; Wang, Y.; Wu, K.; Zhang, Q.; Feng, Y.; Miao, Y.; Yan, Z. Exogenous Application of Chitosan Alleviate Salinity Stress in Lettuce (*Lactuca sativa* L.). *Horticulturae* **2021**, *7*, 342. [CrossRef]
12. Hanaka, A.; Ozimek, E.; Reszczyńska, E.; Jaroszuk-Ściseł, J.; Stolarz, M. Plant Tolerance to Drought Stress in the Presence of Supporting Bacteria and Fungi: An Efficient Strategy in Horticulture 2021. *Horticulturae* **2021**, *7*, 390. [CrossRef]

horticulturae

MDPI

Review

Plant Tolerance to Drought Stress in the Presence of Supporting Bacteria and Fungi: An Efficient Strategy in Horticulture

Agnieszka Hanaka [1,*], Ewa Ozimek [2], Emilia Reszczyńska [1], Jolanta Jaroszuk-Ściseł [2] and Maria Stolarz [1]

[1] Department of Plant Physiology and Biophysics, Faculty of Biology and Biotechnology, Institute of Biological Sciences, Maria Curie-Skłodowska University, Akademicka 19, 20-031 Lublin, Poland; emilia.reszczynska@umcs.pl (E.R.); maria.stolarz@umcs.pl (M.S.)

[2] Department of Industrial and Environmental Microbiology, Faculty of Biology and Biotechnology, Institute of Biological Sciences, Maria Curie-Skłodowska University, Akademicka 19, 20-031 Lublin, Poland; ewa.ozimek@umcs.pl (E.O.); jolanta.jaroszuk-scisel@umcs.pl (J.J.-Ś.)

* Correspondence: agnieszka.hanaka@umcs.pl

Abstract: Increasing temperature leads to intensive water evaporation, contributing to global warming and consequently leading to drought stress. These events are likely to trigger modifications in plant physiology and microbial functioning due to the altered availability of nutrients. Plants exposed to drought have developed different strategies to cope with stress by morphological, physiological, anatomical, and biochemical responses. First, visible changes influence plant biomass and consequently limit the yield of crops. The presented review was undertaken to discuss the impact of climate change with respect to drought stress and its impact on the performance of plants inoculated with plant growth-promoting microorganisms (PGPM). The main challenge for optimal performance of horticultural plants is the application of selected, beneficial microorganisms which actively support plants during drought stress. The most frequently described biochemical mechanisms for plant protection against drought by microorganisms are the production of phytohormones, antioxidants and xeroprotectants, and the induction of plant resistance. Rhizospheric or plant surface-colonizing (rhizoplane) and interior (endophytic) bacteria and fungi appear to be a suitable alternative for drought-stress management. Application of various biopreparations containing PGPM seems to provide hope for a relatively cheap, easy to apply and efficient way of alleviating drought stress in plants, with implications in productivity and food condition.

Keywords: climate change; drought stress; biopreparations; plant stimulation; plant growth-promoting microorganisms

Citation: Hanaka, A.; Ozimek, E.; Reszczyńska, E.; Jaroszuk-Ściseł, J.; Stolarz, M. Plant Tolerance to Drought Stress in the Presence of Supporting Bacteria and Fungi: An Efficient Strategy in Horticulture. *Horticulturae* **2021**, *7*, 390. https://doi.org/10.3390/horticulturae7100390

Academic Editor: Alessandra Francini

Received: 1 September 2021
Accepted: 5 October 2021
Published: 11 October 2021

Publisher's Note: MDPI stays neutral with regard to jurisdictional claims in published maps and institutional affiliations.

1. Introduction

The horticulture system is affected by various abiotic and biotic stresses which directly and indirectly influence soil fertility, plant health and crop yield [1–3]. These stresses result in the loss of soil microbial diversity, soil fertility and availability of nutrients [4]. The condition of the soil under drought strictly corresponds to plant performance, showing consequences in plant morphology, anatomy, physiology, and biochemistry. With reduction in seed germination and seedling growth, plant height, nutrition and biomass are weakened resulting in yield limitation. The huge variety of changes taking place in horticultural plants and the mechanisms of counteracting stress they produce result from a very wide range of horticultural plant species, including types of crops such as those distinguished by the International Society for Horticultural Science (ISHS): (1) tree, bush and perennial fruits, (2) perennial bush and tree nuts, (3) vegetables (roots, tubers, shoots, stems, leaves, fruits and flowers of edible and mainly annual plants), (4) medicinal and aromatic plants, (5) ornamental plants, (6) trees, shrubs, turf and ornamental grasses propagated and produced in nurseries for use in landscaping or for establishing fruit orchards or other crop production units [5]. Facing the current, rapid climate changes, the cultivation of plants is strongly affected by abiotic stresses, which additionally intensify the influence of biotic factors such as pests causing serious plant infections [4]. In this dramatic situation, plant associations with rhizospheric [6,7] and endophytic [8,9] microorganisms colonizing the rhizoplane, rhizosphere and plant tissues should be considered as the main stress relievers [10–14]. Three types of effects of microorganisms associated with plants are distinguished: beneficial, deleterious and neutral ones. Based on the positive effects of microbes, two main groups are listed, plant growth-promoting rhizobacteria (PGPR) or more generally, plant growth-promoting bacteria (PGPB) and plant growth-promoting fungi (PGPF) [14–19]. All mentioned groups of microorganisms can serve as biocontrol agents, biofertilizers, phytostimulators and phytoremediators [2,12,20–22].

The most frequently described biochemical mechanisms of plant protection against drought by microorganisms are the production of phytohormones, antioxidants and xeroprotectants [23]. Trehalose can act as xeroprotectant triggering the plant-defense system to counteract the damage caused by drought. It has been shown that microorganisms with tolerance to desiccation have the ability to protect some plants from drought. It seems to be dependent on the microorganism's ability to regulate the concentration of trehalose in the plant as a signal of drying damage.

In horticultural production, plant–microbe interactions should be considered the main factor of plant growth, protection against abiotic stresses and resistance against adverse conditions [24,25] (e.g., in arid and semiarid areas), and these interactions could also be beneficial in alleviating drought stress in plants [26]. Profound knowledge about the mechanisms of plant–microbe interactions can offer several strategies to increase plant productivity in an environmentally friendly manner [27]. Therefore, in the increasing market for plant growth-promoting products, it is important to develop a successful strategy for microorganism screening [28]. Furthermore, the European Green Deal (EGD), provided by the European Commission in December 2019, is currently focused on the application of natural products in agriculture and horticulture instead of chemical plant-protection products. To cope with this idea, new efficient biological ingredients in the face of changing climate are desired. Nowadays, the most significant consequence of climate change is drought stress [29].

To deal with severe drought stress in the near future, it is strictly necessary to determine the interactions, mechanisms and signaling pathways responsible for increased drought tolerance in terrestrial organisms. The concept of drought and water deficit is difficult to define, but the literature data [30–32] indicate that drought can be defined as a state of the total water capacity being within the range of 12–20% for a period of 16 days. Moreover, the drought state can achieve at least two degrees—mild and severe [33]—while the water deficit [34] refers to the state of water capacity falling below 30%. To handle the drought effect, plants can be supported by both microorganisms inhabiting the rhizoplane (i.e., those adhering to the surface of the roots) and rhizosphere (i.e., living at a further distance within the root secretions) [34,35], as well as endophytic microorganisms inhabiting the inside of the root [36]. The application of plant growth-promoting microorganisms seems to provide hope for a cheap, easy to apply and efficient way of alleviating drought stress in plants with implications in productivity and food condition. The presented review was undertaken to discuss the impact of climate change with respect to drought stress, and to emphasize that modifications in microorganisms composition and their traits should indicate new solutions in the search for efficient compounds of biopreparations supporting plant growth.

2. Climate Change

Global climate change is expected to be considerably critical over the century, leading to influences on various parameters of the environment [17]. Not only atmospheric CO_2 concentrations derived from natural and anthropogenic sources, but also surface temperatures will be increasing gradually, likely from 1.0 to 5.7 °C by the end of this century [37]. Moreover, some regions, such as the Eastern Mediterranean and Middle East (EMME), have been classified as a climate "global hot-spot". In the EMME, the temperature is predicted to increase from 3.5 to 7 °C by the end of the century [38]. Additionally, it is anticipated that rising air temperatures will increase the frequency of extreme weather disasters such as heat waves, drought and heavy precipitation occurrence to a level that has never been monitored before [37]. These strongly temperature-dependent climate changes, combined with water scarcity, will lead to enhanced drought throughout the globe, hurting whole ecosystems and different organisms, including the distribution of plants and microorganisms [17].

In climate studies, calculations concerning crop evapotranspiration are also important [17]. For instance, in South East Europe, the mean annual crop evapotranspiration in the period 1991–2020 reached from 56 mm to 1297 mm, while averages for the future 30 years (between 2021 and 2050), are expected to vary from 59 mm to 1410 mm [17]. These predictions consider the impact of future climate warming. Global warming increases water evaporation and consequently leads to drought stress [39]. High temperature is the crucial factor in melting glaciers and increasing the sea level [8]. The changes in polar and subpolar climate zones also correspond with climate warming [40–42].

Climate change results in altered environmental conditions and negative effects on natural ecosystems, which are likely to trigger modifications in plant physiology [43] and microbial functioning [44] based on the availability of nutrients [4] or signal compounds [2]. It is certain that not only plants, but also plant-associated microorganisms might be remarkably changed in abundance, diversity and activity [44,45]. Both increased temperature and drought may activate correspondent adjustments in plants and microorganisms and their mutual interactions [17]. The adaptational challenges of horticultural plants are not only associated with long-term average climate change, but also with the short-term changes driven by weather extremes and interannual fluctuations [46]. Drought-related cereal production losses are increasing by more than 3% yr^{-1} [46]. In the face of the continuous raise of the world population to an estimated nine billion by 2050 [47], withstanding drought stress according to sustainable agriculture/horticulture is a challenge for the 21st century [48].

3. Plants under Drought Stress

Drought is an uncontrolled stress which affects almost all stages of plant growth and development directly or indirectly [43]. Most of the drought effects on plants are associated with high temperature. Physiological processes occur mostly in temperatures ranging from 0 °C to 40 °C. However, the optimal temperatures for the different stages of growth and development are narrower and strongly depend on the species and ecological origin [1,49].

Plants exposed to drought stress develop numerous responses in different areas, from morphological and physiological mechanisms to anatomical and biochemical or molecular ones [1,39,50] (Figure 1).

Four types of morphological and physiological response strategies to drought stress are highlighted, i.e., tolerance, avoidance, escape and recovery [51] (Figure 2). Tolerance is defined as the plant's ability to resist dehydration using osmoprotectants [52]. Avoidance is based on the undisturbed occurrence of physiological processes (such as stomata regulation, root system development). Escape is the adjustment of the plant's life cycle by shortening of the life cycle to avoid drought stress. Recovery is the ability of a plant to restart growth after the exposure to the extreme drought stress [53].

The morphological features of drought stress include limited seed germination and seedling growth, reduced size, area and number of leaves, restricted number of stomata, reduced number of flowers, disturbed stem and root elongation, impaired plant height, growth, development and yield, and reduced fresh and dry biomass [7,39,50].

Figure 1. Some morphological, anatomical, physiological and biochemical plant responses to drought stress (modified on the basis of [1,39,50]).

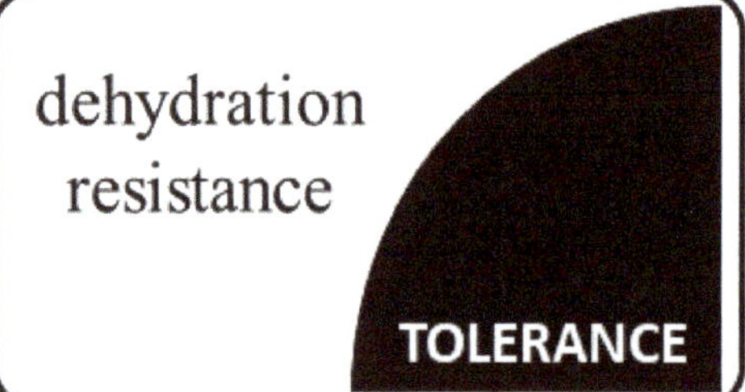
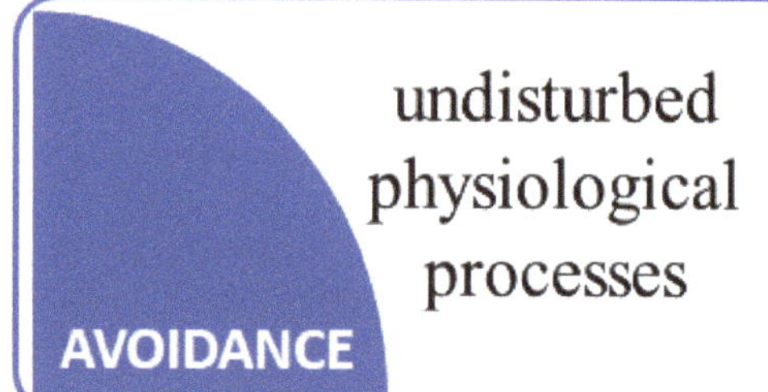
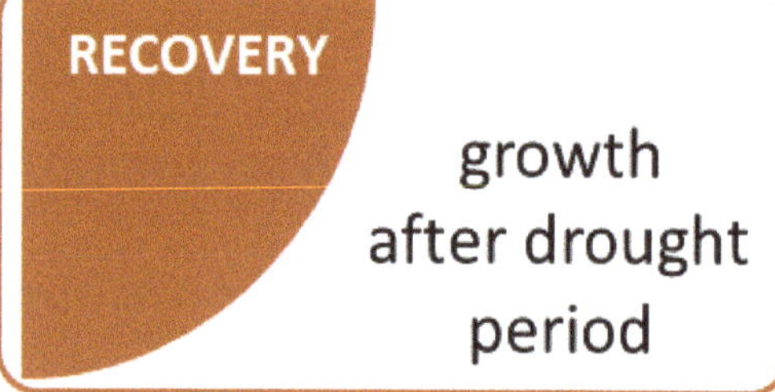

Figure 2. Response strategies to drought stress (modified on the basis of [1,39,50,51]).

In order to adapt to the adverse environment, avoid drought and improve water availability, plants increase the root length and their number [54]. Drought significantly affects the plant's cell elongation and division, its growth and its development, which is mainly caused by the reduction in cellular differentiation, plant growth and yield [50]. The negative effect on the leaf area under the drought condition could be dependent on the reduction in the leaf number, size and longevity, combined with temperature, leaf turgor pressure and assimilation rate [55]. The reduction in plant height and shoot dry weight results in a lower quality of yield [54].

The morphological responses are most frequently combined with anatomical changes in plants exposed to drought, e.g., thickening of cell walls, increased cuticle layer on the leaf surface and improved development of vascular tissues [8,56]. Drought stress results in anatomical changes in the lower and upper epidermis, mesophyll tissue and vascular bundle diameter of leaves [57]. The negative anatomical effects on the leaves are based on a shortage of water supply from the soil, limitations in nutrients uptake, and reduction in photosynthetic rate. Plant hydraulic conductivity is modulated during drought stress leading to the disruption of water flow in the xylem vessels (embolism) or modifications in the vessel size and function [58]. Consequently, the reduced water flow from the root to the shoot causes stomatal closure and transpiration disruption [50].

Drought affects the physiological traits such as the leaf relative water content and water potential, stomatal conductance, transpiration and photosynthetic rates [59,60]. Reduced water content and water conductivity are responsible for the loss of turgidity and limited stomatal conductance resulting in restricted gaseous exchange (the rate of carbon assimilation) [8,61]. Furthermore, climatic conditions, e.g., higher temperature, drought and soil aeration reduce the movement of nutrients in the soil, their uptake by roots and transport in plant tissues [62].

Photosynthesis can be disrupted through the modulation of the electron transport chain and can increase the rate of biochemical reactions catalyzed by different enzymes. Above a certain temperature threshold, enzymes lose their function, influencing the plant tissue tolerance to drought [1,63,64]. Drought stress also affects the translocation of nutrients and the composition of minerals, antioxidants and proteins [39,52]. Under stress conditions, reactive oxygen species (ROS) are highly generated [65,66] causing cell damage and plant necrosis [67]. Additionally, plant hormones and primary and secondary metabolites are modified [1]. Drought is the elicitor that can increase the content of secondary metabolites in plant tissues such as flavonoids, phenolics or more specific molecules, e.g., glycosides and alkaloids [68,69].

Crosstalk between drought and salinity stresses results in secondary stresses such as oxidative and osmotic ones [66]. Drought stress is a major agricultural problem worldwide and almost all of the main agricultural lands are affected by drought stress. The potential mechanisms of drought tolerance include: (1) production of phytohormones (such as indole-3-acetic acid (IAA), cytokinins and abscisic acid (ABA)) (2) synthesis of exopolysaccharides (3) activity of 1-aminocyclopropane-1-carboxylic acid (ACC) deaminase (4) induced systemic tolerance [66,70].

4. Mechanisms of Resistance in Plants

Plants are capable of defending themselves against numerous stress factors, both biotic and abiotic ones, by activating very effective pathways of immunity (Figure 3).

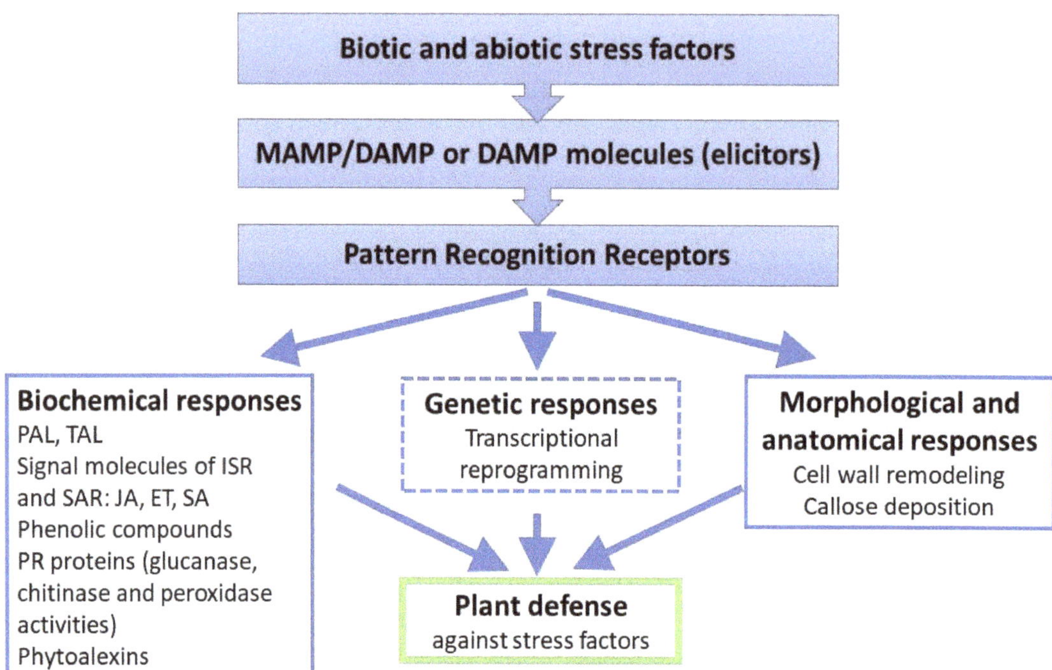

Figure 3. Immune response of plants against stress factors (modified on the basis of [71,72]); PAL—phenylalanine ammonia-lyase; TAL—tyrosine ammonia-lyase; ISR—induced systemic resistance, SAR—systemic acquired resistance; JA—jasmonic acid; ET—ethylene; SA—salicylic acid; PR proteins—pathogenesis related proteins.

The plant might acquire resistance against phytopathogenic infection due to the induction of plant defense responses driven by the very wide range of interactions with above- and below-ground microorganisms [6,73,74]. Several microbial species have displayed plant-priming phenomena. The priming process of plants is typically known in induced systemic resistance (ISR) and systemic acquired resistance (SAR), and microbe-associated molecular pattern (MAMP)-triggered immunity. ISR is mediated through the involvement of phytohormones, e.g., ethylene (ET) and jasmonic acid (JA), and the defense responses against phytopathogenic microorganisms are activated very quickly. SAR and MAMP-triggered immunity are induced as a first line, and unlike ISR, they utilize salicylic acid (SA) as signal substance of the plant resistance pathways [6,73,74]. The nonpathogenic microorganisms and various organic elicitors, mainly derived from microorganisms, act by inducing systemic acquired resistance (SAR) [75]. To elicit defense responses in plants, microorganisms secrete several molecules such as antibiotics (i.e., 2,4-diacetylphloroglucinol, phenazines synthesized by *Pseudomonas* species, and cyclic lipopeptides such as surfactin, synthesized by *Bacillus* strains), volatiles, quorum-sensing signals (N-acyl homoserine lactone of Gram-negative bacteria), proteins and small low-molecular weight compounds [6]. Natural bioactive compounds of microbial, Protist or plant origins with the ability to protect plants against phytopathogens can have fungicidal effects (can kill pathogens) or fungistatic effects (can limit development of phytopathogens), as well as being able to induce plant defense reactions as elicitors [71,76]. Every factor (physical, chemical, biotic, abiotic and their mixture) that induces plant immunity or stimulates the defense mechanisms in a plant is called an elicitor and is defined depending on its origin and molecular structure [68]. MAMP-type molecules are exoelicitors of microbial origin. Pathogen-associated molecular pattern (PAMP)-type molecules are exoelicitors of pathogenic organism origin. Damage/danger-associated molecular pattern (DAMP)-type molecules are endoelicitors of plant origin released during phytopathogen infection or produced under various stresses, [77–79]. Receptor proteins in the plasma membrane-pattern recognition receptors (PRRs) recognize particular molecular patterns of MAMP/PAMP and DAMP molecules [80,81]. The priming or PAMP-triggered, (PTI)-type local immunity, which arises in the absence of virulent pathogens, is due to the rapid onset of intracellular-signaling-pathway activation leading to a very fast and effective defense responses in the plant [82].

In plants, a range of abiotic and biotic elicitors can strengthen tolerance to drought stress, including alginate-derived oligosaccharides, ketoconazole, 2-aminoethanol, ABA, brassinosteroids, and beneficial microorganisms such as *Rhizobium* strains, endo- and exo-mycorrhizal rhizospheric and endophytic nonpathogenic fungi. These elicitors reduced the content of monodehydroascorbate, prevented the accumulation of ROS, increased activities of antioxidant enzymes, and maintained fresh and dry weights, grain yield, and relative water content in a variety of plants in response to drought stress [83]. The term "induced systemic tolerance" (IST) has been suggested for PGPB-induced physical and chemical alterations that result in enhanced tolerance to abiotic stresses [70,84,85].

5. Bacteria Supporting Horticultural Crops

The most promising solution for the future of modern horticulture seems to be the skillful use of biopreparations in the conventional crops and not limiting their use to a narrow range of ecological or organic farming crops. Biopreparations include at least three types of products: (1) biocontrol, or biological plant protection inhibiting directly (antagonism, competition) or indirectly (defense responses) the growth of phytopathogenic fungi or bacteria and other pests such as insects and nematodes, (2) biostimulation, positively affecting the plant development, increasing the plant biomass and yield, and (3) biofertilization, which provides nutrients and enhances plant nutrient uptake [6,29,71], (Table 1). The components of these biopreparations are very diverse, ranging from various inoculum types of microorganisms (either single or consortia of endophytic bacteria, fungi and Protista strains belonging to the plant growth-promoting group), through to metabolites, including phytohormonal and hormonelike substances or parts of microorganism cells, to various metabolites and structural compounds derived from microorganisms, Protista and plants often acting as the plant resistance elicitors [6,86]. Interestingly, the components of biopreparations are composed in such a way that, while performing biocontrol, biofertilizer and biostimulant functions [87–91], they reduce the impact of stresses caused by the numerous and dynamically changing environmental factors. Among these factors are rapid shifts in the temperature and humidity leading to the formation of drought, which reduces the availability of nutrients.

A very common approach is the isolation and application of active microorganisms to similar or the same plant and conditions, e.g., a *Pseudomonas* IACRBru1 strain isolated from *Eruca versicaria* (rucola) tissues improved *Lactuca sativa* (lettuce) biomass (up to 30%) [92]. One of the critical steps for the successful application of microorganisms is their survival and development in the new environment. In drought-stressed soils, the highest efficiency of this inoculation could be achieved using drought-tolerant bacteria isolated from arid soils or drought-resistant plants [93–95]. Bacteria classified as *Bacillus subtilis*, *Bacillus altitudinis*, *Brevibacillus laterosporus* and *Bacillus mojavensis* were isolated from *Cistanthe longiscapa*, a plant native to Atacama Desert in Chile [94]. A consortium of these microorganisms, with various complementary properties such as phosphate solubilization, the ability to grow on N-free culture, IAA, ACC-deaminase, and exopolysaccharide (EPS) synthesis, were applied onto tomato seeds, improving seedlings growth under drought stress.

Table 1. Selected activities of beneficial bacteria under drought-stress conditions in horticultural plant species; H—higher level/content; L—lower level/content; GPX—glutathione peroxidase; CAT—catalase; Chl—chlorophyll; RWC—relative water content; Fv/Fm—quantum efficiency of photosystem II; SOD—superoxide dismutase; MDA—malondialdehyde.

Bacteria	Changes in Plants	Plants	Ref.
Bacillus pumilus	- H total biomass (to 34.9%) - H antioxidant enzyme activities - H flavonoids, polysaccharide and glycyrrhizic acid contents	*Glycyrrhiza uralensis*	[96]
Bacillus sp.	- H roots and shoots fresh and dry weight - H shoot length	*Cucumis sativus*	[97]
Azotobacter chroococcum	- L GPX activity (to 12.5%)	*Mentha pulegium*	[98]
Pseudomonas putida KT2440 (pUCP22:*ots*AB)	- H trehalose [11-fold vs. *P. putida* KT2440 (pUCP) without the gen] - H fresh and dry weight - H fully turgid weight - H RWC	*Capsicum annuum* cv. Maor	[23]
Azospirillum brasilense	- L GPX activity (to 14.7%) - H CAT activity (2.6-fold vs. control)	*Mentha pulegium*	[98]
Variovorax paradoxus 5C-2	- H shoot dry weight - H net photosynthesis - H relative Chl content - L proline content	*Solanum lycopresicum* cv. Boludo F1	[99]
Rhodococcus sp. 4J2A2	xeroprotectant effect of trehalose in preventing the biomolecules	*Solanum esculentum* cv. F144	[23]
Azotobacter chroococcum with *Azospirilum lipofrum*	- H fresh weight - H root growth and length - H total phenolics content in leaves - H peroxidase activity	*Juglans regia*	[100]
Azotobacter chroococcum with *Azospirillum brasilense*	- H RWC (from 64.6% to 72.1%) - H Fv/Fm (from 0.56 to 0.75) - H SOD activity (from 34.7% to 57.2%) - L GPX activity (to 26.9%)	*Mentha pulegium*	[98]
Microbacterium sp. 3J1 or *Arthrobacter koreensis* 5J12A or *Arthrobacter piechaudii* 366-5	- H fresh and dry weight - H turgid weight - H RWC - H roots and stems length	*Capsicum annuum* cv. Maor	[23]
Bacillus cereus AR156 with *Bacillus subtilis* SM21 with *Serratia* sp. XY21	- H proline content in leaves - L MDA content in leaves - L peroxidation in plasmalemma	*Cucumis sativus*	[82]

It is worth emphasizing that among the most frequently mentioned PGPR strains of different genera, e.g., *Pseudomonas, Bacillus, Klebsiella, Azotobacter*, many *Pseudomonas* species show a very high diversity of traits stimulating plant growth. For this reason, many scientific laboratories are looking for such valuable isolates adapted to drought conditions [101–103] (Table 1). *Pseudomonas putida*, isolated by Kumar et al. [101], synthetized IAA, siderophore, ACC-deaminase, formed biofilm and solubilized phosphate. *Pseudomonas aeruginosa* strain, isolated from North East India, additionally showed HCN synthesis and endogenous osmolyte accumulation under the drought condition [103]. Sandhya et al. [102] also selected drought-tolerant *Pseudomonas* spp.: *P. monteilli, P. putida, P. stutzeri, P. syringae* from the rhizosphere of crop plants.

Niu et al. [93] isolated drought-tolerant plant growth-promoting bacteria from *Setaria italica* (foxtail millet) cultivated in arid soils. The bacterial strains identified in the roots and bulk soil (e.g., *Pseudomonas fluorescens* DR7 and DR11, *Pseudomonas migulae* DR35 and *Enterobacter hormaechei* DR16) synthetized ACC-deaminase under drought conditions. All the isolates produced EPS, but IAA activity was confirmed only in DR35 culture. Similarly, *Pseudomonas* sp. isolated from Californian soil exposed to frequent drought also showed significant production of EPS in response to desiccation [104].

Belonging to the PGPR family, *Azospirillum* spp. (Table 1) are a group of free-living soil bacteria mainly known for their ability to fix atmospheric nitrogen but also for releasing phytohormones, enhancing root growth, water and mineral uptake and plant resistance to drought stress [105,106] (Table 1). As a microbial inoculant, *Azospirillum* spp. could be crucial to improve fruit-tree acclimatization when transferred to the *post-vitro* environment [106].

Mariotti et al. [105] revealed that *Azospirillum baldaniorum* cells and their metabolites promote *Ocimum basilicum* cv. Red Rubin (purple basil) growth under the water stress condition. This action was attributed to the synthesis and transport of phytohormones that promoted plant growth and conferred tolerance to the abiotic stress. The plant leaves treated with a relevantly high dose of the filtered culture supernatants of *A. baldaniorum* contained significantly higher concentration of chlorophyll *a* and *b*, total chlorophyll, carotenoids, and anthocyanins. In the presence of these bacteria, in the tissues of purple basil, the concentration of stress-related phytohormones, ABA, JA and SA were higher. *Azospirillum brasilence* accompanied by *Pseudomonas* sp. and *Bacillus lentus* also caused a higher level of chlorophyll content in *Ocimum basilicum* grown under drought stress [107].

Moreover, at the end of the growing season, certain soil species, including soil-borne endophytic microorganisms promoting plant growth (e.g., including *Bacillus, Clostridium* and *Sporolactobacillus* genera), form endospores capable of remaining dormant in the soil. It is extremely important that in adverse environmental conditions (e.g., drought, very high or low temperature or higher amounts of incoming solar radiation) [1,108,109], when spores encounter the appropriate conditions (for example in the next growing season), they survive, germinate and the vegetative cells develop in the soil and are able to inhabit plants [109].

A mixture of three PGPR strains (*Bacillus cereus* AR156, *Bacillus subtilis* SM21, and *Serratia* sp. XY21) (Table 1) stimulated IST in drought stress in cucumber plants by maintaining the root recovery intensity, reducing plasmalemma peroxidation, stabilizing the osmotic potential, increasing photosynthesis efficiency and activities of SOD and cytoplasmic ascorbate peroxidase (APX) in the leaves, without involving the action of ACC deaminase to the lower plant ethylene levels [83].

5.1. Bacillus Species in Drought Stress

Among the features of the soil-aerobic, rod-shaped cells of *Bacillus* species (Table 1) contributing to the biocontrol mechanism is the synthesis and secretion of various antimicrobial peptides and very diverse antibiotics, enzymes, other proteins and organic compounds [110,111]. Inoculation of *Cucumis sativus* (cucumber) with *Bacillus cereus* and *Bacillus subtilis* strains along with *Serratia* sp. induced systemic tolerance to drought stress in plants by maintaining photosynthetic efficiency, root vigor, increasing proline content and enhanced SOD and CAT activities in the leaves [83]. In another experiment, to enhance *Lycopersicon esculentum* (tomato) drought tolerance, *Bacillus cereus* AR156 supernatant was applied. In the treated plants, chlorophyll *a* and *b* contents, as well as the activities of SOD, POD and CAT were increased markedly after culture supernatant application [112].

Plant small heat shock proteins (sHSPs) act as molecular chaperones that prevent irreversible aggregation of denatured proteins [85]. During drought stress, pepper plants inoculated with *Bacillus licheniformis* K11 exhibited enhanced transcription of Cadhn, VA, sHSP, and CaPR-10 genes [113,114]. In the study of Lim and Kim [113], the *Capsicum annuum* (pepper) seedlings were treated with a *Bacillus licheniformis* strain originated from Korean soil. Plants inoculated with drought-tolerant bacteria achieved higher shoot length and dry weight, and the analysis of gene expression in pepper roots indicated higher levels of expression of four genes related to drought and cold stresses. *Bacillus* sp. selected for high levels of cytokine synthesis was introduced into 12-day old *L. sativa* grown in dry soil.

After 3 weeks of seedlings inoculation, the increased amount of cytokinin and higher fresh and dry weights of shoots were confirmed [115].

At the beginning of vegetative season, higher temperature induces microbial metabolism (including releasing of inorganic available P to the soil solution by phosphate solubilizing microorganisms (PSM)) [116]. Gradually, the lack of adequate precipitation, insufficient soil moisture and high temperature decreased the soil microbial activity and the movement of nutrients in the soil [62]. *Bacillus* strains are commonly known to be great phosphate solubilizers [110]. Ying et al. [18] revealed high phosphatase activity of *Bacillus megatherium* and inorganic phosphate solubilization of *Bacillus saryghattati* strains under drought stress. *Bacillus* spp. (*B. cultidtuctinus*, *B. subtilis*, *B. polymyxa* and *B. mojavensis*) isolated from the *Cistanthe longiscapa* rhizosphere grown in the Atacama Desert (Chile) also exhibited phosphate-solubilizing activity [94].

It is worth noting that the activity of phosphate-solubilizing *Bacillus* strains support very energy-consuming processes of nitrogen fixation. Available P is a crucial ingredient of the energy source ATP. It can also replace conventional fertilization. An effective action on N and P uptake by the *Vicia faba* (faba bean) seeds and straws was confirmed after inoculation with the well-known phosphate-solubilizing bacterium *Bacillus megatherium* [117]. After inoculation of the apple trees cv. 'Topaz' with 'Mycostat' (containing *Bacillus subtilis* among strains promoting plant growth) the P root content was the same as in the tissues treated with chemical NPK fertilizer [89]. In addition, when soil moisture declined, the limited diffusion rate of nutrients, particularly P, from the soil matrix into the absorbing surface negatively affected nodulation and biological nitrogen fixation [118].

Plants with symptoms of potassium deficiency show accelerated wilting and lower yield, causing the loss of control of turgor-driven leaf movements [119]. *Bacillus* strains can secrete acidic metabolites (e.g., oxalic, fumaric, lemon, tartaric acids) that dissolve various minerals. Avakyan [120] demonstrated the ability to produce a thick EPS envelope by the strain *Bacillus mucilaginosus*. Secretion of the acidic metabolites by *B. mucilaginosus* cells creates a zone of strong acidification at the soil minerals' surface and allows the dissolution of mineral compounds. The polysaccharides secreted by these microorganisms additionally strongly adsorb SiO_3^{-2} leaving bioavailable K cations for plants in the soil solution [121].

The plant root is involved in the perception and transduction of stress signals via phytoregulators such as ET [122]. The increased level of ET causes premature aging of fruits and vegetables; wilting of flowers and leafy vegetables and defoliation of the mature leaves. Additionally, higher concentrations of ET in the rhizosphere inhibits arbuscular mycorrhizal fungi colonization and the root nodulation of legumes. A *Bacillus subtilis* (LDR2) strain isolated from the rhizosphere of drought-stressed plants, synthetizes ACC deaminase-regulating ET concentration. In the experiment, a LDR2 strain revealed protective mechanisms against the low water availability in soil, and improved *Trigonella* plants' weights (by 56%). Barnawall et al. [122] also demonstrated the enhanced nodulation and arbuscular mycorrhizal fungi colonization in the plants, which caused better nutrient uptake after inoculation of plants with *B. subtilis*.

In the face of climate change, certain future adaptations can be predicted by observing the functioning of organisms in extreme environments. In addition, in natural adverse ecosystems, except for the ability to form spores, microorganisms support plant growth and simultaneously provide an optimal environment for the development of plants tissues [111,117,123], e.g., *Bacillus mojavensis* was isolated from the very extreme environment of the Mojave Desert in California [111]. All the strains belonging to this group are described as endophytic and antagonistic to fungi [124]. The endophytic microorganisms (including both obligate and facultative species) are microbial symbionts residing within plants, mostly influencing host physiology [36,111].

The *B. mojavensis* strain isolated from the soybean plant rhizosphere was a very antagonistic strain, effectively controlling *Rhizoctonia solani*, a pathogenic fungus causing huge harvest losses of horticultural crops [125,126]. The presence of endophytic microorganisms with the biocontrol actions of soil-borne pathogens and the ability to stimulate the growth of cultivated plants from the early stages of its development seems to be a crucial solution for plants under unfavorable climate conditions. The inoculation of soybean seeds with the *Bacillus mojavensis* PB 35(R11) strain enhanced the growth of plant inoculated with *R. solani* (about 30% higher plant fresh weight and over 100% higher plant dry weight) [111]. Moreover, quantitative assays of the PB-35(R11) strain showed HCN, ammonia and siderophore production, as well as phosphate solubilization and chitinase activity. The treatment of seeds gives several advantages for the control of pathogenic fungi as a promising alternative to the use of synthetic pesticides. The endophytic *Bacillus* inoculants are also known for controlling *Fusarium* species, especially *Fusarium verticillioides* [124,127].

5.2. Actinomycetes Species in Drought Stress

A more advantageous strategy is the selection of microorganisms adapted to functioning in the conditions of temporary lack of water, drought, or rapid changes in temperature, because the metabolically active forms of microorganisms may support the growth of sensitive horticultural crops. *Actinomycetes* are Gram-positive, mostly aerobic, saprotrophic bacteria of diverse phenotypes (from cocci to highly differentiated mycelia).

Tangles of filaments grow similarly to filamentous fungi. This pseudomycelial growth (surface, plunge or air) provides penetration of a larger soil volume and into pores of soil, easing access to valuable minerals and simultaneously making them available to plants [128].

The main place of *Actinomycetes* occurrence is the soil (warm and humid or dry), but they are also identified in desert sands, on leaves and in plant tissues [129]. Reproduction of these bacteria occurs by fragmentation of pseudomycelium and spore formation. This group of microorganisms are mostly chemoorganotrophs with the ability to break down difficult decomposing substrates, e.g., cellulose, chitin, steroids, higher fatty acids or aromatic compounds. These activities allow them to survive and outcompete the native microflora in various ecological niches [130]. Lawlor et al. [131] revealed a higher number of *Actinomycetes* colony forming units (CFU) (about 10^6 to 10^7 g^{-1} of dry weight of soil) than CFU of fungi (10^4 to 10^5 g^{-1} of dry weight of soil).

Actinomycetes are known to be producers of bioactive compounds (antibacterial, antifungal), exhibiting great potential in promoting plant growth [129]. Sousa et al. [3] investigated that the *Streptomyces* strains produce siderophores, phytohormones (IAA), and solubilizing phosphate compounds, and exhibit chitinase, xylanase, cellulase, amylase and lipase activities. Additionally, the number of plant growth-promoting *Actinobacteria* is 1.3 times higher than that of the other bacteria [132].

Khamna et al. [133] identified about 30 *Streptomyces* isolates in Thai soil samples collected from the rhizospheres of plants such as *Curcuma magga*, *Zingiber officinale* (ginger), *Ocimum sanctum* (holy basil), and *Cumbopogon citratus* (lemongrass). After 3-day incubation, the *Streptomyces* CMU-H009 strain synthetized the highest concentration of IAA (about 144 μg · mL^{-1}) and its culture filtrates stimulated *Vigna unguiculata* (cowpea) seed germination. El-Tarabily [134] isolated over 60 *Streptomyces* spp. strains from a tomato rhizosphere in the United Arab Emirates and some of them revealed ACC-deaminase and IAA synthesis. The most efficient *Streptomyces filipinensis* 15, *Streptomyces atrovirens* 26 and *Streptomyces albovinaceus* 41 strains increased *Lycopersicon esculentum* (tomato) root and shoot length and dry weight. A higher level of endogenous IAA in the roots and shoots in these plants was also confirmed. *Actinomycetes* exhibited great potential in promoting rice, sorghum [135], tomato [134], maize [136] and soybean seedling growth [137].

It is worth emphasizing that some *Actinomycetes* (e.g., *Frankia* sp.) function in a symbiosis with higher plants, fixing nitrogen, while the plant provides the bacteria with sugars and minerals [136]. Such a favorable relationship has been observed in soybeans, peas, *Elaeagnus umbellata* and *Eleangus angustifolia* (Russian olive) [138].

A variety of activities improving plant development have been indicated in the *Acitnomycetes* species, and their efficiency obtained by adapting to adverse climatic conditions enables them to receive commercial products containing PGPA (plant growth-promoting actinobacteria), such as *Streptomyces lydicus* strains. *Acitnomycetes* species (*Streptomyces kasugaensis*, *Streptomyces griseus* and *Streptomyces cacoi* var. *asoensis*) producing antibacterial and antifungal bioactive compounds are components of biocontrol products applied against plant pathogens [129]. *Actinomycetes* synthesize enzymes such as lysozyme, glucanases, peptide–peptide hydrolases, mannanase and chitinase, which are involved in the lysis of the cell walls of other microorganisms [128]. El Tarabily et al. [134] investigated the promotion and biological control of seedlings and the mature plants of *C. sativus* using endophytic *Actinomycetes* (*Actinoplanes campanulatus*, *Micromonospora chalcea* and *Streptomyces spiralis*). *Pythium aphanidermatum* (oospore-producing soil-borne pathogen) causes seedling and root diseases of cucumber, causing damage to horticultural crops. This experiment proved that *Actinomycetes* colonize the roots of inoculated plants, promoted their growth and reduced the impact of *P. aphanidermatum.* Furthermore, El Tarabily et al. [134] compared the impact of *Actinomycetes* with chemical fungicide (metalaxyl) and demonstrated the possibility of replacing fungicide with plant inoculation with endophytic *Actinomycetes*.

6. Plant Growth-Promoting Fungi in Horticultural Crops

Among microorganisms, fungi can be much more drought-tolerant than bacteria [139] due to a number of mechanisms to overcome drought stress, including osmolytes, thick cell walls, and melanin [140]. Yeast cells are encased in a protective cell wall and cells of filamentous fungi can be connected, allowing water and solutes to flow between them. The filamentous fungi produce extremally long hypha, enabling the extraction of water from remote sites in the soil. Fungal abundance in the soil can increase under drought. They can remain active and even grow under extremely dry conditions. Their resistance to drought allows them to conduct the basic processes of decomposition of polymer compounds and the circulation of C and N [139]. Fungi, bacteria, seaweeds and plants are able to accumulate osmoprotectants, for instance amino acids (e.g., proline, glutamate), carbohydrates (trehalose), sugar alcohols (inositol, mannitol), quaternary ammonium compounds (glycine betaine) and tertiary sulphonium compounds (e.g., dimethylsulphoniopropionate) [141].

Plants use various mechanisms to protect against water deficiency, but some of them are associated with the presence of fungi with special activities (Table 2). Eukaryotic plant endophytes belong mainly to the fungi kingdom [142,143] and the most numerous among these endophytes are Glomeromycota (40%), Ascomycota (31%), Basidiomycota (20%), Zygomycota (0.1%) and unidentified phylla (8%). The Glomeromycota phyllum includes only arbuscular mycorrhizal fungi (AMF), whose species protect against phytopathogens, promote plant growth and counteract diverse stresses (mainly drought and salinity) by activating stress responsive/induced genes in plants. AMF are able to create a symbiosis with many horticultural plants belonging to various families, e.g., *Alliaceae, Apiaceae, Asteraceae, Fabaceae, Solanaceae, Rosaceae,* and *Oleaceae* [144–146]. Arbuscular mycorrhiza (AM) is the endomycorrhizal symbiotic association improving the nutrient uptake and growth of plants which may protect the host plants from pathogens and the harmful effects of drought [144,146,147]. Interactions with AMF, through an extensive network of hyphae, supply the plant with water from distant places. Studies have shown that AMF mainly use plant-derived carbohydrates in symbiosis with plants, and the plant receives access to the bioavailable minerals absorbed by the fungus from the soil (especially phosphorus). Moreover, the hyphae of fungal strains can uptake phosphorus and ammonium ions much more efficiently than the plant roots [148].

Moreover, fungi may influence the hormonal balance of plants by producing phytohormones (auxins, gibberellins) and through tolerance and resistance pathways, which protect the plant against biotic and abiotic factors. Both ectomycorrhizal (EMF, e.g., *Laccaria* spp.)- and endomycorrhizal-AMF fungi (e.g., belonging to the *Glomus, Rhizophagus, Funneliformis* genera) are capable of inducing the ISR resistance pathways involving JA as a signaling substance, or SAR, in which signaling occurs thanks to the SA molecules [149]. The production of auxins by the fungal endophytes increases the growth of plants under stress [150]. After the plant under stress is colonized by endophytes, stress-induced levels of ABA and some genes' expressions (e.g., zeaxanthin epoxidase, 9-cis-epoxycarotenoid dioxygenase 3 and ABA aldehyde oxidase 3) were decreased. Similar effects were achieved in the promotion of plant growth and yield under stress conditions, after exogenous phytohormone application such as gibberellic acid [151]. Sometimes endophytes do not have positive effects on plant growth during drought, but they improve plant recovery after water shortage [152].

The presence of some fungal endophytes (e.g., in the *Nicotiana benthamiana* seedlings) increased the leaf area, chlorophyll content, photosynthetic rate, antioxidative enzyme activities, accumulation of osmoprotectants (sugar, protein and proline) and enhanced expression of drought-related genes [153,154]. On the other hand, in drought conditions, the relative water content in leaves and soluble protein content in the tissues of *Cinnamomum migao* did not change after 120-day inoculation with *Glomus lamellosum* [155] (Table 2). In drought-suffering plants, after endophyte inoculation, a lower level of biomolecule degradation was observed as a consequence of the reduced level of ROS production, e.g., in tomato [156]. In the face of drought stress, inoculation with *Piriformospora indica* (Table 2) mobilized activities of peroxidase (POX), catalase (CAT) and superoxide dismutase (SOD) in the leaves [157]. The endophytic fungal strains of *Ampelomyces* sp. isolated from soil exposed to drought enhanced drought tolerance in tomato [143,158]. Inoculation of tomato seedlings with *Alternaria* spp. strains under drought conditions resulted in the maintenance of the photosynthetic efficiency and effective reduction of ROS accumulation [156].

Table 2. Selected activities of beneficial fungi under drought stress conditions in horticultural plant species; H—higher level/content; L—lower level/content; Chl—chlorophyll; CAT—catalase; SOD—superoxide dismutase; MDA—malondialdehyde.

Fungi	Changes in Plants	Plants	Ref.
Glomus intraradices	- H content of N in roots and shoots - H flower and fruit production - H fruit yield - H dry root mass	*Lycopersicum esculentum*	[159]
Glomus mosesseae	- slow down the reduction of Chl *a+b* - inhibit the decomposition of carotenoids	*Fragaria x ananassa*	[32]
Glomus etunicatum	- H fresh weight - H number of leaves - H content of N, P, Zn in leaves	*Juglans regia*	[100]
Ampelomyces sp.	- H dry weight of root and shoot - H fruit weight - H stress tolerance	*Solanum lycopersicum* var. Better Boy	[158]
Phoma spp.	- H proline content, CAT and SOD activities - H chlorophyll content - H MDA content - H water content in leaves	*Pinus tablaeformis*	[160]
Piriformospora indica	- H root and shoot growth - H lateral root development - H peroxidase, CAT and SOD activities in leaves	*Brassica rapa* subsp. *pekinensis*	[157]
Trichodermaharzianum	- H fresh and dry weight of roots - H osmolyte concentration	*Theobroma cacao*	[160]
Glomus mossae with *Glomus etunicatum*	- H height - H content of N, P, Zn in leaves - L leaves abscission	*Juglans regia*	[100]
Glomus lamellosum or *Glomus etunicatum*	- H fresh and dry weight - H stem fresh weight - H water content in leaves - L MDA content - H CAT activity	*Cinnamomum migao*	[155]

Table 2. *Cont.*

Fungi	Changes in Plants	Plants	Ref.
Glomus mosseae or *G. versiforme* or *G. diaphanum*	- H enzyme activity in soil - L CAT activity in soil - H hyphal density	*Poncirus trifoliata*	[161]
Chaetomium globosum or *Penicillium resedanum*	- H shoot dry weight - H shoot length - H photosynthesis rate	*Capsicum annum*	[151]
Alternaria sp. or *Trichoderma harzianum*	- H root and shoot dry weight	*Solanum lycopersicum* var. Rutger	[156]

The presence of beneficial fungal strains in soil and plants might specifically induce resistance by releasing elicitors belonging to fungal-derived compounds, e.g.: chitin, chitosan, ergosterol, β-glucans [151]. Moreover, among the pathogenic microorganisms, the filamentous fungi are responsible for horticultural crop diseases. It should be also noted that fungal endophytes belonging to PGP (e.g., *Colletotrichum* sp., *Alternaria* sp., *Fusarium* sp. and *Aspergillus* sp.) induce plant resistance and increase plant tolerance to drought, but may also produce mycotoxins in plants [162]. To increase specificity and enhancement of the induction, elicitors can be derived from the nonpathogenic fungi belonging to the same genus as the pathogenic strains causing plant diseases, e.g., *Fusarium* or *Trichoderma* [152]. In grapevine with black-foot disease (*Dactylonectria* and *Cylindrocarpon* genera), the relative abundance of the potential biocontrol agent *Trichoderma* in the root endosphere, rhizosphere, and bulk soil under drought stress (25% irrigation regime) was significantly lower than in control conditions (50–100% irrigation regime) [163]. Moreover, enrichment in AMF *Funneliformis* during drought was observed.

Recent and particularly promising studies have focused on the determination of the effectiveness and reliability of a mixture of bacterial and fungal strains (Table 3).

Table 3. Selected activities of beneficial consortia (bacteria with fungi) under drought stress conditions in horticultural plant species; H—higher level/content; L—lower level/content; CAT—catalase; APX—ascorbate peroxidase; Chl—chlorophyll.

Bacteria with Fungi	Changes in Plants	Plants	Ref.
Pseudomonas fluorescence with *Trichoderma harzianum*	- H growth parameters - H seedling emergency - H root and shoot length - H CAT and APX activities	*Cuminum cyminum*	[164]
Variovorax paradoxus 5C-2 with *Rhizophorus irregularis MULC*	- H shoot dry weight - H net photosynthesis - no change: relative Chl content vs. control - H oxidative damage - L proline content	*Solanum lycopersicum cv.* Boludo F1	[99]

7. Conclusions

The EGD emphasizes sustainable food production by the crucial reduction in the use of pesticides, biocides and chemical mineral fertilizers and increase of organic (ecological) production. Consequently, in many European countries, continuous research has been carried out on natural biopreparations (biocontroling, biofertilizers) containing selected microorganisms with different activities, and/or their metabolites.

The key problems of using biopreparations containing various microorganisms include limiting the possibility of introducing the microbial inoculum to the appropriate conditions and the low repeatability of their activities. This might be due to drought stress during the vegetative period in comparison to microorganisms tested in the optimal conditions.

The resistance of plants that interact with microorganisms in drought conditions is enhanced because it is induced by both abiotic (stress factor) and the biotic (microorganism) elicitors. In drought conditions, many cultivated horticultural plants use their own numerous mechanisms (morphological, physiological, anatomical, biochemical or molecular) to counteract the negative effects and are supported by endophytes that constantly inhabit them, and rhizospheric microorganisms existing in the vicinity of roots.

The use of preparations containing fungal strains, which are more tolerant to drought than bacteria, provides many tolerance mechanisms, and their abundance increases in water-limiting conditions. Fungi, through their specific growth and traits, allow intensive soil exploration, water extraction and penetration of plant tissues influencing the plant and might be more effective compared to bacteria with the same activity.

Studies on the influence of drought stress on horticultural plants have indicated that the application of various microorganisms allows efficient protection of plants, despite our restricted knowledge about these mechanisms of action. Due to such a high variability of the environment, biopreparations should be multicomponent in order to achieve appropriate levels of microorganism cooperation and the final desired effect. The combination of fungal and bacterial strains into one preparation gives even better effectiveness and reliability, allowing us to consider higher crop-specificity, and seems to be particularly promising.

Author Contributions: Conceptualization, A.H.; E.O. and J.J.-Ś.; writing—original draft preparation, E.O.; A.H.; E.R. and J.J.-Ś.; writing—review and editing, A.H.; E.O.; E.R. and M.S.; visualization, A.H.; E.O.; E.R. and M.S. All authors have read and agreed to the published version of the manuscript.

Funding: This research received no external funding.

Institutional Review Board Statement: Not applicable.

Informed Consent Statement: Not applicable.

Conflicts of Interest: The authors declare no conflict of interest.

References

1. Moretti, C.L.; Mattos, L.M.; Calbo, A.G.; Sargent, S.A. Climate changes and potential impacts on postharvest quality of fruit and vegetable crops: A review. *Food Res. Int.* **2010**, *43*, 1824–1832. [CrossRef]
2. Hamid, B.; Zaman, M.; Farooq, S.; Fatima, S.; Sayyed, R.Z.; Baba, Z.A.; Sheikh, T.A.; Reddy, M.S.; El Enshasy, H.; Gafur, A.; et al. Bacterial plant biostimulants: A sustainable way towards improving growth, productivity, and health of crops. *Sustainability* **2021**, *13*, 2856. [CrossRef]
3. Sousa, C.S.; Soares, A.C.F.; Garrido, M.S. Characterization of *Streptomycetes* with potential to promote plant growth and biocontrol. *Sci. Agric.* **2008**, *65*, 50–55. [CrossRef]
4. Shah, A.; Nazari, M.; Antar, M.; Msimbira, L.A.; Naamala, J.; Lyu, D.; Rabileh, M.; Zajonc, J.; Smith, D.L. PGPR in agriculture: A sustainable approach to increasing climate change resilience. *Front. Sustain. Food Syst.* **2021**, *5*, 1–22. [CrossRef]
5. International Society for Horticultural Science. Available online: https://www.ishs.org/defining-horticulture (accessed on 27 September 2021).
6. Vishwakarma, K.; Kumar, N.; Shandilya, C.; Mohapatra, S.; Bhayana, S.; Varma, A. Revisiting plant-microbe interactions and microbial consortia application for enhancing sustainable agriculture: A review. *Front. Microbiol.* **2020**, *11*, 560406. [CrossRef]
7. Khan, N.; Ali, S.; Shahid, M.A.; Mustafa, A.; Sayyed, R.Z.; Curá, J.A. Insights into the interactions among roots, rhizosphere, and rhizobacteria for improving plant growth and tolerance to abiotic stresses: A review. *Cells* **2021**, *10*, 1551. [CrossRef]
8. Ullah, A.; Nisar, M.; Ali, H.; Hazrat, A.; Hayat, K.; Keerio, A.A.; Ihsan, M.; Laiq, M.; Ullah, S.; Fahad, S.; et al. Drought tolerance improvement in plants: An endophytic bacterial approach. *Appl. Microbiol. Biotechnol.* **2019**, *103*, 7385–7397. [CrossRef]
9. Yan, L.; Zhu, J.; Zhao, X.; Shi, J.; Jiang, C.; Shao, D. Beneficial effects of endophytic fungi colonization on plants. *Appl. Microbiol. Biotechnol.* **2019**, *103*, 3327–3340. [CrossRef]
10. Cheng, Y.T.; Zhang, L.; He, S.Y. Plant-microbe interactions facing environmental challenge. *Cell Host Microbe* **2019**, *26*, 183–192. [CrossRef]
11. Kavadia, A.; Omirou, M.; Fasoula, D.; Ioannides, I.M. The importance of microbial inoculants in a climate-changing agriculture in Eastern Mediterranean region. *Atmosphere* **2020**, *11*, 1136. [CrossRef]
12. Malik, A.; Mor, V.S.; Tokas, J.; Punia, H.; Malik, S.; Malik, K.; Sangwan, S.; Tomar, S.; Singh, P.; Singh, N.; et al. Biostimulant-treated seedlings under sustainable agriculture: A global perspective facing climate change. *Agronomy* **2021**, *11*, 14. [CrossRef]
13. Porter, S.S.; Bantay, R.; Friel, C.A.; Garoutte, A.; Gdanetz, K.; Ibarreta, K.; Moore, B.M.; Shetty, P.; Siler, E.; Friesen, M.L. Beneficial microbes ameliorate abiotic and biotic sources of stress on plants. *Funct. Ecol.* **2020**, *34*, 2075–2086. [CrossRef]

14. Fincheira, P.; Quiroz, A.; Tortella, G.; Diez, M.C.; Rubilar, O. Current advances in plant-microbe communication *via* volatile organic compounds as an innovative strategy to improve plant growth. *Microbiol. Res.* **2021**, *247*, 126726. [CrossRef] [PubMed]

15. Hanaka, A.; Ozimek, E.; Majewska, M.; Rysiak, A.; Jaroszuk-Ściseł, J. Physiological diversity of Spitsbergen soil microbial communities suggests their potential as plant growth-promoting bacteria. *Int. J. Mol. Sci.* **2019**, *20*, 1207. [CrossRef] [PubMed]

16. Hanaka, A.; Nowak, A.; Plak, A.; Dresler, S.; Ozimek, E.; Jaroszuk-Ściseł, J.; Wójciak-Kosior, M.; Sowa, I. Bacterial isolate inhabiting Spitsbergen soil modifies the physiological response of *Phaseolus coccineus* in control conditions and under exogenous application of methyl jasmonate and copper excess. *Int. J. Mol. Sci.* **2019**, *20*, 1909. [CrossRef] [PubMed]

17. Compant, S.; Van Der Heijden, M.G.A.; Sessitsch, A. Climate change effects on beneficial plant-microorganism interactions. *FEMS Microbiol. Ecol.* **2010**, *73*, 197–214. [CrossRef]

18. Ma, Y.; Vosátka, M.; Freitas, H. Editorial: Beneficial microbes alleviate climatic stresses in plants. *Front. Plant Sci.* **2019**, *10*, 595. [CrossRef] [PubMed]

19. Vimal, S.R.; Singh, J.S.; Arora, N.K.; Singh, S. Soil-plant-microbe interactions in stressed agriculture management: A review. *Pedosphere* **2017**, *27*, 177–192. [CrossRef]

20. Gupta, S.; Seth, R.; Sharma, A. *Plant Growth-Promoting Rhizobacteria Play a Role as Phytostimulators for Sustainable Agriculture*; Choudhary, D., Varma, A., Tuteja, N., Eds.; Springer: Singapore, 2016.

21. Ma, Y.; Oliveira, R.S.; Freitas, H.; Zhang, C. Biochemical and molecular mechanisms of plant-microbe-metal interactions: Relevance for phytoremediation. *Front. Plant Sci.* **2016**, *7*, 918. [CrossRef]

22. Pirttilä, A.M.; Tabas, H.M.P.; Baruah, N.; Koskimäki, J.J. Biofertilizers and biocontrol agents for agriculture: How to identify and develop new potent microbial strains and traits. *Microorganisms* **2021**, *9*, 817. [CrossRef]

23. Vílchez, J.I.; García-Fontana, C.; Román-Naranjo, D.; González-López, J.; Manzanera, M. Plant drought tolerance enhancement by trehalose production of desiccation-tolerant microorganisms. *Front. Microbiol.* **2016**, *7*, 1577. [CrossRef]

24. Finkel, O.M.; Castrillo, G.; Herrera Paredes, S.; Salas González, I.; Dangl, J.L. Understanding and exploiting plant beneficial microbes. *Curr. Opin. Plant Biol.* **2017**, *38*, 155–163. [CrossRef] [PubMed]

25. Meena, K.K.; Sorty, A.M.; Bitla, U.M.; Choudhary, K.; Gupta, P.; Pareek, A.; Singh, D.P.; Prabha, R.; Sahu, P.K.; Gupta, V.K.; et al. Abiotic stress responses and microbe-mediated mitigation in plants: The omics strategies. *Front. Plant Sci.* **2017**, *8*, 172. [CrossRef] [PubMed]

26. Anjum, S.A.; Ashraf, U.; Zohaib, A.; Tanveer, M.; Naeem, M.; Ali, I.; Tabassum, T.; Nazir, U. Growth and developmental responses of crop plants under drought stress: A review. *Zemdirb. Agric.* **2017**, *104*, 267–276. [CrossRef]

27. Sharma, M.; Sudheer, S.; Usmani, Z.; Rani, R.; Gupta, P. Deciphering the omics of plant-microbe interaction: Perspectives and new insights. *Curr. Genom.* **2020**, *21*, 343–362. [CrossRef]

28. Vasseur-Coronado, M.; du Boulois, H.D.; Pertot, I.; Puopolo, G. Selection of plant growth promoting rhizobacteria sharing suitable features to be commercially developed as biostimulant products. *Microbiol. Res.* **2021**, *245*, 1–10. [CrossRef]

29. Pylak, M.; Oszust, K.; Frąc, M. Review report on the role of bioproducts, biopreparations, biostimulants and microbial inoculants in organic production of fruit. *Rev. Environ. Sci. Biotechnol.* **2019**, *18*, 597–616. [CrossRef]

30. Kränzlein, M.; Geilfus, C.-M.; Franzisky, B.L.; Zhang, X.; Wimmer, M.A.; Zörb, C. Physiological responses of contrasting maize (*Zea mays* L.) hybrids to repeated drought. *J. Plant Growth Regul.* **2021**. [CrossRef]

31. Jamil, M.; Ahamd, M.; Anwar, F.; Zahir, Z.A.; Kharal, M.A.; Nazli, F. Inducing drought tolerance in wheat through combined use of L-tryptophan and *Pseudomonas fluorescens*. *Pak. J. Agric. Sci.* **2018**, *55*, 331–337. [CrossRef]

32. Yin, B.; Wang, Y.; Liu, P.; Hu, J.; Zhen, W. Effects of vesicular-arbuscular mycorrhiza on the protective system in strawberry leaves under drought stress. *Front. Agric. China* **2010**, *4*, 165–169. [CrossRef]

33. Hone, H.; Mann, R.; Yang, G.; Kaur, J.; Tannenbaum, I.; Li, T.; Spangenberg, G.; Sawbridge, T. Profiling, isolation and characterisation of beneficial microbes from the seed microbiomes of drought tolerant wheat. *Sci. Rep.* **2021**, *11*, 11916. [CrossRef]

34. Zapata, T.; Galindo, D.M.; Corrales-Ducuara, A.R.; Ocampo-Ibáñez, I.D. The diversity of culture-dependent gram-negative *Rhizobacteria* associated with manihot esculenta crantz plants subjected to water-deficit stress. *Diversity* **2021**, *13*, 366. [CrossRef]

35. Mustafa, S.; Kabir, S.; Shabbir, U.; Batool, R. Plant growth promoting rhizobacteria in sustainable agriculture: From theoretical to pragmatic approach. *Symbiosis* **2019**, *78*, 115–123. [CrossRef]

36. Ozimek, E.; Hanaka, A. *Mortierella* species as the plant growth-promoting fungi present in the agricultural soils. *Agriculture* **2021**, *11*, 7. [CrossRef]

37. IPCC Climate Change. *The Physical Science Basis 2021*; IPCC Climate Change: Geneva, Switzerland, 2021.

38. Lelieveld, J.; Hadjinicolaou, P.; Kostopoulou, E.; Chenoweth, J.; El Maayar, M.; Giannakopoulos, C.; Hannides, C.; Lange, M.A.; Tanarhte, M.; Tyrlis, E.; et al. Climate change and impacts in the Eastern Mediterranean and the Middle East. *Clim. Chang.* **2012**, *114*, 667–687. [CrossRef] [PubMed]

39. Ilyas, M.; Nisar, M.; Khan, N.; Hazrat, A.; Khan, A.H.; Hayat, K.; Fahad, S.; Khan, A.; Ullah, A. Drought tolerance strategies in plants: A mechanistic approach. *J. Plant Growth Regul.* **2021**, *40*, 926–944. [CrossRef]

40. Hanaka, A.; Plak, A.; Zagórski, P.; Ozimek, E.; Rysiak, A.; Majewska, M.; Jaroszuk-Ściseł, J. Relationships between the properties of Spitsbergen soil, number and biodiversity of rhizosphere microorganisms, and heavy metal concentration in selected plant species. *Plant Soil* **2019**, *436*, 49–69. [CrossRef]

41. Mędrek, K.; Gluza, A.; Siwek, K.; Zagórski, P. The meteorological conditions on the Calypsobyen in summer 2014 on the background of multiyear 1986–2011. *Probl. Klim. Polar.* **2014**, *24*, 37–50. (In Polish)

42. Franczak, Ł.; Kociuba, W.; Gajek, G. Runoff variability in the Scott River (SW Spitsbergen) in summer seasons 2012–2013 in comparison with the period 1986–2009. *QuaGeo* **2016**, *35*, 39–50. [CrossRef]

43. Hu, Y.; Xie, G.; Jiang, X.; Shao, K.; Tang, X.; Gao, G. The relationships between the free-living and particle-attached bacterial communities in response to elevated eutrophication. *Front. Microbiol.* **2020**, *11*, 423. [CrossRef]

44. Schimel, J.P. Life in dry soils: Effects of drought on soil microbial communities and processes. *Annu. Rev. Ecol. Evol. Syst.* **2018**, *49*, 409–432. [CrossRef]

45. Manzanera, M. Dealing with water stress and microbial preservation. *Environ. Microbiol.* **2021**, *23*, 3351–3359. [CrossRef]

46. Brás, T.A.; Seixas, J.; Carvalhais, N.; Jagermeyr, J. Severity of drought and heatwave crop losses tripled over the last five decades in Europe. *Environ. Res. Lett.* **2021**, *16*, 065012. [CrossRef]

47. Singh, R.; Parihar, P.; Singh, M.; Bajguz, A.; Kumar, J.; Singh, S.; Singh, V.P.; Prasad, S.M. Uncovering potential applications of cyanobacteria and algal metabolites in biology, agriculture and medicine: Current status and future prospects. *Front. Microbiol.* **2017**, *8*, 515. [CrossRef]

48. Camaille, M.; Fabre, N.; Clément, C.; Barka, E.A. Advances in wheat physiology in response to drought and the role of plant growth promoting rhizobacteria to trigger drought tolerance. *Microorganisms* **2021**, *9*, 687. [CrossRef]

49. Rysiak, A.; Dresler, S.; Hanaka, A.; Hawrylak-Nowak, B.; Strzemski, M.; Kováčik, J.; Sowa, I.; Latalski, M.; Wójciak, M. High temperature alters secondary metabolites and photosynthetic efficiency in *Heracleum sosnowskyi*. *Int. J. Mol. Sci.* **2021**, *22*, 4756. [CrossRef]

50. Abdelaal, K.; Alkahtani, M.; Attia, K.; Hafez, Y.; Király, L.; Künstler, A. The role of plant growth-promoting bacteria in alleviating the adverse effects of drought on plants. *Biology* **2021**, *10*, 520. [CrossRef] [PubMed]

51. Fang, Y.; Xiong, L. General mechanisms of drought response and their application in drought resistance improvement in plants. *Cell. Mol. Life Sci.* **2015**, *72*, 673–689. [CrossRef] [PubMed]

52. Luo, W.; Xu, C.; Ma, W.; Yue, X.; Liang, X.; Zuo, X.; Knapp, A.K.; Smith, M.D.; Sardans, J.; Dijkstra, F.A.; et al. Effects of extreme drought on plant nutrient uptake and resorption in rhizomatous vs bunchgrass-dominated grasslands. *Oecologia* **2018**, *188*, 633–643. [CrossRef] [PubMed]

53. Manavalan, L.P.; Guttikonda, S.K.; Phan Tran, L.S.; Nguyen, H.T. Physiological and molecular approaches to improve drought resistance in soybean. *Plant Cell Physiol.* **2009**, *50*, 1260–1276. [CrossRef] [PubMed]

54. Shao, H.B.; Chu, L.Y.; Jaleel, C.A.; Zhao, C.X. Water-deficit stress-induced anatomical changes in higher plants. *Comptes Rendus Biol.* **2008**, *331*, 215–225. [CrossRef]

55. Reddy, A.R.; Chaitanya, K.V.; Vivekanandan, M. Drought-induced responses of photosynthesis and antioxidant metabolism in higher plants. *J. Plant Physiol.* **2004**, *161*, 1189–1202. [CrossRef]

56. Abdelaal, K.A.A.; Hafez, Y.M.; El-Afry, M.M.; Tantawy, D.S.; Alshaal, T. Effect of some osmoregulators on photosynthesis, lipid peroxidation, antioxidative capacity, and productivity of barley (*Hordeum vulgare* L.) under water deficit stress. *Environ. Sci. Pollut. Res.* **2018**, *25*, 30199–30211. [CrossRef] [PubMed]

57. Hafez, Y.; Attia, K.; Alamery, S.; Ghazy, A.; Al-Doss, A.; Ibrahim, E.; Rashwan, E.; El-Maghraby, L.; Awad, A.; Abdelaal, K. Beneficial effects of biochar and chitosan on antioxidative capacity, osmolytes accumulation, and anatomical characters of water-stressed barley plants. *Agronomy* **2020**, *10*, 630. [CrossRef]

58. Hargravei, K.R.; Kolb, K.J.; Ewers, F.W.; Davis, S.D. Conduit diameter and drought-induced embolism in *Salvia mellifera* Greene (*Labiatae*). *New Phytol.* **1994**, *126*, 695–705. [CrossRef]

59. Liu, F.; Jensen, C.R.; Shahanzari, A.; Andersen, M.N.; Jacobsen, S.E. ABA regulated stomatal control and photosynthetic water use efficiency of potato (*Solanum tuberosum* L.) during progressive soil drying. *Plant Sci.* **2005**, *168*, 831–836. [CrossRef]

60. Ullah, A.; Mushtaq, H.; Fahad, S.; Hakim; Shah, A.; Chaudhary, H.J. Plant growth promoting potential of bacterial endophytes in novel association with *Olea ferruginea* and *Withania coagulans*. *Microbiology* **2017**, *86*, 119–127. [CrossRef]

61. Hussain, H.A.; Hussain, S.; Khaliq, A.; Ashraf, U.; Anjum, S.A.; Men, S.; Wang, L. Chilling and drought stresses in crop plants: Implications, cross talk, and potential management opportunities. *Front. Plant Sci.* **2018**, *9*, 393. [CrossRef]

62. Ragel, P.; Raddatz, N.; Leidi, E.O.; Quintero, F.J.; Pardo, J.M. Regulation of K^+ nutrition in plants. *Front. Plant Sci.* **2019**, *10*, 281. [CrossRef]

63. Wang, Z.; Li, G.; Sun, H.; Ma, L.; Guo, Y.; Zhao, Z.; Gao, H.; Mei, L. Effects of drought stress on photosynthesis and photosynthetic electron transport chain in young apple tree leaves. *Biol. Open* **2018**, *7*. [CrossRef]

64. Zhang, Y.B.; Yang, S.L.; Dao, J.M.; Deng, J.; Shahzad, A.N.; Fan, X.; Li, R.D.; Quan, Y.J.; Bukhari, S.A.H.; Zeng, Z.H. Drought-induced alterations in photosynthetic, ultrastructural and biochemical traits of contrasting sugarcane genotypes. *PLoS ONE* **2020**, *15*, e0235845. [CrossRef]

65. Cruz de Calvadio, M.H. Drought stress and reactive oxygen species. *Plant Signal. Behav.* **2008**, *3*, 156–165. [CrossRef]

66. Verma, G.; Srivastava, D.; Tiwari, P.; Chakrabarty, D. Reactive oxygen, nitrogen and sulfur species in plants: Production, metabolism, signaling and defense mechanisms. ROS modulation in crop plants under drought stress. In *Reactive Oxygen, Nitrogen and Sulfur Species in Plants: Production, Metabolism, Signaling and Defense Mechanisms*; Hasanuzzaman, M., Fotopoulos, V., Nahar, K., Fujita, M., Eds.; Wiley: Hoboken, NJ, USA, 2019; pp. 311–336.

67. Petrov, V.; Hille, J.; Mueller-Roeber, B.; Gechev, T.S. ROS-mediated abiotic stress-induced programmed cell death in plants. *Front. Plant Sci.* **2015**, *6*, 1–16. [CrossRef]

68. Thakur, M.; Sohal, B.S. Role of elicitors in inducing resistance in plants against pathogen infection: A review. *ISRN Biochem.* **2013**, *2013*, 1–10. [CrossRef]
69. Ali, S.; Khan, N. Delineation of mechanistic approaches employed by plant growth promoting microorganisms for improving drought stress tolerance in plants. *Microbiol. Res.* **2021**, *249*, 126771. [CrossRef] [PubMed]
70. Vardharajula, S.; Ali, S.Z.; Grover, M.; Reddy, G.; Bandi, V. Drought-tolerant plant growth promoting *Bacillus* spp.: Effect on growth, osmolytes, and antioxidant status of maize under drought stress. *J. Plant Interact.* **2011**, *6*, 1–14. [CrossRef]
71. Jamiołkowska, A. Natural compounds as elicitors of plant resistance against diseases and new biocontrol strategies. *Agronomy* **2020**, *10*, 173. [CrossRef]
72. Walters, D.; Newton, A.; Lyon, G. (Eds.) *Induced Resistance for Plant Defence. A Sustainable Approach to Crop Protection*; Blackwell Publishing: Oxford, UK, 2007; ISBN 978-1-4051-3447-7.
73. Conrath, U.; Beckers, G.J.; Flors, V.; Garcia-Augustin, P.; Jakab, G.; Mauch, F.; Newman, M.-A.; Pieterse, C.M.J.; Poinssot, B.; Pozo, M.J.; et al. Priming: Getting ready for battle. *Mol. Plant Microbe Interact.* **2006**, *19*, 1062–1071. [CrossRef] [PubMed]
74. Martinez-Medina, A.; Flors, V.; Heil, M.; Mauch-Mani, B.; Pieterse, C.M.; Pozo, M.J.; Ton, J. Recognizing plant defense priming. *Trends Plant Sci.* **2016**, *21*, 818–822. [CrossRef]
75. Koziara, W.; Sulewska, H.; Panasiewicz, K. Effect of resistance stimulator application to some agricultural crops. *J. Res. Appl. Agric. Eng.* **2006**, *51*, 82–86.
76. Babosha, A.V. Changes in lectin activity in plants treated with resistance inducers. *Plant Physiol.* **2004**, *31*, 51–55. [CrossRef]
77. Jones, J.D.G.; Dangl, J.L. The plant immune system. *Nature* **2006**, *444*, 323–329. [CrossRef] [PubMed]
78. Boller, T.; Felix, G. A renaissance of elicitors: Perception of microbe-associated molecular patterns and danger signals by pattern-recognition receptors. *Annu. Rev. Plant Biol.* **2009**, *60*, 379–407. [CrossRef] [PubMed]
79. Henry, G.; Thonart, P.; Ongena, M. PAMPs, MAMPs, DAMPs and others: An update on the diversity of plant immunity elicitors. *Biotechnol. Agron. Soc. Environ.* **2012**, *16*, 257–268.
80. Schwessinger, B.; Ronald, P.C. Plant innate immunity: Perception of conserved microbial signatures. *Annu. Rev. Plant Biol.* **2012**, *63*, 451–482. [CrossRef]
81. Ranf, S. Sensing of molecular patterns through cell surface immune receptors. *Curr. Opin. Plant Biol.* **2017**, *38*, 68–77. [CrossRef]
82. Bigeard, J.; Colcombet, J.; Hirt, H. Signaling mechanisms in pattern-triggered immunity (PTI). *Mol. Plant* **2015**, *8*, 521–539. [CrossRef]
83. Wang, C.J.; Yang, W.; Wang, C.; Gu, C.; Niu, D.D.; Liu, H.X.; Wang, Y.P.; Guo, J.H. Induction of drought tolerance in cucumber plants by a consortium of three plant growth-promoting *Rhizobacterium* strains. *PLoS ONE* **2012**, *7*, e52565. [CrossRef]
84. Shrivastava, P.; Kumar, R. Soil salinity: A serious environmental issue and plant growth promoting bacteria as one of the tools for its alleviation. *Saudi J. Biol. Sci.* **2015**, *22*, 123–131. [CrossRef]
85. Singh, H.B.; Keswani, C.; Reddy, M.S.; Sansinenea, E.; García-Estrada, C. *Secondary Metabolites of Plant Growth Promoting Rhizomicroorganisms: Discovery and Applications*; Springer: Singapore, 2019; ISBN 9789811358623.
86. Oszust, K.; Pylak, M.; Frąc, M. *Trichoderma*-based biopreparation with prebiotics supplementation for the naturalization of raspberry plant rhizosphere. *J. Mol. Sci.* **2021**, *22*, 6356. [CrossRef]
87. Timmusk, S.; Behers, L.; Muthoni, J.; Muraya, A.; Aronsson, A.C. Perspectives and challenges of microbial application for crop improvement. *Front. Plant Sci.* **2017**, *8*, 49. [CrossRef]
88. Sokolova, M.G.; Akimova, G.; Vaishlya, O.; Vedernikova, A. Physiological research of efficiency of biologically safe bacterial fertilizers. *J. Manuf. Technol. Manag.* **2010**, *21*, 956–970. [CrossRef]
89. Abd El-Gleel Mosa, W.F.; Paszt, L.S.; Frąc, M.; Trzciński, P.; Treder, W.; Klamkowski, K. The role of biofertilizers in improving vegetative growth, yield and fruit quality of apple. *Hortic. Sci.* **2018**, *45*, 173–180. [CrossRef]
90. Anli, M.; Baslam, M.; Tahiri, A.; Raklami, A.; Symanczik, S.; Boutasknit, A.; Ait-El-Mokhtar, M.; Ben-Laouane, R.; Toubali, S.; Ait Rahou, Y.; et al. Biofertilizers as strategies to improve photosynthetic apparatus, growth, and drought stress tolerance in the date palm. *Front. Plant Sci.* **2020**, *11*, 516818. [CrossRef] [PubMed]
91. Aseri, G.K.; Jain, N.; Panwar, J.; Rao, A.V.; Meghwal, P.R. Biofertilizers improve plant growth, fruit yield, nutrition, metabolism and rhizosphere enzyme activities of pomegranate (*Punica granatum* L.) in Indian Thar Desert. *Sci. Hortic.* **2008**, *117*, 130–135. [CrossRef]
92. Cipriano, M.A.P.; Lupatini, M.; Lopes-Santos, L.; da Silva, M.J.; Roesch, L.F.W.; Destéfano, S.A.L.; Freitas, S.S.; Kuramae, E.E. Lettuce and rhizosphere microbiome responses to growth promoting *Pseudomonas* species under field conditions. *FEMS Microbiol. Ecol.* **2016**, *92*, fiw197. [CrossRef]
93. Niu, X.; Song, L.; Xiao, Y.; Ge, W. Drought-tolerant plant growth-promoting rhizobacteria associated with foxtail millet in a semi-arid and their potential in alleviating drought stress. *Front. Microbiol.* **2018**, *8*, 2580. [CrossRef] [PubMed]
94. Astorga-Eló, M.; Gonzalez, S.; Acuña, J.J.; Sadowsky, M.J.; Jorquera, M.A. *Rhizobacteria* from 'flowering desert' events contribute to the mitigation of water scarcity stress during tomato seedling germination and growth. *Sci. Rep.* **2021**, *11*, 13745. [CrossRef]
95. Milet, A.; Chaouche, N.K.; Dehimat, L.; Kaki, A.A.; Mounira, K.A.; Philippe, T. Flow cytometry approach for studying the interaction between *Bacillus mojavensis* and *Alternaria alternata*. *Afr. J. Biotechnol.* **2016**, *15*, 1417–1428. [CrossRef]
96. Xie, Z.; Chu, Y.; Zhang, W.; Lang, D.; Zhang, X. *Bacillus pumilus* alleviates drought stress and increases metabolite accumulation in *Glycyrrhiza uralensis* Fisch. *Environ. Exp. Bot.* **2019**, *158*, 99–106. [CrossRef]

97. Li, Y.; Shi, H.; Zhang, H.; Chen, S. Amelioration of drought effects in wheat and cucumber by the combined application of super absorbent polymer and potential biofertilizer. *PeerJ* **2019**, *7*, e6073. [CrossRef]
98. Asghari, B.; Khademian, R.; Sedaghati, B. Plant growth promoting rhizobacteria (PGPR) confer drought resistance and stimulate biosynthesis of secondary metabolites in pennyroyal (*Mentha pulegium* L.) under water shortage condition. *Sci. Hortic.* **2020**, *263*, 109132. [CrossRef]
99. Calvo-Polanco, M.; Sánchez-Romera, B.; Aroca, R.; Asins, M.J.; Declerck, S.; Dodd, I.C.; Martínez-Andújar, C.; Albacete, A.; Ruiz-Lozano, J.M. Exploring the use of recombinant inbred lines in combination with beneficial microbial inoculants (AM fungus and PGPR) to improve drought stress tolerance in tomato. *Environ. Exp. Bot.* **2016**, *131*, 47–57. [CrossRef]
100. Behrooz, A.; Vahdati, K.; Rejali, F.; Lotfi, M.; Sarikhani, S.; Leslie, C. Arbuscular mycorrhiza and plant growth-promoting bacteria alleviate drought stress in walnut. *HortScience* **2019**, *54*, 1087–1092. [CrossRef]
101. Kumar, M.; Mishra, S.; Dixit, V.; Kumar, M.; Agarwal, L.; Chauhan, P.S.; Nautiyal, C.S. Synergistic effect of *Pseudomonas putida* and *Bacillus amyloliquefaciens* ameliorates drought stress in chickpea (*Cicer arietinum* L.). *Plant Signal. Behav.* **2016**, *11*. [CrossRef] [PubMed]
102. Sandhya, V.; Ali, S.Z.; Grover, M.; Reddy, G.; Venkateswarlu, B. Effect of plant growth promoting *Pseudomonas* spp. on compatible solutes, antioxidant status and plant growth of maize under drought stress. *Plant Growth Regul.* **2010**, *62*, 21–30. [CrossRef]
103. Sarma, R.K.; Saikia, R. Alleviation of drought stress in mung bean by strain *Pseudomonas aeruginosa* GGRJ21. *Plant Soil* **2014**, *377*, 111–126. [CrossRef]
104. Roberson, E.B.; Firestone, M.K. Relationship between desiccation and exopolysaccharide production in a soil *Pseudomonas* sp. *Appl. Environ. Microbiol.* **1992**, *58*, 1284–1291. [CrossRef] [PubMed]
105. Mariotti, L.; Scartazza, A.; Curadi, M.; Picciarelli, P.; Toffanin, A. *Azospirillum baldaniorum* sp245 induces physiological responses to alleviate the adverse effects of drought stress in purple basil. *Plants* **2021**, *10*, 1141. [CrossRef]
106. Vettori, L.; Russo, A.; Felici, C.; Fiaschi, G.; Morini, S.; Toffanin, A. Improving micropropagation: Effect of *Azospirillum brasilense* Sp245 on acclimatization of rootstocks of fruit tree. *J. Plant Interact.* **2010**, *5*, 249–259. [CrossRef]
107. Heidari, M.; Mousavinik, S.M.; Golpayegani, A. Plant growth promoting Rhizobacteria (PGPR) effect on physiological parameters and mineral uptake in basil (*Ociumum basilicum* L.) under water stress. *ARPN J. Agric. Biol. Sci.* **2011**, *6*, 6–11.
108. Jones, P.D.; Lister, D.H.; Jaggard, K.W.; Pidgeon, J.D. Future climate impact on the productivity of sugar beet (*Beta vulgaris* L.) in Europe. *Clim. Chang.* **2003**, *58*, 98–103. [CrossRef]
109. Gauvry, E.; Mathot, A.-G.; Leguérinel, I.; Couvert, O.; Postollec, F.; Broussolle, V.; Coroller, L. Knowledge of the physiology of spore-forming bacteria can explain the origin of spores in the food environment. *Res. Microbiol.* **2016**, *168*, 369–378. [CrossRef]
110. Radhakrishnan, R.; Hashem, A.; Abd Allah, E.F. *Bacillus*: A biological tool for crop improvement through bio-molecular changes in adverse environments. *Front. Physiol.* **2017**, *8*, 667. [CrossRef] [PubMed]
111. Prajakta, B.M.; Suvarna, P.P.; Raghvendra, S.P.; Alok, R.R. Potential biocontrol and superlative plant growth promoting activity of indigenous *Bacillus mojavensis* PB-35 (R11) of soybean (*Glycine max*) rhizosphere. *SN Appl. Sci.* **2019**, *1*, 1143. [CrossRef]
112. Wang, C.; Guo, Y.; Wang, C.; Liu, H.; Niu, D.; Wang, Y.; Guo, J. Enhancement of tomato (*Lycopersicon esculentum*) tolerance to drought stress by plant-growth-promoting rhizobacterium (PGPR) *Bacillus cereus* AR. *J. Agric. Biotechnol.* **2012**, *20*, 1097–1105.
113. Lim, J.H.; Kim, S.D. Induction of drought stress resistance by multi-functional PGPR *Bacillus licheniformis* K11 in pepper. *Plant Pathol. J.* **2013**, *29*, 201–208. [CrossRef]
114. Kaushal, M.; Wani, S.P. Plant-growth-promoting rhizobacteria: Drought stress alleviators to ameliorate crop production in drylands. *Ann. Microbiol.* **2016**, *66*, 35–42. [CrossRef]
115. Arkhipova, T.N.; Prinsen, E.; Veselov, S.U.; Martinenko, E.V.; Melentiev, A.I.; Kudoyarova, G.R. Cytokinin producing bacteria enhance plant growth in drying soil. *Plant Soil* **2007**, *292*, 305–315. [CrossRef]
116. Kramer, S.; Green, D.M. Acid and alkaline phosphatase dynamics and their relationship to soil microclimate in a semiarid woodland. *Soil Biol. Biochem.* **2000**, *32*, 179–188. [CrossRef]
117. Omer, A.M. Bioformulations of *Bacillus* spores for using as biofertilizer biovalorization of olive mill waste water for the production of natural biofertilizers and antioxidants view project isolation and identification of fungi and bacteria from different Egyptian e. *Life Sci. J.* **2010**, *7*, 1097–8135.
118. O'Hara, G.W.; Boonkerd, N.; Dilworth, M.J. Mineral constraints to nitrogen fixation. *Plant Soil* **1988**, *108*, 93–110. [CrossRef]
119. Xu, X.; Du, X.; Wang, F.; Sha, J.; Chen, Q.; Tian, G.; Zhu, Z.; Ge, S.; Jiang, Y. Effects of potassium levels on plant growth, accumulation and distribution of carbon, and nitrate metabolism in apple Dwarf Rootstock seedlings. *Front. Plant Sci.* **2020**, *11*, 904. [CrossRef] [PubMed]
120. Avakyan, Z.A. Silicon compounds in solution bacteria quartz degradation. *Mikrobiologiya* **1984**, *54*, 301–307.
121. Malinovskaya, I.M.; Kosenko, L.V.; Votselko, S.K.; Podgorsky, V.S. The role of *Bacillus mucilaginosus* polysaccharide in the destruction of silicate minerals. *Mikrobiologiya* **1990**, *59*, 70–78.
122. Barnawal, D.; Maji, D.; Bharti, N.; Chanotiya, C.S.; Kalra, A. ACC deaminase-containing *Bacillus subtilis* reduces stress ethylene-induced damage and improves mycorrhizal colonization and rhizobial nodulation in *Trigonella foenum-graecum* under drought stress. *J. Plant Growth Regul.* **2013**, *32*, 809–822. [CrossRef]
123. Malusá, E.; Sas-Paszt, L.; Ciesielka, J. Technologies for beneficial microorganisms inocula used as biofertilizers. *Sci. World J.* **2012**, 1–12. [CrossRef]

124. Roberts, M.S.; Nakamura, L.K.; Cohan, F.M. *Bacillus mojavensis* sp. nov., distinguishable from *Bacillus subtilis* by sexual isolation, divergence in DNA sequence, and differences in fatty acid composition. *Int. J. Syst. Bacteriol.* **1994**, *44*, 256–264. [CrossRef] [PubMed]

125. Cao, L.; Qiu, Z.; You, J.; Tan, H.; Zhou, S. Isolation and characterization of endophytic *Streptomyces* strains from surface-sterilized tomato (*Lycopersicon esculentum*) roots. *Lett. Appl. Microbiol.* **2004**, *39*, 425–430. [CrossRef]

126. Moussa, T.A. Studies on biological control of sugar beet pathogen *Rhizoctonia solani* Kühn. *J. Biol. Sci.* **2002**, *2*, 800–804.

127. Snook, M.E.; Mitchell, T.; Hinton, D.M.; Bacon, C.W. Isolation and characterization of Leu 7-surfactin from the endophytic bacterium *Bacillus mojavensis* RRC 101, a biocontrol agent for *Fusarium verticillioides*. *J. Agric. Food Chem.* **2009**, *57*, 4287–4292. [CrossRef]

128. Romano-Armada, N.; Yañez-Yazlle, M.F.; Irazusta, V.P.; Rajal, V.B.; Moraga, N.B. Potential of bioremediation and PGP traits in *Streptomyces* as strategies for bio-reclamation of salt-affected soils for agriculture. *Pathogens* **2020**, *9*, 117. [CrossRef] [PubMed]

129. Hamedi, J.; Mohammadipanah, F. Biotechnological application and taxonomical distribution of plant growth promoting actinobacteria. *J. Ind. Microbiol. Biotechnol.* **2015**, *42*, 157–171. [CrossRef]

130. Chitraselvi, R.P.E. *Actinomycetes*: Dependable tool for sustainable agriculture. *Curr. Investig. Agric. Curr. Res.* **2018**, *1*, 128–130. [CrossRef]

131. Lawlor, K.; Knight, B.P.; Barbosa-Jefferson, V.L.; Lane, P.W.; Lilley, A.K.; Paton, G.I.; McGrath, S.P.; O'Flaherty, S.M.; Hirsch, P.R. Comparison of methods to investigate microbial populations in soils under different agricultural management. *FEMS Microbiol. Ecol.* **2000**, *33*, 129–137. [CrossRef] [PubMed]

132. Pierzynski, G.M.; Vance, G.F.; Sims, J.T. *Soils and Environmental Quality*; Taylor and Francis: Boca Raton, FL, USA, 2005.

133. Khamna, S.; Yokota, A.; Peberdy, J.F.; Lumyong, S. Indole-3-acetic acid production by *Streptomyces* sp. isolated from some Thai medicinal plant rhizosphere soils. *Eur. Asian J. Biosci.* **2010**, *4*, 23–32. [CrossRef]

134. El-Tarabily, K.A. Promotion of tomato (*Lycopersicon esculentum* Mill.) plant growth by rhizosphere competent 1-aminocyclopropane-1-carboxylic acid deaminase-producing *Streptomycete Actinomycetes*. *Plant Soil* **2008**, *308*, 161–174. [CrossRef]

135. Gopalakrishnan, S.; Srinivas, V.; Vidya, M.S.; Rathore, A. Plant growth-promoting activities of *Streptomyces* spp. in sorghum and rice. *Springerplus* **2013**, *2*, 1–8. [CrossRef]

136. El-Sayed, S.F.; Hassan, H.A.; El-Mogy, M.M. Impact of bio- and organic fertilizers on potato yield, quality and tuber weight loss after harvest. *Potato Res.* **2015**, *58*, 67–81. [CrossRef]

137. Wahyudi, A.T.; Priyanto, J.A.; Afrista, R.; Kurniati, D.; Astuti, R.I.; Akhdiya, A. Plant growth promoting activity of *Actinomycetes* isolated from soybean rhizosphere. *Online J. Biol. Sci.* **2019**, *19*, 1–8. [CrossRef]

138. Tredici, P. Del Nitrogen fixation: The story of the *Frankia* symbiosis. *Arnold Arbor.* **1995**, *55*, 26–31.

139. Treseder, K.K.; Berlemont, R.; Allison, S.D.; Martiny, A.C. Drought increases the frequencies of fungal functional genes related to carbon and nitrogen acquisition. *PLoS ONE* **2018**, *13*, e0206441. [CrossRef] [PubMed]

140. Treseder, K.K.; Lennon, J.T. Fungal traits that drive ecosystem dynamics on land. *Microbiol. Mol. Biol. Rev.* **2015**, *79*, 243–262. [CrossRef] [PubMed]

141. Jiménez-Arias, D.; García-Machado, F.J.; Morales-Sierra, S.; García-García, A.L.; Herrera, A.J.; Valdés, F.; Luis, J.C.; Borges, A.A. A beginner's guide to osmoprotection by biostimulants. *Plants* **2021**, *10*, 363. [CrossRef] [PubMed]

142. Hardoim, P.R.; van Overbeek, L.S.; Berg, G.; Pirttilä, A.M.; Compant, S.; Campisano, A.; Döring, M.; Sessitsch, A. The hidden world within plants: Ecological and evolutionary considerations for defining functioning of microbial endophytes. *Microbiol. Mol. Biol. Rev.* **2015**, *79*, 293–320. [CrossRef]

143. Verma, H.; Kumar, D.; Kumar, V.; Kumari, M.; Singh, S.K.; Sharma, V.K.; Droby, S.; Santoyo, G.; White, J.F.; Kumar, A. The potential application of endophytes in management of stress from drought and salinity in crop plants. *Microorganisms* **2021**, *9*, 1729. [CrossRef]

144. Bizos, G.; Papatheodorou, E.M.; Chatzistathis, T.; Ntalli, N.; Aschonitis, V.G.; Monokrousos, N. The role of microbial inoculants on plant protection, growth stimulation, and crop productivity of the olive tree (*Olea europea* L.). *Plants* **2020**, *9*, 743. [CrossRef]

145. Bona, E.; Lingua, G.; Manassero, P.; Cantamessa, S.; Marsano, F.; Todeschini, V.; Copetta, A.; D'Agostino, G.; Massa, N.; Avidano, L.; et al. AM fungi and PGP *Pseudomonads* increase flowering, fruit production, and vitamin content in strawberry grown at low nitrogen and phosphorus levels. *Mycorrhiza* **2015**, *25*, 181–193. [CrossRef]

146. Adamec, S.; Andrejiová, A. Mycorrhiza and stress tolerance of vegetables: A review. *Acta Hortic. Regiotect.* **2018**, *21*, 30–35. [CrossRef]

147. Marulanda, A.; Barea, J.M.; Azcón, R. Stimulation of plant growth and drought tolerance by native microorganisms (AM fungi and bacteria) from dry environments: Mechanisms related to bacterial effectiveness. *J. Plant Growth Regul.* **2009**, *28*, 115–124. [CrossRef]

148. Matias, S.R.; Pagano, M.C.; Muzzi, F.C.; Oliveira, C.A.; Carneiro, A.A.; Horta, S.N.; Scotti, M.R. Effect of rhizobia, mycorrhizal fungi and phosphate-solubilizing microorganisms in the rhizosphere of native plants used to recover an iron ore area in Brazil. *Eur. J. Soil Biol.* **2009**, *45*, 259–266. [CrossRef]

149. Dreischhoff, S.; Das, I.S.; Jakobi, M.; Kasper, K.; Polle, A. Local responses and systemic induced resistance mediated by ectomycorrhizal fungi. *Front. Plant Sci.* **2020**, *11*, 590063. [CrossRef] [PubMed]

150. De Battista, J.P.; Bacon, C.W.; Severson, R.; Plattner, R.D. Indole acetic acid production by the fungal endophyte of Tall Fescue. *Agron. J.* **1990**, *82*, 878–880. [CrossRef]

151. Khan, A.L.; Waqas, M.; Khan, A.R.; Hussain, J.; Kang, S.M.; Gilani, S.A.; Hamayun, M.; Shin, J.H.; Kamran, M.; Al-Harrasi, A.; et al. Fungal endophyte *Penicillium janthinellum* LK5 improves growth of ABA-deficient tomato under salinity. *World J. Microbiol. Biotechnol.* **2013**, *29*, 2133–2144. [CrossRef] [PubMed]
152. Ren, A.; Clayy, K. Impact of a horizontally transmitted endophyte, *Balansia henningsiana*, on growth and drought tolerance of *Panicum rigidulum. Int. J. Plant Sci.* **2009**, *170*, 599–608. [CrossRef]
153. Dastogeer, K.M.G.; Li, H.; Sivasithamparam, K.; Jones, M.G.K.; Wylie, S.J. Fungal endophytes and a virus confer drought tolerance to *Nicotiana benthamiana* plants through modulating osmolytes, antioxidant enzymes and expression of host drought responsive genes. *Environ. Exp. Bot.* **2018**, *149*, 95–108. [CrossRef]
154. Davies, F.T.; Potter, J.R.; Linderman Davies, R.G.; Ft, R.; Linderman, R. Drought resistance of mycorrhizal pepper plants independent of leaf P concentration-response in gas exchange and water relations. *Physiol. Plant.* **1993**, *87*, 45–53. [CrossRef]
155. Liao, X.; Chen, J.; Guan, R.; Liu, J.; Sun, Q. Two arbuscular mycorrhizal fungi alleviates drought stress and improves plant growth in *Cinnamomum migao* seedlings. *Mycobiology* **2021**, *49*, 396–405. [CrossRef]
156. Azad, K.; Kaminskyj, S. A fungal endophyte strategy for mitigating the effect of salt and drought stress on plant growth. *Symbiosis* **2016**, *68*, 73–78. [CrossRef]
157. Sun, C.; Johnson, J.M.; Cai, D.; Sherameti, I.; Oelmüller, R.; Lou, B. *Piriformospora indica* confers drought tolerance in Chinese cabbage leaves by stimulating antioxidant enzymes, the expression of drought-related genes and the plastid-localized CAS protein. *J. Plant Physiol.* **2010**, *167*, 1009–1017. [CrossRef]
158. Morsy, M.; Cleckler, B.; Armuelles-Millican, H. Fungal endophytes promote tomato growth and enhance drought and salt tolerance. *Plants* **2020**, *9*, 877. [CrossRef]
159. Subramanian, K.S.; Santhanakrishnan, P.; Balasubramanian, P. Responses of field grown tomato plants to arbuscular mycorrhizal fungal colonization under varying intensities of drought stress. *Sci. Hortic.* **2006**, *107*, 245–253. [CrossRef]
160. Bae, H.; Sicher, R.C.; Kim, M.S.; Kim, S.H.; Strem, M.D.; Melnick, R.L.; Bailey, B.A. The beneficial endophyte *Trichoderma hamatum* isolate DIS 219b promotes growth and delays the onset of the drought response in *Theobroma cacao. J. Exp. Bot.* **2009**, *60*, 3279–3295. [CrossRef]
161. Wu, Q.S.; Xia, R.X.; Zou, Y.N. Improved soil structure and citrus growth after inoculation with three arbuscular mycorrhizal fungi under drought stress. *Eur. J. Soil Biol.* **2008**, *44*, 122–128. [CrossRef]
162. Thirumalai, E.; Venkatachalam, A.; Suryanarayanan, T.S. Fungal endophytes of betel leaves: The need to study mycotoxin-producing endophytes in leafy vegetables. *Sydowia* **2020**, *73*, 83–88. [CrossRef]
163. Carbone, M.J.; Alaniz, S.; Mondino, P.; Gelabert, M.; Eichmeier, A.; Tekielska, D.; Bujanda, R.; Gramaje, D. Drought influences fungal community dynamics in the grapevine rhizosphere and root microbiome. *J. Fungi* **2021**, *7*, 686. [CrossRef]
164. Piri, R.; Moradi, A.; Balouchi, H.; Salehi, A. Improvement of cumin (*Cuminum cyminum*) seed performance under drought stress by seed coating and biopriming. *Sci. Hortic.* **2019**, *257*, 108667. [CrossRef]

horticulturae

Article

Investigating Evolution and Balance of Grape Sugars and Organic Acids in Some New Pathogen-Resistant White Grapevine Varieties

Tommaso Frioni *, Cecilia Squeri, Filippo Del Zozzo, Paolo Guadagna, Matteo Gatti, Alberto Vercesi and Stefano Poni

Department of Sustainable Crops Production, Università Cattolica del Sacro Cuore, 29122 Piacenza, Italy; cecilia.squeri@unicatt.it (C.S.); filippo.delzozzo@unicatt.it (F.D.Z.); paolo.guadagna@unicatt.it (P.G.); matteo.gatti@unicatt.it (M.G.); alberto.vercesi@unicatt.it (A.V.); stefano.poni@unicatt.it (S.P.)
* Correspondence: tommaso.frioni@unicatt.it; Tel.: +39-05-2359-9384

Citation: Frioni, T.; Squeri, C.; Del Zozzo, F.; Guadagna, P.; Gatti, M.; Vercesi, A.; Poni, S. Investigating Evolution and Balance of Grape Sugars and Organic Acids in Some New Pathogen-Resistant White Grapevine Varieties. *Horticulturae* **2021**, *7*, 229. https://doi.org/ 10.3390/horticulturae7080229

Academic Editors: Agnieszka Hanaka, Jolanta Jaroszuk-Ściseł and Małgorzata Majewska

Received: 13 July 2021
Accepted: 2 August 2021
Published: 6 August 2021

Publisher's Note: MDPI stays neutral with regard to jurisdictional claims in published maps and institutional affiliations.

Abstract: Breeding technologies exploiting marker-assisted selection have accelerated the selection of new cross-bred pathogen-resistant grapevine varieties. Several genotypes have been patented and admitted to cultivation; however, while their tolerance to fungal diseases has been the object of several in vitro and field studies, their productive and fruit composition traits during ripening are still poorly explored, especially in warm sites. In this study, five white pathogen-resistant varieties (PRV) listed as UD 80–100, Soreli, UD 30–080, Sauvignon Rytos, Sauvignon Kretos were tested over two consecutive seasons in a site with a seasonal heat accumulation of about 2000 growing degree days (GDDs), and their performances were compared to two *Vitis vinifera* L. traditional varieties, Ortrugo and Sauvignon Blanc. Berries were weekly sampled from pre-veraison until harvest to determine total soluble solids (TSS) and titratable acidity (TA) dynamics. All tested PRV exhibited an earlier onset of veraison and a faster sugar accumulation, as compared to Ortrugo and Sauvignon Blanc, especially in 2019. At harvest, Sauvignon Blanc was the cultivar showing the highest titratable acidity (8.8 g/L). Ortrugo and PRV showed very low TA (about 4.7 g/L), with the exception of Sauvignon Rytos (6.5 g/L). However, data disclose that Sauvignon Rytos higher acidity at harvest relies on higher tartrate (+1.1 to +2.2 g/L, as compared to other PRV), whereas in Sauvignon Blanc, high TA at harvest is due to either tartaric (+1 g/L, compared to PRV) and malic (+2.5 g/L, compared to PRV) acid retention. Overall, Sauvignon Rytos is the most suited PRV to be grown in a warm climate, where retaining adequate acidity at harvest is crucial to produce high-quality white wines. Nevertheless, canopy and ripening management strategies must be significantly adjusted, as compared to the standard practice employed for the parental Sauvignon Blanc.

Keywords: *Vitis* spp.; piwi cultivars; disease-resistant varieties; malic acid; ripening; fruit composition; downy mildew

1. Introduction

Warming trends are severely endangering viticulture and its sustainability [1,2]. In many wine regions, the incidence of drought and the rise of temperature are affecting vineyard performances by compromising vine physiology, compressing phenology, and boosting grapes' metabolism. In white varieties, especially when intended for sparkling wine, accelerated ripening leads to excessive sugars and inadequate acidity and aromas at harvest, resulting in unbalanced wines [1,3]. Moreover, the advancement of veraison increases the susceptibility of grapes to sunburn and dehydration phenomena by exposing ripening berries to the hottest days of the year, when the evaporative demand is maximum [2]. As a consequence, several wine regions are seeking either new cultural practices viable to decompress sugar accumulation and acidity decrease, or late-ripening cultivars, especially when the target hits white and/or sparkling wines [1,4–6]. In addition, contrary

29

to what one might think, fungal diseases are a serious issue also in warm and hot regions. Pathogen spread varies according to the seasonal weather evolution; the increase of average temperatures broadens the time window available for pathogens to complete additional biological cycles within one year [7]. Consequently, numerous pesticides applications are still frequently needed during the season to control downy mildew (DM) and powdery mildew (PM), even in warm climates. For instance, in northern Italy, where average heat accumulation easily exceeds >2000 growing degree days (GDDs), several pesticides residues have been detected in the groundwater [8].

Pathogen-resistant grapevine varieties (PRV) are complex interspecific hybrids obtained by multiple crossings of resistant *Vitis* spp. accessions (mostly *V. amurensis*, *V. rupestris*, and *V. berlandieri*) with selected cultivars of *Vitis vinifera* which, conversely, lacks genetic resistance towards *Plasmopara viticola* (Pv) and *Erisyphae necator* (En). In most Mediterranean grape growing regions, adoption of PRV has been hindered for many decades. Hindrance to PRV was not merely due to their botanical origins, yet to the fact that non-*vinifera* genotypes often carry undesired or atypical flavour, and wines are not appreciated [9–12].

Over the last decades, however, the increasing concerns of consumers about ecological issues and product safety, the rapid surge of organic (or environmentally certified) wines, and the spread of new awareness and values in modern society, have led to a renovated interest in PRV. Moreover, new breeding technologies have opened up new frontiers [9,11]. For instance, in Italy, a few breeding programmes have originated new pathogen-resistant varieties by marker-assisted selection (MAS) sorted out within progenies obtained by crossing *Vitis vinifera* cultivars with complex hybrids having genetic resistance to DM and PM. MAS has considerably shortened the time needed to identify the crossings carrying one or more quantitative trait loci (QTL), inducing higher tolerance to Pv and En. Within this pool of individuals, the genotypes having the most similar must biochemical composition to the *V. vinifera* parentals were identified. Currently, several selected resistant cultivars are available for growers [11,13–15]. At the same time, also national and European regulations changed and, from 2009 to date, many hybrids PRV have been admitted for cultivation.

A fairly high number of papers have described the tolerance to DM and PM of these new PRV under various environmental conditions. Depending upon the number and types of QTLs, these PRV exhibit a different degree of tolerance to Pv and En; nonetheless, these varieties have been confirmed to be more tolerant than traditional *V. vinifera* [9,14–19].

Conversely, the literature shows a paucity of scientific papers evaluating the productive performances and fruit composition of such new PRV. Poni et al. [20] have reported that potted vines of the red resistant accession UD 72–096 (Sangiovese x Bianca) had comparable yield with a standard susceptible Sangiovese clone but, when harvested on the same date, UD 72–096 showed significantly higher TSS and TA than Sangiovese. Moreover, UD 72–096 grapes had higher concentrations of skin acylated and coumarated anthocyanins, lower mono-glucosidic, and higher di-glucosidic anthocyanin forms, as well as lower quercetin 3-O-glucoside concentration, as compared to the susceptible Sangiovese. In north-eastern Spain, the resistant genotype Sauvignon Kretos (Sauvignon Blanc x Kozma 20-3) exhibited a similar yield and must composition to the parental Sauvignon Blanc. In this case, the two varieties were picked at different dates, but the date of harvest of Sauvignon Blanc was not provided [21].

To the best of our knowledge, no other scientific papers have tested fruit ripening kinetics or enological parameters of new PRV. Several technical reports suggest that these varietals are poor in di-glucoside anthocyanin forms and free from furaneol and other undesired volatile compounds typical of non-*vinifera* cultivars [22,23]. Analytical and sensorial traits of musts and wines are overall promising and confirm that these varietals, if adequately managed in the field and the winery, can produce wines comparable to those obtained from the *V. vinifera* parentals [22–25].

A shared result of these technical reports is the earlier annual cycle of PRV [22]. The high heritability of earliness traits, when crossing grapevine cultivars, is something well known, and some PRV originating from pioneer breeding programmes, such as Bronner, Solaris, or Souvignier Gris, show accelerated ripening and fast sugar accumulation [9,22,26–31]. This can be linked to the locations and the times of the selection of these genotypes, obtained by breeding programmes set in cool climates, and well before warming trends affected Mediterranean wine regions. PRV introduction in warm regions, where global warming has already caused a significant compression of the annual growth cycle and a consistent advancement of phenological stages, seems to be an additional matter of concern.

The aim of this work was to evaluate vine performances and fruit-ripening dynamics of five new white PRV, in a region where local viticulture is suffering the negative effects of warming trends on grape biochemical balance. The resistant genotypes were compared to Sauvignon Blanc and Ortrugo, two of the most cultivated white *Vitis vinifera* genotypes in the area. Our hypothesis was that the tested PRV might have a different degree of suitability to warm and hot sites according to three specific traits, namely, (i) onset of veraison time, (ii) malic acid degradation rates, and (iii) maintenance of a minimum acid pool at harvest.

2. Materials and Methods

2.1. Experimental Site and Treatment Layout

The study was carried out for two years (2019–2020) in a varietal collection located at Vicobarone (Ziano Piacentino, Italy, 44°59′31.7″ N 9°21′27.8″ E, 268 m a.s.l.). In the vineyard, 5 recently obtained white PRV were planted in 2016. The PRV present in the collection were obtained by crossing a *V. vinifera* parental (namely Sauvignon Blanc or Friulano) with an interspecific *Vitis* hybrid (namely, Kozma 20-3 or Bianca) able to confer resistance to PM and DM [32,33]. The five PRV were UD 80–100 (Friulano x Bianca), Soreli (Friulano x Kozma 20-3), UD 30–080 (Sauvignon Blanc x Kozma 20-3), Sauvignon Kretos (Sauvignon Blanc x Kozma 20-3), and Sauvignon Rytos (Sauvignon Blanc x Bianca) (Figure S1). To date, Sauvignon Rytos, Sauvignon Kretos, and Soreli are already admitted for cultivation in Italy, whereas UD 80–100 and UD 30–080 are still under evaluation. Ortrugo (VCR245) and Sauvignon Blanc (R3) vines, planted in the same year nearby these 5 PRV, were selected as *V. vinifera* references. Ortrugo was chosen since it is the most common white variety in the region and also is susceptible to summer temperatures in terms of fast acidity loss [6]. Sauvignon Blanc was instead included in the study since it was one of the parentals of three PRV among the five planted in the vineyard. All the cultivars, grafted on SO4 rootstock, were planted at 2.4 m × 0.8 m spacing (between row and within row distance, respectively) for a resulting density of 5125 plants/hectare. The vineyard has a soft slope (about 6°) and an east-facing aspect, with rows following E–W orientation. Vines were trained to a unilateral Guyot with about 10 nodes on the primary horizontal cane and two more on a spur left for annual cane renewal.

Each cultivar was present in one row of 80 m, encompassing 100 vines. The vineyard was divided into three uniform blocks along the rows. Nine test vines per varietal (three vines per cultivar per block) were randomly chosen in 2019, tagged, and then maintained also for the following season. These selected vines were used for detailed assessment of vegetative growth, yield components, and grape composition at harvest. The vineyard is typically non-irrigated, whereas fertilisation was uniform across all the vineyard surfaces and conducted based on local sustainable practices. Vines were trimmed once shoots outgrew 20 cm above the top wire. In order to prevent the spread of pathogens on both PRV and reference varieties, control of diseases was differentiated based on the degree of tolerance of the different genotypes. Details of pest management layout are provided in Table S1. In both seasons, none of the experimental vines showed symptoms of DM and PM at harvest. The minimum, mean, and maximum daily air temperature (°C) and daily rainfall (mm) from 1 January (DOY 1) to 31 December (DOY 365) were recorded in each season by a nearby weather station. Cumulative GDDs were then calculated according to Winkler [34].

2.2. Phenological Stages, Vegetative Growth, and Yield Components

In both years, bud break (BBCH09) and the onset of veraison (BBCH81) were assessed on each tagged vine according to Lorenz et al. [35]. Each season, in late spring (end of May–beginning of June), the number of inflorescences bore on each shoot was recorded according to the position of the shoot onto the horizontal cane. Total vine fruitfulness was then calculated as total inflorescences/total shoots ratio for the entire vine and for basal nodes (base node + count nodes 1 and 2).

All the experimental vines were picked on the same day, when a berry total soluble solids concentration of about 20 °Brix was achieved for Ortrugo, according to the optimal ripening threshold for this cultivar identified by Gatti et al. [36]. At harvest, tagged vines were individually picked, the mass of grapes was weighed, and the total bunch number per vine was counted. The average bunch weight was then calculated. Concurrently, three representative bunches per vine, usually inserted on basal, median, and apical cane portions, were taken to the laboratory. On each bunch, the number of berries was counted, and the mass of berries was weighed. Rachis length was measured, and bunch compactness was expressed as the ratio of total berry mass to rachis length. Berries were crushed, and the obtained must was then used for technological maturity and organic acids determination (see next paragraphs).

At harvest, the leaves inserted at nodes 3, 6, 9, 12, 15, 18, and 21 of the distal shoot of each tagged vine were collected with all the leaves from two lateral shoots developing below the trimming cut. The area of each leaf was measured with an LI-3000A leaf area meter (LI-COR Biosciences, Lincoln, NE, USA). Immediately after leaf fall, the number of nodes per cane and the number of nodes of each lateral cane were counted. The final leaf area was then estimated from the main and lateral shoots per vine on the basis of node counts and leaf-blade areas. Total vine leaf area was calculated as a sum of the two components. Leaf area to yield ratio (LA/Y) was finally calculated by dividing the total leaf area and yield of each tagged vine.

2.3. Grape Composition

Each year, from veraison (TSS ~4.5 to 5 °Brix) until harvest, three 100-berry samples were taken weekly from untagged vines of each varietal. These samples were not taken from the tagged vines so that the natural dynamic of grape ripening would not be altered due to the progressive reduction of the pending yield. During sampling, it was assured that the removed berries were taken from bunches located on both sides of the row and, within each bunch, the top, median, and bottom portions were also represented. In 2020, untagged Ortrugo and Sauvignon Blanc were not picked, and two additional post-harvest samplings were conducted. Sampled berries were brought to the laboratory, weighed, and crushed to obtain a must. Musts were analysed immediately for TSS using a temperature-compensated desk refractometer, whereas pH and TA were measured by titration with 0.1 N NaOH to a pH 8.2 end point and expressed as g/L of tartaric acid equivalents. TSS/TA ratio at harvest was then calculated.

TSS accumulation rates (°Brix day^{-1}) were calculated from pre-veraison to harvest dividing the difference in TSS between two subsequent samplings by the number of elapsed days. The same procedure was carried out based on malic acid concentration in order to calculate degradation rates (g/L day^{-1}).

For better readability of data, when appropriate, seasonal trends and correlations of specific parameters were graphed separately for PRV having Sauvignon Blanc as *V. vinifera parental* (UD 30–080, Sauvignon Kretos, and Sauvignon Rytos, compared with Sauvignon Blanc) and for PRV obtained crossing Friulano (UD 80–100 and Soreli, compared with Ortrugo).

2.4. HPLC Analysis

To assess tartaric and malic acid concentrations in all samples taken seasonally and at harvest, an aliquot of the must was diluted four times, then filtered through a 0.22 μm polypropylene syringe for high-performance liquid chromatography (HPLC) analysis and transferred to auto-sampler vials. All solvents were of HPLC grade. Water Milli-Q quality, acetonitrile, and methanol were obtained from VWR. L-(+)-tartaric acid and L-(-)-malic acid standards were purchased from Sigma-Aldrich. The chromatographic method was developed using an Agilent 1260 Infinity Quaternary LC (Agilent Technology) consisting of a G1311B/C quaternary pump with an inline degassing unit, G1329B autosampler, G1330B thermostat, G1316B thermostated column compartment, and a G4212B diode array detector (DAD) fitted with a 10 mm path, 1 µL volume Max-Light cartridge flow cell. The instrument was controlled using the Agilent Chemstation software version A.01.05. The organic acids analysis used an Allure Organic Acid Column, 300×4.6 mm, 5 µm (Restek). Separation was performed in isocratic conditions using water, pH-adjusted to 2.5 using ortho-phosphoric acid, at a flow rate of 0.8 mL/min. The column temperature was maintained at 30 ± 0.1 °C, and 15 µL of the sample was injected. The elution was monitored at 200 to 700 nm and detected by UV–Vis absorption with DAD at 210 nm. Organic acids were identified using authentic standards, and quantification was based on peak areas and performed by external calibration with standards. The ratio between tartaric acid and malic acid (HT/HM) was then calculated.

2.5. Statistical Analysis

Vine performance data were subjected to a two-way analysis of variance (ANOVA) using IBM SPSS 25 (IBM, Chicago, IL, USA). Treatment comparison was performed using the Student–Neuman–Keuls test at $p \leq 0.05$. Year $\times$ treatment interaction was partitioned only when the F test was significant.

Repeated measures of the same parameters (TSS, pH, TA, tartaric acid, malic acid) taken at different dates along the season were analysed with the repeated-measure analysis of variance (ANOVA) routine embedded in IBM SPSS Statistics 25. Equality of variances of the differences between all possible pairs of within-subject conditions was assessed via Mauchly's sphericity test. The least squared (LS) mean method at $p \leq 0.05$ was used for multiple comparisons within dates.

Data about grapes TSS, TA, malic acid, and tartaric acid progression were also re-elaborated to predict fruit composition at the thresholds of TSS = 20 °Brix and, separately, TA = 7.5 g/L, for all the tested varieties. The balance among malate, tartrate, and TA or TSS for grapes harvested at those thresholds was compared by a one-way analysis of variance (ANOVA) using IBM SPSS 25 (IBM, Chicago, IL, USA).

The correlations existing between TSS and TA values of grapes sampled during the season were subjected to regression analysis, using SigmaPlot 11 (Systat Software Inc., San Jose, CA, USA).

3. Results

3.1. Weather Conditions and Phenology

At the experimental site, 2019 was marked by a wet and cold spring (Figure 1A). From 1 April to the end of May, 274 mm of rain and only 260 GDDs were recorded, vs. 62 mm and 369 GDDs recorded in the same period of 2020 (Figure 1B). Then, summer 2019 was conversely hotter than summer 2020, so that at the end of August, a comparable amount of GDDs were accumulated from 1 April in the two years (1568 GDDs in 2019, 1571 GDDs in 2020). From 1 April to 31 October, 2007 GDDs were accumulated in 2019 vs. 1917 GDDs in 2020.

All the genotypes tested in the trial had a similar bud-break date (BBCH09) in both seasons, with cv. Ortrugo showing a delay of a few days, as compared to other varieties (Table 1). Sauvignon Kretos was the cultivar showing the earliest onset of veraison (BBCH81) on both seasons (between DOYs 195–197). The onset of veraison was anticipated in 2020 in Sauvignon Rytos, Sauvignon Blanc, and Ortrugo by 8 days, compared to the previous year, whereas other genotypes did not exhibit a similar variation between seasons.

Considering only resistant varieties, UD 30–080 was the cultivar exhibiting, in both years, the later onset of veraison.

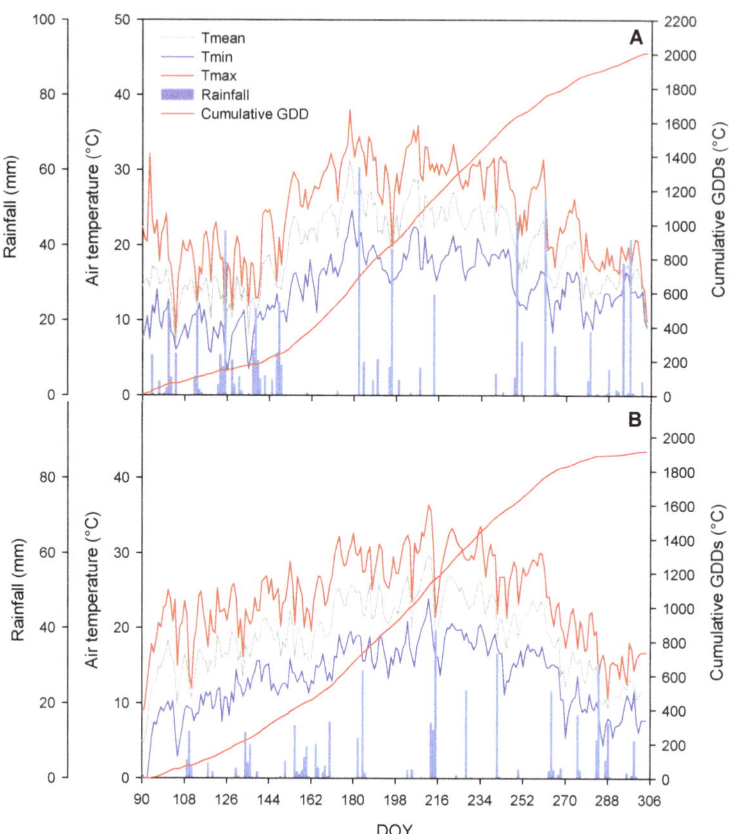

Figure 1. Seasonal daily trends of minimum temperature (Tmin), mean temperature (Tmean), maximum temperature (Tmax), rainfall (blue bars), and heat accumulation (cumulative GDDs) calculated following Winkler [34] in 2019 (**A**) and 2020 (**B**). DOY = day of the year.

Table 1. Date of achievement of main phenological stages and key ripening thresholds for five pathogen-resistant grapevine genotypes and two non-resistant *Vitis vinifera* cultivars in 2019 and 2020.

	UD 80–100	Soreli	UD 30–080	Cultivar Sauvignon Kretos	Sauvignon Rytos	Sauvignon Blanc	Ortrugo
2019				(DOY) [1]			
BBCH09 [2]	92	90	92	93	92	92	94
BBCH81 [2]	203	203	210	197	210	217	217
TSStr [3]	224	231	231	224	231	252	245
Matr [3]	224	224	231	217	231	252	238
2020							
BBCH09 [2]	90	90	92	92	92	92	95
BBCH81 [2]	202	202	209	195	202	209	209
TSStr [3]	213	223	237	213	223	252	252
Matr [3]	223	213	223	213	213	252	223

[1] DOY = day of the year. [2] Phenological stages were identified according to Lorenz et al. [35]. [3] TSStr = achievement of grapes total soluble solids threshold of 20 ± 1 °Brix; MAtr = achievement of grapes malic acid threshold of 2.5 ± 0.5 g/L.

3.2. Vegetative Growth and Shoot Fruitfulness

UD 30–080 had a significantly higher main shoot leaf area/vine at the end of the season, as compared to Sauvignon Rytos and UD 80–100, with other genotypes scoring intermediate values (Table 2). Lateral shoot leaf area/vine was unaffected by the genotype, so that differences in total vine leaf area mostly tracked those in the main shoot leaf area.

Table 2. Vegetative growth and shoot fruitfulness of five pathogen-resistant grapevine genotypes and two non-resistant *Vitis vinifera* cultivars in 2019 and 2020.

Cultivar	Main Shoots Leaf Area	Lateral Shoots Leaf Area	Vine Total Leaf Area	Shoot Fruitfulness	Shoot Fruitfulness Nodes 0–2
	(m^2/Vine)	(m^2/Vine)	(m^2/Vine)	(N. Inflorescences/Shoot)	(N. Inflorescences/Shoot)
UD 80–100	2.38 b [2]	0.35	2.73 ab	1.36 b	1.16 ab
Soreli	2.96 ab	0.32	3.27 ab	1.76 a	1.51 a
UD 30–080	3.17 a	0.24	3.41 a	1.47 b	1.20 ab
Sauvignon Kretos	3.01 ab	0.23	3.24 ab	1.58 ab	1.19 ab
Sauvignon Rytos	2.27 b	0.12	2.39 b	1.73 a	1.44 a
Sauvignon Blanc	3.12 ab	0.45	3.55 a	1.23 bc	0.75 b
Ortrugo	2.95 ab	0.39	3.34 a	0.93 c	0.33 c
2019	3.12	0.38 a	3.50 a	1.62 a	1.26 b
2020	2.58	0.24 b	2.82 b	1.17 b	1.02 a
V [1]	** [3]	ns	*	***	***
Y	ns	*	**	***	***
V×Y	ns	ns	**	**	ns

[1] V = variety; Y = year. [2] Means within columns noted by different letters are different by SNK test. [3] *, ** and *** indicate significant difference per $p \leq 0.05$, $p \leq 0.01$ and $p \leq 0.001$, respectively. ns: not significant.

Soreli and Sauvignon Rytos exhibited the highest shoot fruitfulness (1.76 and 1.73 inflorescences per shoot, respectively). UD 80–100 and UD 80–030 showed a slightly lower shoot fruitfulness (Table 2), with Sauvignon Kretos setting at intermediate levels. Sauvignon Blanc and Ortrugo were the two cultivars having the lowest number of inflorescences per shoot (1.23 and 0.93, respectively). This trend was substantially confirmed when considering only basal nodes' fruitfulness (i.e., base bud until count node 2).

3.3. Yield, Bunch Morphology, and Vine Balance

Harvest was performed on DOY 245 in 2019 and on DOY 238 in 2020. Soreli was the variety having the highest yield per vine (3.53 kg). All the remaining cultivars exhibited a significantly lower yield (from 2.08 to 2.43 kg/vine), with the sole exception of Sauvignon Blanc (Table 3). The high productivity of Soreli was associated with the high number of bunches per vine (30). However, the number of bunches per vine in Sauvignon Rytos and Sauvignon Kretos was not significantly lower than Soreli; rather, these two genotypes had a lower bunch weight and number of berries per bunch (−29% and −23%, respectively). Ortrugo was the variety with the lower number of bunches per vine (13), paralleled by the highest bunch weight (181 g) and number of berries per bunch (170). Sauvignon Rytos had small bunches (85 g) with a low number of berries per bunch (83). All PRV had medium bunch compactness, significantly lower than Ortrugo.

UD 30–080 showed the highest LA/Y ratio (1.64 m^2/kg), whereas all the other cultivars had a lower LA/Y ratio (from 0.92 to 1.33 m^2/kg), except for Ortrugo, which set at intermediate levels (1.42 m^2/kg).

Table 3. Yield, vine balance, and bunch morphology of five pathogen-resistant grapevine genotypes and two non-resistant *Vitis vinifera* cultivars in 2019 and 2020.

Cultivar	Yield	Bunches Per vine	Leaf Area to Yield Ratio	Bunch Weight	Berry Mass	Berries Per Bunch	Bunch Compactness
	(kg/vine)	(n.)	(m^2/kg)	(g)	(g)	(n.)	(g/cm)
UD 80–100	2.31 b [2]	22 b	1.18 b	110 ab	1.17 ab	106 b	12.4 b
Soreli	3.53 a	30 a	0.92 b	120 ab	1.15 ab	104 b	11.1 b
UD 30–080	2.08 b	22 b	1.64 a	103 bc	1.36 a	76 c	10.4 b
Sauvignon Kretos	2.43 b	25 ab	1.33 b	107 ab	1.33 ab	80 c	10.1 b
Sauvignon Rytos	2.41 b	29 a	0.99 b	85 c	1.02 b	83 c	9.3 b
Sauvignon Blanc	2.86 ab	20 b	1.24 b	143 ab	1.45 a	99 bc	13.5 ab
Ortrugo	2.35 b	13 c	1.42 ab	181 a	1.06 b	170 a	17.1 a
2019	2.28 b	25 a	1.61	92 b	1.11 a	83 b	10.1 b
2020	2.82 a	20 b	1.40	141 a	1.32 b	126 a	13.3 a
V [1]	*** [3]	***	**	***	***	***	***
Y	***	***	ns	***	*	***	*
VxY	ns	ns	ns	**	**	*	ns

[1] V = variety; Y = year. [2] Means within columns noted by different letters are different by SNK test. [3] *, ** and *** indicate significant difference per $p \leq 0.05$, $p \leq 0.01$ and $p \leq 0.001$, respectively. ns: not significant.

3.4. Grapes' TSS, pH, and TA during Ripening

All PRV showed earlier berry sugar accumulation than Ortrugo and Sauvignon Blanc (Figure 2). In both years, Sauvignon Kretos had a significantly higher TSS than any of the other genotypes at any sampling date. UD 30–080 and Sauvignon Rytos had lower sugars than the other PRV right after veraison. However, later on, both genotypes showed a faster TSS accumulation rate, reducing the gap with other PRV. The threshold of 20 °Brix was achieved much earlier by the PRV in both years (Table 1). Overall, Sauvignon Blanc and Ortrugo lagged behind PRV by 15 days in 2019, and by approximately 10 days in 2020.

Similarly, must pH in PRV was constantly higher in both years (Figure 2C,D). In Sauvignon Rytos, only the dynamic of pH was more similar to the one of Ortrugo and Sauvignon Blanc, especially in 2020.

Figure 2E,F shows the early loss of TA by UD 80–100, Soreli and Sauvignon Kretos. In both years, Sauvignon Rytos maintained higher TA than other PRV until the end of the season. Ortrugo and Sauvignon Blanc showed a delayed acidity loss than Sauvignon Rytos, especially in 2019; however, Ortrugo crossed values lower than 5 g/L (on DOY 245 in 2019, on DOY 230 in 2020), whilst Sauvignon Blanc maintained a TA higher than 8 g/L until the end of the season. Over the last sampling dates, in 2019, Sauvignon Rytos tracked Ortrugo in terms of TA, whereas, in 2020, it set at intermediate levels between Ortrugo (and PRV) and Sauvignon Blanc.

All correlations between TSS and TA fit a quadratic model for any of the tested cultivars (Figure 3). The model shows that in 2019 Sauvignon Blanc, Sauvignon Kretos and UD 30–080 had similar TA for any TSS level, up to the threshold of 15 °Brix (Figure 3A). At higher TSS, Sauvignon Kretos and UD 30–080 had a TA lower than Sauvignon Rytos and Sauvignon Blanc. In 2020 (Figure 3B), conversely, Sauvignon Blanc had lower TA for any TSS below 15 °Brix, whereas, above this threshold, Sauvignon Blanc, Sauvignon Rytos, and Sauvignon Kretos grouped together and only UD 30–080 had lower TA. In both years, Ortrugo and Soreli had quite low TA for any TSS level, whereas, conversely, UD 80–100 exhibited the opposite behaviour (Figure 3C,D).

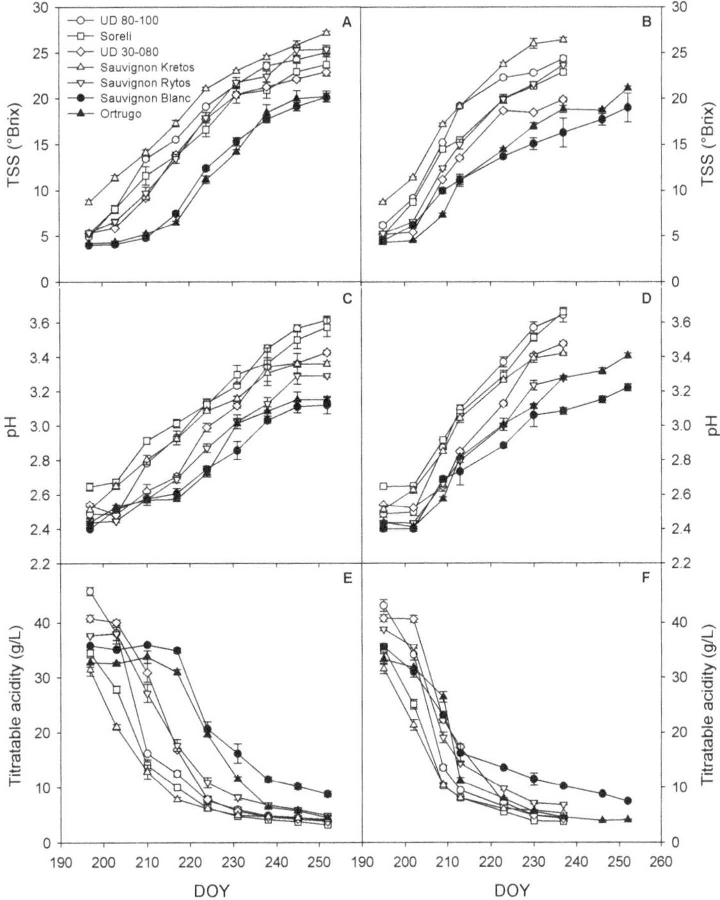

Figure 2. Seasonal evolution of grapes total soluble solids (TSS, panels **A,B**), pH (**C,D**) and titratable acidity (**E,F**) in 2019 (**A,C,E**) and 2020 (**B,D,F**) for 5 pathogens-resistant varieties (white symbols) and 2 reference *V. vinifera* cultivars (black symbols). Each point represents the average of three replicates ± SE. DOY = day of year.

3.5. Trends for Grapes Organic Acids Concentration

In both years, UD 80–100 was the variety showing the highest malic acid concentration pre-veraison (about 35 g/L). Sauvignon Kretos had the earliest decrease of malic acid, achieving the threshold of 2.5 g/L on DOY 217 in 2019 and on DOY 213 in 2020 (Figure 4A,B, Table 1). In 2019 all the PRV had an earlier peak of malic acid and onset of its degradation, as compared to Ortrugo and Sauvignon Blanc. In 2020, this was confirmed as related to Sauvignon Blanc, even if the gap was ostensibly narrower, whereas in Ortrugo the trend of malic acid degradation was comparable to the one exhibited by Soreli and Sauvignon Rytos. In both years, at the end of the season, Ortrugo reached a minimum malic acid concentration comparable to those of PRV (below 1 g/L), whereas Sauvignon Blanc maintained significantly higher malic acid concentrations (above 2.5 g/L).

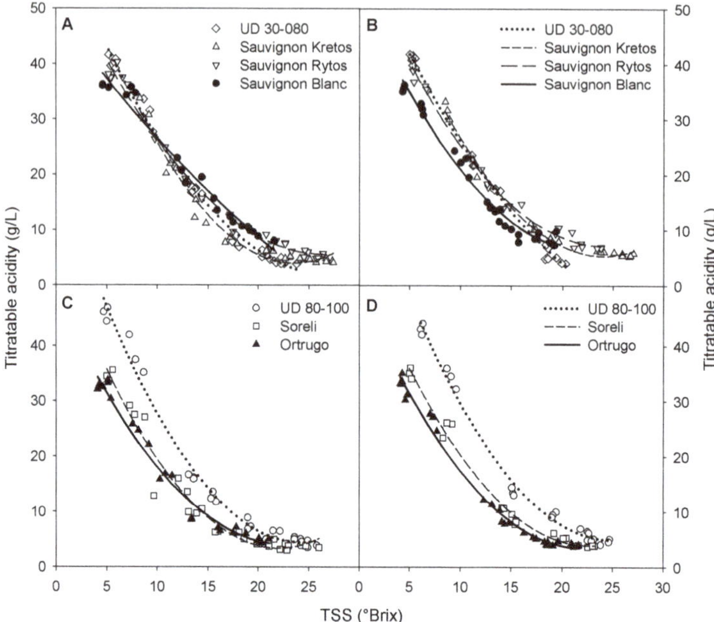

Figure 3. Seasonal variation of titratable acidity expressed as a function of total soluble solids (TSS) in 2019 (**A,C**), and 2020 (**B,D**), for 5 pathogens-resistant varieties (white symbols) and 2 reference *V. vinifera* cultivars (black symbols). Data were fit to the following equations: UD 30–080 2019 $y = 70.07 - 5.62x + 0.12x^2$, $R^2 = 0.994$; UD 30–080 2020 $y = 60.68 - 4.10x + 0.06x^2$, $R^2 = 0.996$; Sauvignon Kretos 2019 $y = 67.86 - 5.21x + 0.11x^2$, $R^2 = 0.976$; Sauvignon Kretos 2020 $y = 63.80 - 4.54x + 0.08x^2$, $R^2 = 0.991$; Sauvignon Rytos 2019 $y = 58.94 - 4.03x + 0.08x^2$, $R^2 = 0.991$; Sauvignon Rytos 2020 $y = 59.89 - 4.39x + 0.09x^2$, $R^2 = 0.993$; Sauvignon Blanc 2019 $y = 49.32 - 2.53x + 0.02x^2$, $R^2 = 0.971$; Sauvignon Blanc 2020 $y = 53.47 - 4.20x + 0.09x^2$, $R^2 = 0.980$; UD 80–100 2019 $y = 72.20 - 5.60x + 0.12x^2$, $R^2 = 0.989$; UD 80–100 2020 $y = 73.80 - 5.53x + 0.11x^2$, $R^2 = 0.992$; Soreli 2019 $y = 57.47.86 - 4.91x + 0.11x^2$, $R^2 = 0.966$; Soreli 2020 $y = 56.61 - 4.66x + 0.10x^2$, $R^2 = 0.989$; Ortrugo 2019 $y = 49.13 - 4.00x + 0.09x^2$, $R^2 = 0.983$; Ortrugo 2020 $y = 50.06 - 4.23x + 0.10x^2$, $R^2 = 0.991$. All the correlations listed were significant per $p < 0.05$.

In both years, Sauvignon Rytos had the highest grapes tartaric acid concentration during ripening, if considering only PRV (Figure 4C,D). In 2019, Ortrugo and Sauvignon Blanc had lower tartaric acid than Sauvignon Rytos after veraison, even if later in the season Sauvignon Blanc maintained significantly higher tartaric acid than Ortrugo and Sauvignon Rytos (approximately +3 g/L). In 2020, conversely, Sauvignon Blanc had a similar decrease of tartaric acid to the one exhibited by Ortrugo, whereas in Sauvignon, Rytos tartaric acid concentration was consistently higher (+3.3 g/L on DOY 237).

3.6. Sugar Accumulation and Malic Acid Degradation Rates

In 2019, UD-30–080, Sauvignon Kretos and Sauvignon Rytos exhibited relatively high TSS accumulation rates (Figure 5A), ranging between 0.40 and 0.65 °Brix day^{-1} until DOY 231. Conversely, Sauvignon Blanc jumped from 0.1 to 0.4 TSS day^{-1} on DOY 217 and, after peaking at 0.65 TSS day^{-1} on DOY 224, it decreased, together with the PRV. On the other hand, Ortrugo did not exceed 0.6 °Brix day^{-1}, although it maintained higher sugar accumulation rates late in the season (Figure 5C). In 2020 (Figure 5B,D), PRV had a higher peak of TSS accumulation rates (0.8–1.0 °Brix day^{-1}), as compared to the previous year, yet occurring approximately at the same DOYs (209–223). Conversely, in Sauvignon

Blanc and Ortrugo, maximum TSS accumulation rates occurred much earlier than in 2019 (approximately 15–20 days). Moreover, in Ortrugo, the maximum TSS accumulation rate, recorded on DOY 214, reached the peak of 1.4 °Brix day^{-1}, before declining to values comprised between 0.2 and 0.4 °Brix day^{-1} for the rest of the season.

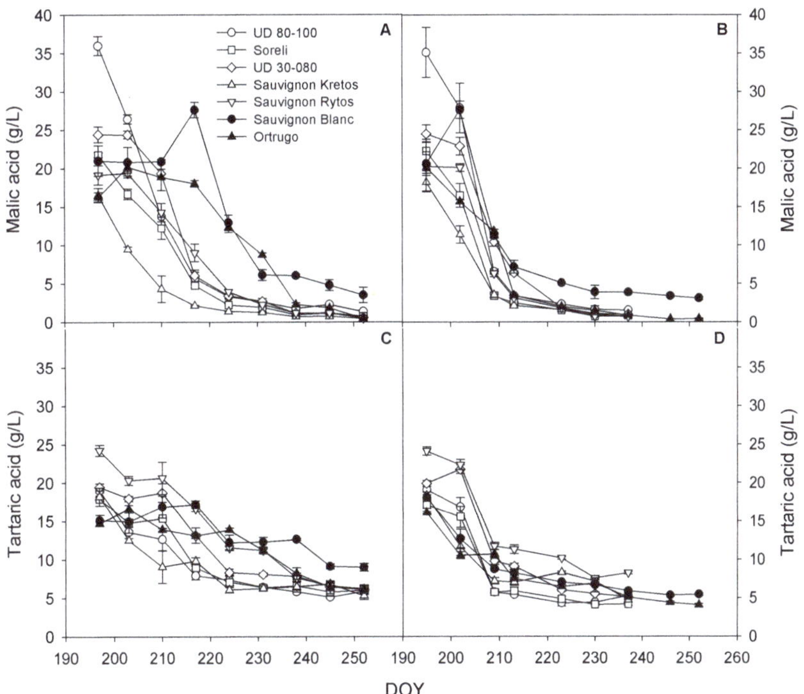

Figure 4. Seasonal evolution of grapes malic acid (panels **A,B**) and tartaric acid (**C,D**) in 2019 (**A,C**) and 2020 (**B,D**) for 5 pathogens-resistant varieties (white symbols) and 2 reference *V. vinifera* cultivars (black symbols). Each point represents the average of three replicates ± SE. DOY = day of year.

Sauvignon Kretos was also the cultivar showing the earliest peak of malic acid degradation rate, in both years (Figure 5G,H). In 2019, Sauvignon Blanc was the variety showing the most delayed onset of malic acid degradation, with a sudden peak of 2.1 g/L day^{-1} occurring only on DOY 224. UD 30–080 and Sauvignon Rytos set at intermediate levels between Sauvignon Blanc and Sauvignon Kretos, in terms of timing and magnitude of malic acid degradation rates increase.

In 2020, UD 30–080 and Sauvignon Rytos had a dynamic of the degradation of malic acid comparable to the one exhibited by Sauvignon Blanc, even if this latter peaked at 2.4 g/L day^{-1} vs. 1.6–2.0 g/L day^{-1} exhibited by UD 30–080 and Sauvignon Rytos. UD 80–100 and Soreli had a malic-acid-degradation-rate dynamic comparable to other PRV (Figure 5I,J). In 2020, UD 80–100 scored, on DOY 209, the highest malic acid degradation rate recorded during the experiment (3.1 g/L day^{-1}).

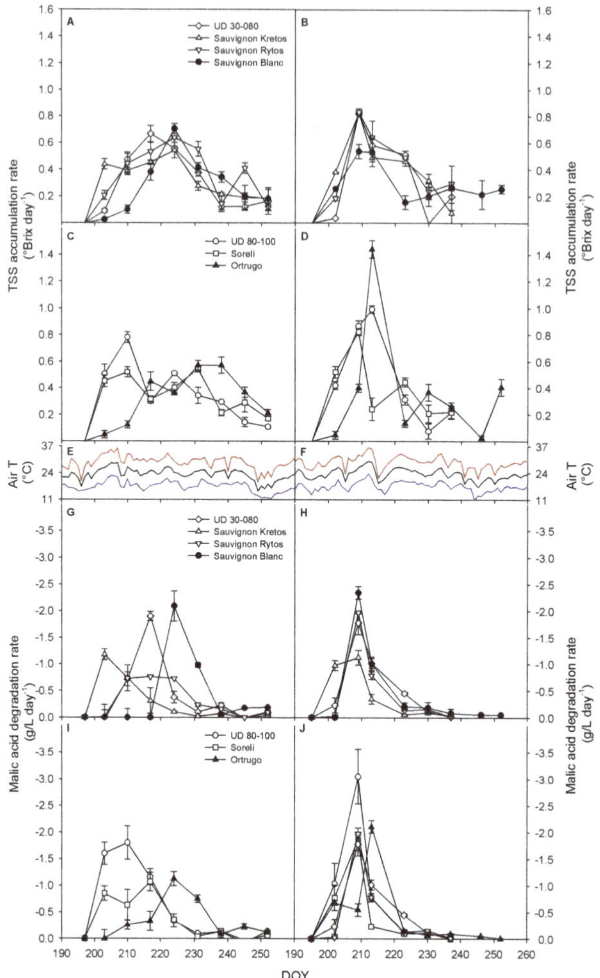

Figure 5. Seasonal evolution of grapes total soluble solids accumulation rates (panels **A–D**), malic acid degradation rate (panels **G–J**) and minimum temperature (blue line), mean temperature (grey line), and maximum temperatures (red lines) of the period (panels **E,F**) in 2019 (**A,C,E,G,I**) and 2020 (**B,D,F,H,J**) for 5 pathogens-resistant varieties (white symbols) and 2 reference *V. vinifera* cultivars (black symbols). Each point represents the average of three replicates ± SE. DOY = day of year. T = temperature.

3.7. Fruit Composition at Harvest

At harvest, all the PRV had higher TSS than Ortrugo and Sauvignon Blanc (from +2.5 °Brix in Soreli to +5.5 °Brix in Sauvignon Kretos), with the sole exception of UD 30–080 (Table 4). Sauvignon Blanc was the variety showing the higher TA at harvest (8.81 g/L). Ortrugo, UD 30–080, Soreli, UD 80–100 had quite low TA, comprised between 4.08 and 5.07 g/L. By contrast, Sauvignon Rytos maintained a TA of 6.50 g/L, halfway between this group of cultivars and Sauvignon Blanc. The high TA of Sauvignon Blanc is linked to a high malic acid concentration (3.45 g/L). Conversely, in Sauvignon Rytos, the malic acid concentration at harvest was similar to those of cultivars having low TA. Sauvignon Rytos had, instead, the highest tartaric acid concentration (7.44 g/L), together with Sauvignon Blanc.

Table 4. Fruit composition at harvest of five pathogen-resistant grapevine genotypes and two non-resistant *Vitis vinifera* cultivars in 2019 and 2020.

Cultivar	TSS	pH	TA	TSS/TA	Malic Acid	Tartaric Acid	HT/HM
	(°Brix)		(g/L)		(g/L)	(g/L)	
UD 80–100	22.6 ab [2]	3.43 a	5.07 c	4.46 a	1.38 b	5.34 bc	3.87 d
Soreli	22.0 b	3.42 a	4.08 d	5.39 a	0.90 c	4.67 c	5.19 c
UD 30–080	20.1 bc	3.32 ab	4.66 cd	4.29 ab	1.04 bc	5.35 bc	5.14 c
Sauvignon Kretos	24.5 a	3.33 ab	5.07 c	4.83 a	0.75 c	6.36 b	8.48 a
Sauvignon Rytos	23.4 ab	3.23 b	6.50 b	3.60 b	0.89 c	7.44 a	8.36 a
Sauvignon Blanc	19.0 c	3.06 c	8.81 a	2.16 c	3.45 a	7.61 a	2.21 e
Ortrugo	19.5 c	3.15 bc	4.95 cd	3.94 b	0.81 c	5.61 bc	6.93 b
2019	21.6	3.26	4.56	4.74	1.24	6.15	4.96
2020	21.4	3.35	5.59	3.83	1.39	5.92	4.26
V [1]	*** [3]	***	***	***	***	***	***
Y	ns	ns	ns	ns	ns	ns	ns
VxY	ns	ns	ns	ns	ns	ns	ns

[1] V = variety; Y = year. [2] Means within columns noted by different letters are different by SNK test. [3] *, ** and *** indicate significant difference per $p \leq 0.05$, $p \leq 0.01$ and $p \leq 0.001$, respectively. ns: not significant.

Sauvignon Blanc was the variety scoring the lowest TSS/TA and HT/HM ratio (2.16 and 2.21, respectively). Sauvignon Rytos had a TSS/TA ratio of 3.60, comparable to Ortrugo, and an HT/HM of 8.36, similar to Sauvignon Kretos (8.48). Ortrugo had a lower HT/HM than Sauvignon Rytos and Sauvignon Kretos (6.93), yet higher than all the other PRV.

Figure 6A shows that, if the tested varieties had been harvested at TSS = 20 °Brix, the acid pool of Sauvignon Rytos would be mostly contributed by tartaric acid, whereas the acidity of Sauvignon Blanc would be also driven by higher malic acid. Sauvignon Kretos and UD 30–080 would exhibit comparable tartaric acid to Sauvignon Blanc but lower malic acid, resulting in a lower TA. Figure 6B shows the low TA of Ortrugo and Soreli at 20 °Brix, linked to a very low malic acid concentration and relatively low tartaric acid level. Conversely, UD 80–100 would have retained a good balance between organic acids and total acids.

Simulated harvest at TA of 7 g/L reveals that both Sauvignon Kretos and Sauvignon Rytos would have shown higher TSS than Sauvignon Blanc, but the TA would be again linked to high tartaric acid and low malic acid, compared to Sauvignon Blanc (Figure 6C). Ortrugo and Soreli would exhibit instead very low TSS, whereas UD 80–100 shows again a good balance among TSS, malic acid, and tartaric acid (Figure 6D).

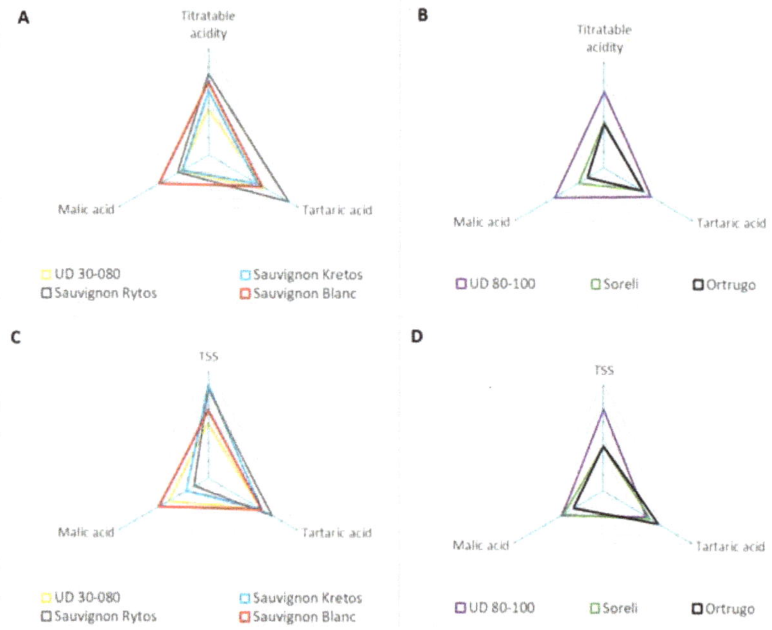

Figure 6. Balance among titratable acidity and organic acids in grapes of 5 pathogens-resistant varieties and 2 reference *V. vinifera* cultivars, at the total soluble solids threshold of about 20 °Brix (panels **A,B**) and balance among total soluble solids and organic acids in grapes of the same varieties, at the titratable acidity threshold of about 7 g/L (panels **C,D**). Scaling for titratable acidity and tartaric acid is 0–12 g/L, for malic acid is 0–6 g/L, and for total soluble solids is 12–24 °Brix. Data pooled over two seasons.

4. Discussion

Compared to Ortrugo and Sauvignon Blanc (Table 2), all tested PRV had a medium-to-high yield per vine linked to a higher shoot fruitfulness. Nowadays, remunerative productivity is a necessary requirement for new genotypes and it should be associated with desired grape quality and low to moderate susceptibility to biotic and abiotic stress [2]. Moreover, high basal bud fruitfulness allows the implementation of spur-pruning systems prone to the mechanisation of winter operations [2,37,38].

Although bud break of all tested genotypes occurred within a quite narrow time window (i.e., 4–5 days), the unseasonal low temperatures recorded in 2019 between DOYs 100–160 likely pushed back veraison in Ortrugo and Sauvignon Blanc, whereas PRV were substantially unaffected by this weather trend. This suggests that PRV have lower requirements in terms of heat accumulation to shift from vegetative activity to a prioritised reproductive activity. This trait could indicate, at the same time, a weak point or an advantage. In fact, if the advancement of phenological stages might indeed increase grapes susceptibility to sunburn and biochemical unbalances in several traditional wine districts [1,39], the PRV early veraison observed in 2019 renders these varieties more suitable to exploit cool climates or areas located at higher altitudes [40–42].

Overall, TSS accumulation had an earlier onset and faster pace in PRV varieties, as compared to Ortrugo and Sauvignon Blanc. The two reference varieties lagged behind PRV by 10 to 20 days in terms of TSS accumulation (Figure 2A,B), and sugar accumulation never caught up to the final TSS of any of the PRV in both seasons. Moreover, the time interval when TSS accumulation rates were higher than 0.3 °Brix day^{-1} was much longer in PRV than in Ortrugo and Sauvignon Blanc, especially in 2020 (Figure 5A–D). While Ortrugo and

Sauvignon Blanc are well known for their slow sugar accumulation and low maximum sugar thresholds [42,43], such fast sugar accumulation dynamic exhibited especially by Sauvignon Kretos and UD 80–100 could easily lead to excessive wine alcohol content and unbalanced aroma, especially in white wines [1]. Our results are overall similar to the findings of Poni et al. [20], who compared Sangiovese and UD 72–096 (a Sangiovese x Bianca PRV).

Interestingly, the fostered sugar accumulation found in Sauvignon Kretos and some others PRV was not associated with a higher LA/Y ratio when compared to Sauvignon Blanc and Ortrugo. The only variety showing a higher LA/Y ratio was UD 30–080 (1.64 m^2/kg) that, concurrently, was the PRV having the most delayed TSS pattern in both years and the lowest TSS at harvest (Figure 2A,B and Table 3). Therefore, the fast sugar accumulation observed in PRV was likely due to the actual genotype efficiency in accumulating sugars and not to differential source/sink relations among cultivars. Moreover, our data suggest that canopy management of PRV should substantially differ from that of *V. vinifera* cultivars such as Ortrugo and Sauvignon Blanc. For instance, Gatti et al. [44] demonstrated the effectiveness of bunch thinning in promoting faster sugar accumulation making it possible to target the enological standards required for producing Ortrugo sparkling wines (TSS of 20–21 °Brix and TA of 6.5 g/L). Accordingly, if, in the two reference cultivars, a good choice could be increasing the LA/Y ratio in order to promote TSS concentration corresponding to optimal TA, then in the PRV tested in this trial, the best strategy should be, on the contrary, to reduce LA/Y ratio trying to slow down sugar accumulation rates [37].

In ripening fruits, acidity is driven by the relative changes in the abundance of organic acids [4]. Malic acid is one of the main substrates for respiration in grapevine berries, and it is maximum pre-veraison and minimum at the end of the season [45,46]. Malic acid respiration rates are mainly driven by temperatures and the abundance of the substrate [47]. However, each genotype has a specific pattern in malic acid degradation rates and, perhaps more importantly, a different maximum and minimum organic acid concentration [48,49]. For instance, our study shows that UD 80–100 boasts the highest pre-veraison malic acid concentration among the tested varieties and that all PRV retain very low malic acid at high TSS concentrations (Figure 4A,B). The dynamic of malate degradation rates (Figure 5G–J) reveals that daily loss of malic acid is a complex function of the onset of veraison time, the pool of pre-veraison malic acid, and air temperature. In fact, in 2019, Sauvignon Blanc and Ortrugo grapes, due to their later onset of veraison, somewhat mitigated the loss of malic acid under the high temperatures recorded between DOYs 200 and 210. Contrariwise, on the same dates, all PRV had already crossed veraison and suffered a drastic malic acid loss. However, in 2019, maximum malic acid loss per day was exhibited by Sauvignon Blanc immediately after veraison, confirming that air temperature is not the only factor driving degradation rates. In 2020, when the onset of veraison time was similar for all tested cultivars, a stronger role was likely played by the availability of malate for respiration. In fact, genotypes showing the highest malic acid degradation rates corresponded with those varieties showing the highest pre-veraison malic acid pool (Figures 4B and 5H,J). Moreover, our data demonstrate that malic acid degradation rates are not the main factor determining final malic acid at the end of the season. Sauvignon Blanc was indeed the variety that exhibited in both years the more intense malic acid degradation rates and the highest minimum malic acid concentration rates at the end of the season (Figure 4A,B and Figure 5G,H).

If the primary goal is screening for varieties capable to maintain adequate acidity in a hot region, two cultivars seem to be of some interest among the five tested PRV: UD 80–100 and Sauvignon Rytos. UD 80–100 shows adequate acidity at high TSS levels and optimal balance between malic and tartaric acid (Figure 3C,D and Figure 6C,D, and Table 4); however, its sugar accumulation is excessively fast, and grapes should be harvested very early (Figure 2A,B and Figure 5C,D). Sauvignon Rytos was the only PRV showing higher TA than Ortrugo in 2020, whereas, in 2019, these two varieties showed very similar malate concentration from DOY 208 to the end of the season (Figure 2A,B). Ortrugo is a variety that suffers a fast acidity loss after veraison to very low values, often not compatible with white and sparkling wine-making standards [6,44]. This is also confirmed by our data, showing poor balance among sugars, acidity, and malic vs. tartaric acid in Ortrugo, either if harvested at optimal TSS or at optimal TA thresholds (Figure 6B,D). However, Table 1 and Figure 2 show that in 2019, Ortrugo had a delayed veraison and TA loss, together with higher minimum acidity, compared to data recorded in the subsequent season. Notably, in Sauvignon Rytos, as compared to Ortrugo, optimal acidity matched higher TSS concentration, meaning that harvest might be anticipated with no detrimental effects on sugars (Table 4 and Figure 6C,D). However, data also support that Sauvignon Rytos acidity cannot be even close to that of Sauvignon Blanc, a variety that is known for the high acidity retained late in the season [50], a trait confirmed by our results (Figure 2E,F). If our data suggest that, in our conditions, Sauvignon Rytos was the PRV more capable to retain some acidity, Figure 4 reveals a relevant difference between Sauvignon Rytos and its parental Sauvignon Blanc: the former holds high acidity mostly due to the high concentration of tartaric acid, with a very limited contribution of malic acid; Sauvignon Blanc acidity at harvest relies instead on both malate and tartrate high concentration late at the end of the season (Figure 4A,B and Table 4). Different from malic acid, tartaric acid is not subjected to respiration, and changes in its abundance are mainly due to dilution and salts formation with K+ [4,45,47]. Interestingly, QTL, responding to grapes' cation mobility and organic acid metabolism, and catabolism have been identified, and today, several breeding programmes based on MAS or new breeding technologies are in progress to obtain new varieties less responsive to organic acid depletion under high temperatures [51–55].

Comparing balance between organic acids at optimal TSS or TA confirms that the difference in the ratio between organic acid concentration in Sauvignon Rytos and Sauvignon Blanc grapes is consistent, either choosing optimal TSS or optimal TA as the key parameter for placing harvest time (Figure 6A,C). This is quite essential information because it suggests that acidity preservation in these two varieties should be pursued based upon contrasting strategies: in Sauvignon Blanc, in order to preserve malic acid from respiration, grapes should be protected from excessive radiation and temperature, choosing trellis system shielding bunches from direct radiation and selecting appropriate aspects and locations [42,56–58]. Contrariwise, in Sauvignon Rytos actions should aim at limiting berry dilution, for instance, by calibrating adequate competition for water and nutrients, and by carefully managing K+ availability and nutrition in order to contrast potassium salts formation [47].

5. Conclusions

To the best of our knowledge, ripening patterns and organic acid depletion rates in new PRV were never studied in detail. Even if many technical reports suggest that wines obtained from these varieties could have similar aromatic traits to their *Vitis vinifera* parentals, our data demonstrate that, in regions with significant heat summation during the summer (~2000 GDDs), PRV exhibit anticipated veraison, as compared to early ripening *V. vinifera* cultivars, and that their organic acids balance and sugar accumulation rates largely differ from those observed in *vinifera* genotypes. Sauvignon Kretos, Sauvignon Rytos, Soreli, and UD 80–100 exhibited an early and fast grape sugar accumulation. Among the tested PRV genotypes, Sauvignon Rytos was the only one capable of maintaining higher titratable acidity at harvest, due to a significant tartaric acid pool. Conversely, in the Sauvignon Blanc parental, the high acidity at harvest was linked to high final malic acid concentration, resulting in a different acidic balance.

Our work also suggests that all new PRV should be subjected to different canopy and ripening management strategies, as compared to the two reference *V. vinifera* cultivars, given the faster sugar accumulation rates at comparable levels of LA/Y ratio.

Overall, our data indicate that PRV could perform better in north-facing hillsides, in cooler climates, or at higher altitudes, where their good resistance to mildews could match an adequate grapes' biochemical balance.

Supplementary Materials: The following are available online at https://www.mdpi.com/article/10.3390/horticulturae7080229/s1, Table S1: Integrated pest management protocol applied for pathogen-resistant varieties (PRV) and non-resistant cultivars, Figure S1: Ripe clusters of the five resistant cultivars evaluated in the study and their parentals with respective pedigree.

Author Contributions: Conceptualisation, T.F. and S.P.; methodology, S.P. and M.G.; investigation, T.F., C.S., F.D.Z. and P.G.; resources, S.P.; data curation, T.F.; writing—original draft preparation, T.F.; writing—review and editing, M.G., S.P., A.V., C.S, F.D.Z. and P.G.; funding acquisition, S.P. All authors have read and agreed to the published version of the manuscript.

Funding: Research partially supported by the ValorInVitis project, funded by the Emilia Romagna Region under the RDP program (PSR Emilia Romagna 2014–2020 Mis. 16.1.01 FA 2A), Grant No. 5004320, and by Catholic University of Sacred Heart D1 funds.

Institutional Review Board Statement: Not applicable.

Informed Consent Statement: Not applicable.

Data Availability Statement: The data presented in this study are available on request from the corresponding author.

Acknowledgments: The authors would like to thank Cantina Sociale di Vicobarone, Vivai Cooperativi Rauscedo, and Roberto Miravalle, for establishing the experimental vineyard; Stefano Carrà and Luca Ampeli, for the technical support in the vineyard; Maria Giulia Parisi, Lily Ronney, and Alessandra Garavani, for the technical help; Alice Richards, for revising the written English manuscript; Roberto Miravalle, Marco Profumo, Mauro Mazzocchi, Claudio Gazzola, Andrea Pradelli, and Andrea Illari, for topic discussion.

Conflicts of Interest: The authors declare no conflict of interest.

References

1. Palliotti, A.; Tombesi, S.; Silvestroni, O.; Lanari, V.; Gatti, M.; Poni, S. Changes in vineyard establishment and canopy management urged by earlier climate-related grape ripening: A review. *Sci. Hortic.* **2014**, *178*, 43–54. [CrossRef]
2. Poni, S.; Gatti, M.; Palliotti, A.; Dai, Z.; Duchêne, E.; Truong, T.T.; Ferrara, G.; Matarrese, A.M.S.; Gallotta, A.; Bellincontro, A.; et al. Grapevine quality: A multiple choice issue. *Sci. Hortic.* **2018**, *234*, 445–462. [CrossRef]
3. Bernardo, S.; Dinis, L.T.; Machado, N.; Moutinho-Pereira, J. Grapevine abiotic stress assessment and search for sustainable adaptation strategies in Mediterranean-like climates. A review. *Agron. Sustain. Dev.* **2018**, *38*, 66. [CrossRef]
4. Duchêne, E.; Huard, F.; Dumas, V.; Schneider, C.; Merdinoglu, D. The challenge of adapting grapevine varieties to climate change. *Clim. Res.* **2010**, *41*, 193–204. [CrossRef]
5. Moran, M.A.; Sadras, V.O.; Petrie, P.R. Late pruning and carry-over effects on phenology, yield components and berry traits in Shiraz. *Aust. J. Grape Wine Res.* **2017**, *23*, 390–398. [CrossRef]
6. Frioni, T.; Bertoloni, G.; Squeri, C.; Garavani, A.; Ronney, L.; Poni, S.; Gatti, M. Biodiversity of Local *Vitis vinifera* L. Germplasm: A Powerful Tool Toward Adaptation to Global Warming and Desired Grape Composition. *Front. Plant Sci.* **2020**, *11*, 608. [CrossRef] [PubMed]
7. Garrett, K.A.; Nita, M.; De Wolf, E.D.; Esker, P.D.; Gomez-Montano, L.; Sparks, A.H. Plant pathogens as indicators of climate change. In *Climate Change*; Elsevier: Amsterdam, The Netherlands, 2021; pp. 499–513.
8. Zambito Marsala, R.; Capri, E.; Russo, E.; Bisagni, M.; Colla, R.; Lucini, L.; Gallo, A.; Suciu, N.A. First evaluation of pesticides occurrence in groundwater of Tidone Valley, an area with intensive viticulture. *Sci. Total Environ.* **2020**, *736*, 139730. [CrossRef] [PubMed]
9. Pedneault, K.; Provost, C. Fungus resistant grape varieties as a suitable alternative for organic wine production: Benefits, limits, and challenges. *Sci. Hortic.* **2016**, *208*, 57–77. [CrossRef]
10. Töpfer, R.; Hausmann, L.; Harst, M.; Maul, E.; Zyprian, E.; Eibach, R. New horizons for grapevine breeding. *Fruit Veg. Cereal Sci. Biotechnol.* **2011**, *5*, 79–100.
11. Montaigne, E.; Coelho, A.; Khefifi, L. Economic issues and perspectives on innovation in new resistant grapevine varieties in France. *Wine Econ. Policy* **2016**, *5*, 73–77. [CrossRef]
12. Pertot, I.; Caffi, T.; Rossi, V.; Mugnai, L.; Hoffmann, C.; Grando, M.S.; Gary, C.; Lafond, D.; Duso, C.; Thiery, D.; et al. A critical review of plant protection tools for reducing pesticide use on grapevine and new perspectives for the implementation of IPM in viticulture. *Crop Prot.* **2017**, *97*, 70–84. [CrossRef]
13. Bavaresco, L.; Gardiman, M.; Brancadoro, L.; Espen, L.; Failla, O.; Scienza, A.; Vezzulli, S.; Zulini, L.; Velasco, R.; Stefanini, M.; et al. Grapevine breeding programs in Italy. In *Grapevine Breeding Programs for the Wine Industry*; Elsevier Inc.: Amsterdam, The Netherlands, 2015; pp. 135–157. ISBN 978-1-78242-080-4.

14. Di Gaspero, G.; Foria, S. Molecular grapevine breeding techniques. In *Grapevine Breeding Programs for the Wine Industry*; Elsevier Inc.: Amsterdam, The Netherlands, 2015; pp. 23–37. ISBN 978-1-78242-080-4.
15. Vezzulli, S.; Dolzani, C.; Migliaro, D.; Banchi, E.; Stedile, T.; Zatelli, A.; Dallaserra, M.; Clementi, S.; Dorigatti, C.; Velasco, R.; et al. The Fondazione Edmund Mach grapevine breeding program for downy and powdery mildew resistances: Toward a green viticulture. *Acta Hortic.* **2019**, *1248*, 109–113. [CrossRef]
16. Bove, F.; Bavaresco, L.; Caffi, T.; Rossi, V. Assessment of Resistance Components for Improved Phenotyping of Grapevine Varieties Resistant to Downy Mildew. *Front. Plant Sci.* **2019**, *10*, 1559. [CrossRef]
17. Casagrande, K.; Falginella, L.; Castellarin, S.D.; Testolin, R.; Di Gaspero, G. Defence responses in Rpv3-dependent resistance to grapevine downy mildew. *Planta* **2011**, *234*, 1097–1109. [CrossRef] [PubMed]
18. Vezzulli, S.; Vecchione, A.; Stefanini, M.; Zulini, L. Downy mildew resistance evaluation in 28 grapevine hybrids promising for breeding programs in Trentino region (Italy). *Eur. J. Plant Pathol.* **2018**, *150*, 485–495. [CrossRef]
19. Zanghelini, J.A.; Bogo, A.; Dal Vesco, L.L.; Gomes, B.R.; Mecabô, C.V.; Herpich, C.H.; Welter, L.J. Response of PIWI grapevine cultivars to downy mildew in highland region of southern Brazil. *Eur. J. Plant Pathol.* **2019**, *154*, 1051–1058. [CrossRef]
20. Poni, S.; Chiari, G.; Caffi, T.; Bove, F.; Tombesi, S.; Moncalvo, A.; Gatti, M. Canopy physiology, vine performance and host-pathogen interaction in a fungi resistant cv. Sangiovese x Bianca accession vs. a susceptible clone. *Sci. Rep.* **2017**, *7*, 1–14. [CrossRef]
21. Casanova-Gascón, J.; Ferrer-Martín, C.; Bernad-Eustaquio, A.; Elbaile-Mur, A.; Ayuso-Rodríguez, J.M.; Torres-Sánchez, S.; Jarne-Casasús, A.; Martín-Ramos, P. Behavior of vine varieties resistant to fungal diseases in the Somontano region. *Agronomy* **2019**, *9*, 738. [CrossRef]
22. Zambon, Y.; Khavizova, A.; Colautti, M.; Georgofili, E.S. Varietà di Vite Resistenti Alle Malattie e Rame: Opportunità e Limiti. In *Riflessioni Sull'uso del Rame per la Protezione Delle Piante*; Accademia dei Georgofili, Quaderni 2019-III, Società Editrice Fiorentina: Firenze, Italy, 2020.
23. Gratl, V.; Sturm, S.; Zini, E.; Letschka, T.; Stefanini, M.; Vezzulli, S.; Stuppner, H. Comprehensive polyphenolic profiling in promising resistant grapevine hybrids including 17 novel breeds in northern Italy. *J. Sci. Food Agric.* **2021**, *101*, 2380–2388. [CrossRef]
24. Gelmetti, A.; Roman, T.; Bottura, M.; Stefanini, M.; Pedò, S. Performance agronomiche di viti resistenti. Primi risultati di un progetto poliennale. *Vite e Vino* **2019**, *2*, 54–61.
25. Nicolini, G.; Barp, L.; Roman, T.; Larcher, R.; Malacarne, M.; Bottura, M.; Tait, F.; Battisti, F.; Mereles, M.S.; Battistella, R. Resistenti bianchi e rossi: Primi dati da esperienze trentine sulla concentrazione nei vini di shikimico e flavonoidi. *L'Enologo* **2018**, *3*, 89–93.
26. Ruehl, E.; Schmid, J.; Eibach, R.; Töpfer, R. Grapevine breeding programmes in Germany. In *Grapevine Breeding Programs for the Wine Industry*; Elsevier Inc.: Amsterdam, The Netherlands, 2015; pp. 77–101. ISBN 978-1-78242-080-4.
27. Frioni, T.; Green, A.; Emling, J.E.; Zhuang, S.; Palliotti, A.; Sivilotti, P.; Falchi, R.; Sabbatini, P. Impact of spring freeze on yield, vine performance and fruit quality of Vitis interspecific hybrid Marquette. *Sci. Hortic.* **2017**, *219*, 302–309. [CrossRef]
28. Teissedre, P.L. Composition of grape and wine from resistant vine varieties. *OENO One* **2018**, *52*, 197–203. [CrossRef]
29. Pedò, S.; Bottura, M.; Porro, D. Development, yield potential and nutritional aspects of resistant grapevine varieties in Trentino Alto Adige. *BIO Web Conf.* **2019**, *13*, 02004. [CrossRef]
30. Smith, M.S.; Centinari, M. Young grapevines exhibit interspecific differences in hydraulic response to freeze stress but not in recovery. *Planta* **2019**, *250*, 495–505. [CrossRef]
31. Hed, B.; Centinari, M. Gibberellin application improved bunch rot control of vignoles grape, but response to mechanical defoliation varied between training systems. *Plant Dis.* **2021**, *105*, 339–345. [CrossRef]
32. Tóth-Lencsés, A.K.; Kozma, P.; Szőke, A.; Kerekes, A.; Veres, A.; Kiss, E. Parentage analysis in Hungarian grapevine cultivars of "Seibel'-'Seyve-Villard" origin. *Vitis* **2015**, *54*, 27–29.
33. Venuti, S.; Copetti, D.; Foria, S.; Falginella, L.; Hoffmann, S.; Bellin, D.; Cindrić, P.; Kozma, P.; Scalabrin, S.; Morgante, M.; et al. Historical Introgression of the Downy Mildew Resistance Gene Rpv12 from the Asian Species Vitis amurensis into Grapevine Varieties. *PLoS ONE* **2013**, *8*, e61228. [CrossRef] [PubMed]
34. Winkler, A. *General Viticulture*; University of California Press: Berkeley, CA, USA, 1965.
35. Lorenz, D.H.; Eichhorn, K.W.; Bleiholder, H.; Klose, R.; Meier, U.; Weber, E. Growth Stages of the Grapevine: Phenological growth stages of the grapevine (*Vitis vinifera* L. ssp. vinifera)—Codes and descriptions according to the extended BBCH scale. *Aust. J. Grape Wine Res.* **1995**, *1*, 100–103. [CrossRef]
36. Gatti, M.; Garavani, A.; Krajecz, K.; Ughini, V.; Parisi, M.G.; Frioni, T.; Poni, S. Mechanical mid-shoot leaf removal on ortrugo (*Vitis vinifera* L.) at pre-or mid-veraison alters fruit growth and maturation. *Am. J. Enol. Vitic.* **2019**, *70*, 88–97. [CrossRef]
37. Kliewer, W.M.; Dokoozlian, N.K. Leaf Area/Crop Weight Ratios of Grapevines: Influence on Fruit Composition and Wine Quality. *Am. J. Enol. Vitic.* **2005**, *56*, 170–181.
38. Poni, S.; Tombesi, S.; Palliotti, A.; Ughini, V.; Gatti, M. Mechanical winter pruning of grapevine: Physiological bases and applications. *Sci. Hortic.* **2016**, *204*, 88–98. [CrossRef]
39. Jarvis, C.; Darbyshire, R.; Goodwin, I.; Barlow, E.W.R.; Eckard, R. Advancement of winegrape maturity continuing for winegrowing regions in Australia with variable evidence of compression of the harvest period. *Aust. J. Grape Wine Res.* **2018**, *25*, 101–108. [CrossRef]
40. Gladstones, J. *Viticulture and Environment*; Winetitles: Adelaide, Australia, 1992.

41. Kemp, B.; Pedneault, K.; Pickering, G.; Usher, K.; Willwerth, J. Red Winemaking in Cool Climates. In *Red Wine Technology*; Elsevier: Amsterdam, The Netherlands, 2018; pp. 341–356. ISBN 978-0-12814-400-8.

42. Frioni, T.; Bronzoni, V.; Moncalvo, A.; Poni, S.; Gatti, M. Evaluation of local minor cultivars and marginal areas to improve wines and increase the sustainability of the district 'Colli Piacentini'. *Acta Hortic.* **2020**, *1276*, 111–118. [CrossRef]

43. Nistor, E.; Dobrei, A.G.; Dobrei, A.; Camen, D. Growing Season Climate Variability and its Influence on Sauvignon Blanc and Pinot Gris Berries and Wine Quality: Study Case in Romania (2005–2015). *S. Afr. J. Enol. Vitic.* **2018**, *39*, 196–207. [CrossRef]

44. Gatti, M.; Garavani, A.; Cantatore, A.; Parisi, M.G.; Bobeica, N.; Merli, M.C.; Vercesi, A.; Poni, S. Interactions of summer pruning techniques and vine performance in the white *Vitis vinifera* cv. Ortrugo. *Aust. J. Grape Wine Res.* **2015**, *21*, 80–89. [CrossRef]

45. Hale, C.R. Synthesis of Organic Acids in the Fruit of the Grape. *Nature* **1962**, *195*, 917–918. [CrossRef]

46. Sweetman, C.; Deluc, L.G.; Cramer, G.R.; Ford, C.M.; Soole, K.L. Regulation of malate metabolism in grape berry and other developing fruits. *Phytochemistry* **2009**, *70*, 1329–1344. [CrossRef]

47. Ford, C.M. The Biochemistry of organic acids in the Grape. In *The Biochemistry of the Grape Berry*; Geròs, H., Chaves, M.M., Delrot, S., Eds.; Bentham Books: Dubai, United Arab Emirates, 2012; pp. 67–88.

48. Famiani, F.; Farinelli, D.; Frioni, T.; Palliotti, A.; Battistelli, A.; Moscatello, S.; Walker, R.P. Malate as substrate for catabolism and gluconeogenesis during ripening in the pericarp of different grape cultivars. *Biol. Plant.* **2016**, *60*, 155–162. [CrossRef]

49. Famiani, F.; Battistelli, A.; Moscatello, S.; Cruz-Castillo, J.G.; Walker, R.P. The organic acids that are accumulated in the flesh of fruits: Occurrence, metabolism and factors affecting their contents—A review. *Rev. Chapingo. Ser. Hortic.* **2015**, *21*, 97–128. [CrossRef]

50. Ewart, A.J.W.; Gawel, R.; Thistlewood, S.P.; McCarthy, M.G. Evaluation of must composition and wine quality of six clones of *Vitis vinifera* cv. Sauvignon Blanc. *J. Exp. Agric.* **1993**, *33*, 945–951. [CrossRef]

51. Costantini, L.; Battilana, J.; Lamaj, F.; Fanizza, G.; Grando, M.S. Berry and phenology-related traits in grapevine (*Vitis vinifera* L.): From Quantitative Trait Loci to underlying genes. *BMC Plant Biol.* **2008**, *8*, 1–17. [CrossRef]

52. Duchêne, É. How can grapevine genetics contribute to the adaptation to climate change? *OENO One* **2016**, *50*, 113–124. [CrossRef]

53. Bayo-Canha, A.; Costantini, L.; Fernández-Fernández, J.I.; Martínez-Cutillas, A.; Ruiz-García, L. QTLs Related to Berry Acidity Identified in a Wine Grapevine Population Grown in Warm Weather. *Plant Mol. Biol. Report.* **2019**, *37*, 157–169. [CrossRef]

54. Bigard, A.; Berhe, D.T.; Maoddi, E.; Sire, Y.; Boursiquot, J.M.; Ojeda, H.; Péros, J.P.; Doligez, A.; Romieu, C.; Torregrosa, L. *Vitis vinifera* L. fruit diversity to breed varieties anticipating climate changes. *Front. Plant Sci.* **2018**, *9*, 455. [CrossRef]

55. Bigard, A.; Romieu, C.; Sire, Y.; Torregrosa, L. *Vitis* vinifera L. Diversity for Cations and Acidity Is Suitable for Breeding Fruits Coping With Climate Warming. *Front. Plant Sci.* **2020**, *11*, 1175. [CrossRef] [PubMed]

56. Bernizzoni, F.; Gatti, M.; Civardi, S.; Poni, S. Long-term Performance of Barbera Grown under Different Training Systems and Within-Row Vine Spacings. *Am. J. Enol. Vitic.* **2009**, *60*, 339–348.

57. Reynolds, A.G.; Vanden Heuvel, J.E. Influence of Grapevine Training Systems on Vine Growth and Fruit Composition: A Review. *Am. J. Enol. Vitic.* **2009**, *60*, 251–268.

58. Dinis, L.T.; Malheiro, A.C.; Luzio, A.; Fraga, H.; Ferreira, H.; Gonçalves, I.; Pinto, G.; Correia, C.M.; Moutinho-Pereira, J. Improvement of grapevine physiology and yield under summer stress by kaolin-foliar application: Water relations, photosynthesis and oxidative damage. *Photosynthetica* **2018**, *56*, 641–651. [CrossRef]

MDPI

Article

The Use of Halophytic Companion Plant (*Portulaca oleracea* L.) on Some Growth, Fruit, and Biochemical Parameters of Strawberry Plants under Salt Stress

Sema Karakas [1,*], Ibrahim Bolat [2] and Murat Dikilitas [3]

[1] Department of Soil Science and Plant Nutrition, Faculty of Agriculture, Harran University, Sanliurfa 63300, Turkey

[2] Department of Horticulture, Faculty of Agriculture, Harran University, Sanliurfa 63300, Turkey; ibolat@harran.edu.tr

[3] Department of Plant Protection, Faculty of Agriculture, Harran University, Sanliurfa 63300, Turkey; m.dikilitas@harran.edu.tr

* Correspondence: skarakas@harran.edu.tr

Abstract: Strawberry is a salt-sensitive plant adversely affected by slightly or moderately saline conditions. The growth, fruit, and biochemical parameters of strawberry plants grown under NaCl (0, 30, 60, and 90 mmol L^{-1}) conditions with or without a halophytic companion plant (*Portulaca oleracea* L.) were elucidated in a pot experiment. Salt stress negatively affected the growth, physiological (stomatal conductance and electrolyte leakage), and biochemical parameters such as chlorophyll contents (chl-*a* and chl-*b*); proline, hydrogen peroxide, malondialdehyde, catalase, and peroxidase enzyme activities; total soluble solids; and lycopene and vitamin C contents, as well as the mineral uptake, of strawberry plants. The companionship of *P. oleracea* increased fresh weight, dry weight, and fruit average weight, as well as the total fruit yield of strawberry plants along with improvements of physiological and biochemical parameters. This study showed that the cultivation of *P. oleracea* with strawberry plants under salt stress conditions effectively increased strawberry fruit yield and quality. Therefore, we suggest that approaches towards the use of *P. oleracea* could be an environmentally friendly method that should be commonly practiced where salinity is of great concern.

Keywords: abiotic stress; strawberry; companion plants; phytoremediation

Citation: Karakas, S.; Bolat, I.; Dikilitas, M. The Use of Halophytic Companion Plant (*Portulaca oleracea* L.) on Some Growth, Fruit, and Biochemical Parameters of Strawberry Plants under Salt Stress. *Horticulturae* 2021, 7, 63. https://doi.org/10.3390/horticulturae7040063

Academic Editors: Agnieszka Hanaka, Jolanta Jaroszuk-Ściseł and Małgorzata Majewska

Received: 10 February 2021
Accepted: 23 March 2021
Published: 26 March 2021

1. Introduction

Salinity is one of the most devastating environmental problems limiting crop productivity and quality in many regions of the world. This problem is more prevalent in arid and semi-arid climatic regions. It affects approximately 20% of the cultivated and 50% of irrigated agricultural lands [1,2]. It has now been estimated that 1.5 million hectares of lands have been lost every year due to salinity problems. If salinization goes with this trend, nearly 50% of cultivable lands will be lost by the mid-point of this century [3,4].

Salinity negatively affects plant growth in terms of osmotic, ionic, and nutrient imbalance [5]. These disorders cause oxidative stress on plants. If plants cannot get enough water under high salt stress, turgor pressure significantly decreases, and thus the closure of the stomata of plants becomes inevitable to conserve water [6]. This significantly affects the photosynthetic capacity of plants. Ionic toxicity, on the other hand, inhibits cellular metabolism and biochemical pathways. For example, Na^+ ions at the root cell disturb enzymatic activities and inhibit the uptake of other minerals such as K^+ and Ca^{++} [7]. The high accumulation of Na^+ and Cl^- ions result in many morphological, physiological, molecular, and biochemical pathways in plants. Due to disturbed mechanisms in plants, NaCl stress leads to the development of leaf chlorosis and necrosis, as well as the loss of quality in crops. As a consequence, the assimilation of carbohydrates and sugars allocated

49

for fruit development is significantly reduced due to stress development and defense mechanisms [8,9].

Plants can be divided into halophytes and glycophytes as responses to salinity stress. Most glycophytes are salt-sensitive, even at low concentrations, while halophytes are highly salt-tolerant plants, which enables them to survive and thrive in extremely saline environments [10,11]. Salt ions have to be taken up by halophytes and deposited in the vacuoles of leaf or root tissues or in separate organelles. In general, salt secretion takes place through the shedding of salty leaves and salt glands or specialized leaf cells [12]. Most halophytes are able to survive by maintaining negative water potential under extreme salt concentrations. Therefore, a true halophyte is considered to maintain its viability and complete the life cycle at NaCl levels between 200 and 1000 mmol L^{-1}. These concentrations are very close to the concentrations of seawater level. Some halophytes, on the other hand, tolerate much higher concentrations of NaCl [10,13]. Halophytes such as *Atriplex* spp., *Chenopodium* spp., *Portulaca* spp., *Suaeda* spp., and *Salsola* spp. uptake salt ions through their roots and store them in their leaves. It is quite possible that these plants could be used as companion plants with crop plants, especially salt-sensitive glycophytes, to reduce the negative effects of NaCl through the uptake of toxic ions [14,15]. For example, *Portulaca oleracea* L. (purslane) (which is a member of Portulacaceae), is a drought- and salt-tolerant annual plant. The plant is a promising crop species in saline–alkali soils [16,17]. Moreover, *P. oleracea* could effectively absorb salts from soil media to remediate saline–alkali soils [18]. Previous studies have investigated the effects of salinity on *P. oleracea* growth. For example, Grieve and Suarez [19] evaluated *P. oleracea* responses with saline irrigation, and they showed that the plant could survive at a salinity of 28.5 dS m^{-1}. The authors further elucidated the performances of salt-tolerant halophyte species of *P. oleracea* and *Salsola soda* against increasing NaCl levels. They reported that *P. oleracea* and *S. soda* seeds were effectively germinated between 250 and 350 mmol L^{-1} NaCl levels [20].

Strawberry is an economically important fruit crop that is globally cultivated. It belongs to the *Fragaria* genus in Rosaceae family, which contains 23 species [21,22]. The popularity of strawberry fruit crops is increasing in the world due to increasing consumption. Its popularity is also increasing along with the generation of new varieties. Strawberry cultivation has therefore become an important greenhouse and open field crop in the Mediterranean area, although drought and salinity have played significant roles in limiting crop production [23,24].

Strawberry is considered to be sensitive to NaCl salinity due to increased osmotic pressure and Na$^+$ or Cl$^-$ ion toxicity. NaCl salinity not only reduces the crop yield but also deteriorates the quality parameters in many crops, including strawberries [25,26].

In the present study, we elucidated the effects of different NaCl concentrations on strawberry plants grown with or without the halophytic companion plant *P. oleracea* L. to remediate the physiological and biochemical parameters of strawberry plants.

2. Materials and Methods

2.1. Experimental Design and the Growth of Plants

The experiment was performed between September 2018 and January 2019 in a semi-controlled greenhouse at the University of Harran, Sanliurfa, Turkey. Fresh strawberry (Rubygem variety) plants were grown alone or in combination with *P. oleracea* seedlings in 8-L pots containing peat (Klasmann TS 1) under natural light conditions. Peat is a very porous substrate with an excellent water capacity. Its slow degradation rate, high porosity, and high-water holding capacity makes it one of the most commonly used growth medium, especially for saline-related studies in vegetables and ornamental plants [27,28]. It has low nutrient values, so it is highly unlikely to affect the mineral uptake of macro and micronutrient elements.

The average temperatures for day and night were 35 ± 2/28 ± 2 °C during the course of the experiment. The trial was carried out in a randomized block design. The first group of strawberry plants grown under differing NaCl conditions (0, 30, 60, and

90 mmol L^{-1}) was designated as S_0, S_{30}, S_{60}, and S_{90}; the second group of plants grown with *P. oleracea* under the same NaCl conditions was designated as SP_0, SP_{30}, SP_{60}, and SP_{90}. Treatments in each group were replicated five times. Seedlings were individually transplanted to the pots. Strawberry seedlings following one week of establishment growth in pots were accompanied with *P. oleracea* seeds that were sown and germinated at a rate of 25 companion plants per pot. After germinations (five weeks), the pots were irrigated with or without salt to the full pot capacity throughout the treatment period (twelve weeks). The plants were fertigated with Hoagland's nutrient solution once a week. The experimental trial from the very beginning of obtaining strawberry seedlings to the end of harvest took five months. The plants were harvested at the optimum stage of physiological maturity for the evaluation of salinity responses (Figure 1).

Figure 1. Strawberry plants were grown with or without *Portulaca oleracea* under different NaCl conditions.

2.2. Plant Growth and Fruit Properties

Strawberry fruits were harvested when 90% of the fruit surface had reached a fully red color. At the end of the experimental period, total fruit weights were determined and the average fruit yield was calculated.

Plant crown and root fresh weight (Fwt) were analyzed immediately after the harvest. The dry weight (Dwt) of plant organs was determined following the drying of plant samples at 70 °C until a constant weight.

Total soluble solids (TSS) were assessed from the fruit juice with a hand refractometer. The results are expressed in percent (%) Catania et al. [29].

The lycopene content of strawberry fruits was assessed according to the method of Barrett and Anthon [30] with minor modifications [31]. One gram of strawberry fruit was homogenized with 10 mL of an ethanol:hexane solution (4:3). The mixture was then centrifuged at 10,000× g for 10 min at room temperature. The supernatant (100 µL) was added to 7 mL of the ethanol:hexane solution (4:3) mixture and vortexed. After 1 h of incubation at room temperature, 1 mL of H_2O was added to the tubes and vortexed. The tubes were incubated in the dark to form two different phases. The top phase was taken and read at 503 nm against a hexane blank with a UV microplate spectrophotometer (Epoch, SN: 1611187, Winooski, VT, USA). The lycopene content was calculated according to the following formula (Equation (1)).

$$\mu g \text{ Lycopene g}^{-1} \text{ Fwt} = \frac{A_{503} \times 2.7}{172 \times 0.1} \times 537 \tag{1}$$

where 537 g/mole is the molecular weight of lycopene, 2.7 mL is the volume of the hexane layer, 172 $mmol^{-1}$ is the extinction coefficient for lycopene in hexane, and 0.1 g is the weight of the strawberry.

The vitamin C content of strawberry fruits was assessed according to the method of Oz [32] with small modifications [31]. Strawberry fruits (5 g) were extracted with 25 mL of oxalic acid. The mixture was centrifuged at 10,000× g for 10 min. Then, 1 mL of this mixture was added to 7 mL of a 1% oxalic acid solution and 8 mL of a dye reagent. The dye reagent was prepared according to the recipe of [31]. The mixture was filtered through Whatman No.2 filter paper and diluted to 100 mL with deionized H_2O. Then, 25 mL of this solution were taken and diluted to 500 mL with deionized H_2O, vortexed, and kept at 4 °C until use. The mixture was once more vortexed before measurement at 518 nm against the oxalic acid and dye mixture with a UV microplate spectrophotometer (Epoch, SN: 1611187, Winooski, VT, USA).

Electrolyte leakage (EL) was assessed following the method of Lutts et al. [33] using leaf discs for each treatment. Fully expanded young leaves were cleaned three times with deionized H_2O to remove dust and surface-adhered electrolytes. Leaf discs were placed in closed vials containing 10 mL of H_2O and incubated at 25 °C on a rotary shaker for 24 h; subsequently, the electrical conductivity of the solution (EC_1) was measured. The final electrical conductivity (EC_2) was determined following the autoclaving of the leaf samples at 120 °C for 20 min. Leaf samples were then equilibrated at 25 °C, and the EL was calculated as follows (Equation (2)).

$$EL\ (\%) = \frac{EC1}{EC2} \times 100 \tag{2}$$

Stomatal conductivity (SC) was determined on the youngest fully expanded leaves on the upper branches of the strawberry plants with a leaf promoter (SC−1) at midday. Measurements were performed by clamping the leaves in the leaf chamber. The actual vapor flux from the leaf through the stomata is expressed as mmol $m^{-2}\ s^{-1}$, following the work of Karlidag et al. [34].

2.3. Biochemical Parameters

Strawberry plant leaf chlorophyll content (Chl-*a* and Chl-*b*) was extracted following the method of Arnon [35] with minor modifications [31]. A sample of the fresh leaf (0.5 g) was homogenized with 10 mL of an acetone:water (80/20, *v/v*) mixture and filtered through Whatman No.2 filter paper and then put into the dark tubes. The Chl-*a* and Chl-*b* contents of the filtrate was measured with a UV microplate spectrophotometer (Epoch, SN: 1611187, Winooski, VT, USA) at 663 and 645 nm, respectively, against an 80% acetone blank. The findings were expressed as mg L^{-1} and calculated as mg g^{-1} Fwt.

The proline concentration was determined following the method of Bates et al. [36] with minor modifications [31]. Leaf samples (0.5 g) were extracted with 10 mL of 3% *w/v* sulphosalicylic acid using a mortar and a pestle. The extract was filtered through Whatman No.2 filter paper. Then, the 2 mL filtrate was mixed with 2 mL of acid–ninhydrin in a test tube and boiled at 100 °C for 1 h. The reaction was terminated in an ice bath. Then, the mixture was extracted using 5 mL of toluene. The tubes were vortexed for 20 s and then left for 20 min at room temperature to achieve two layers of separation. The organic phase was collected, and the absorbance of the extracts was read at 515 nm using a toluene blank. Proline concentration was made from a standard curve using L-proline (Sigma-Aldrich, Taufkirchen-Germany). The results are expressed as μmol g^{-1} Fwt.

Hydrogen peroxide (H_2O_2) levels were assessed following the method of Velikova et al. [37] with small modifications [38]. Leaf samples (0.5 g) were extracted with 5 mL of 0.1% (*w:v*) trichloroacetic acid (TCA). The extract was centrifuged at 12,000$\times$ *g* at 4 °C for 15 min, and the supernatant (0.5 mL) was added to 0.5 mL of a 10 mmol L^{-1} potassium phosphate buffer (pH 7.0) and 1 mL of a 1 mol L^{-1} potassium iodide. The absorbance was read at 390 nm in a UV microplate spectrophotometer (Epoch, SN: 1611187, Winooski, VT, USA). The H_2O_2 content was expressed as μmol g^{-1} Fwt.

The malondialdehyde (MDA) content was assessed following the method of Sairam and Sexena [39] with minor modifications [38]. The leaf samples (0.5 g) were extracted with 10 mL of a 0.1% (*w/v*) TCA solution. The extract was centrifuged at 10,000$\times$ *g* for 5 min. Four milliliters of 20% *v/v* TCA containing 0.5% *v/v* thiobarbituric acid (TBA) was added to 1 mL of the supernatant. The mixture was kept in boiling water for 30 min, and then the reaction was stopped in an ice bath. The mixture was once more centrifuged at 10,000$\times$ *g* for 5 min and then read in a UV microplate spectrophotometer (Epoch, SN: 1611187, Winooski, VT, USA) at 532 and 600 nm. The MDA content was calculated and expressed as nmol g^{-1} Fwt (Equation (3)).

$$\text{MDA}\left(\text{nmol g}^{-1}\right) = \frac{\text{Extract volume (ml)} \times \left[(A_{532} - A_{600})/\left(155\,\text{mM}^{-1}\,\text{cm}^{-1}\right)\right]}{\text{Sample amaunt (g)}} \times 10^3 \tag{3}$$

Catalase enzyme activity (CAT; Enzyme Code. 1.11.1.6) was determined following the method of Milosevic and Slusarenko [40] with minor modifications [38]. Leaf samples (0.5 g) were extracted with 10 mL of a 50 mmol L^{-1} Na-phosphate buffer solution, and then 50 mL of the extract were added to a 2.95 mL reaction mixture (50 mmol L^{-1} Na-phosphate buffer, 10 mmol L^{-1} H_2O_2, and 4 mmol L^{-1} Na_2EDTA) and read with a UV microplate spectrophotometer (Epoch, SN: 1611187, Winooski, VT, USA) at 240 nm for 30 s. One CAT unit (U) was defined as an increase in absorbance of 0.1 at 240 nm. The activity is expressed as enzyme unite mg^{-1} Fwt.

Peroxidase enzyme activity (POX; Enzyme Code. 1.11.1.7) was assayed following the method of Cvikrova et al. [41] with minor modifications [38]. For the analysis, 100 mL of the homogenate (obtained as above) was added to 3 mL of the reaction mixture (50 mmol L^{-1} Na-phosphate, 5 mmol L^{-1} H_2O_2, 13 mmol L^{-1} guaiacols, and pH 6.5). Activity was defined by the range of change in absorbance at 470 nm with a UV microplate spectrophotometer (Epoch, SN: 1611187, Winooski, VT, USA). One unit of POX was defined as a change of 0.1 absorbance unit per minute at 470 nm. Activity is expressed as enzyme unit mg^{-1} Fwt.

Leaf mineral (K^+, Na^+, Ca^{2+}, Mg^{2+}, and Cl^-) contents were determined according to the procedure made by Chapman and Pratt [42] with minor modifications [31]. Dry plant samples (0.5 g) were ground in porcelain crucibles. The porcelain crucibles were placed into a muffle furnace, and the temperature was gradually increased up to 500 °C. Following cooling, the ash was dissolved in 5 mL of 2 N hydrochloric acid. After 30 min, the volume was made up to 50 mL with distilled H_2O, and the supernatant was filtered through Whatman No.42 filter paper. The resulting supernatant containing Na^+, K^+, Ca^{+2}, and Mg^{+2} ions were assessed by Inductively Coupled Plasma (ICP, Perkin Elmer). Chloride was determined using ion chromatography after the filtration through Whatman No.42 filter paper.

Duncan's multiple range test (DMRT) was used to evaluate the data using SPSS 22 (ANOVA test) at a significance level of $p \leq 0.05$ using. Data are presented as a mean value $\pm$ with standard error.

3. Results

Strawberry plant growth, fruit properties, biochemical parameters, and mineral contents were significantly affected by all salinity levels. The crown fresh and dry weights of strawberry plants in saline conditions were significantly lower in plants grown alone in saline conditions when compared to those of plants grown in combination with *P. oleracea* under the same conditions. For example, the crown fresh weights of the plants were 55.16, 37.62, and 35.16 g plant^{-1} grown alone in S_{30}, S_{60}, and S_{90} mmol L^{-1} NaCl conditions, respectively. When plants were grown in combinations with *P. oleracea*, their conditions were significantly improved at all NaCl conditions. The fresh weights of plants increased to 64.38, 44.76, and 44.49 plant^{-1} at the SP_{30}, SP_{60}, and SP_{90} mmol L^{-1} NaCl conditions, respectively (Table 1). Similar improvements were recorded for the dry weights of plants (Table 1). In general, the combination of companion plants (*P. oleracea*) was found to be effective in increasing the Fwt and Dwt under each NaCl condition.

Table 1. Growth and physiological parameters of strawberry plants grown alone or in combination with *P. oleracea* at differing NaCl levels.

Treatments	Crown Fwt (g Plant^{-1})	Crown Dwt (g Plant^{-1})	EL (%)	SC (mmol m^{-2} s^{-1})
S_0	85.32 ± 4.17 [a]	18.39 ± 0.70 [a]	11.90 ± 0.77 [e]	241.98 ± 4.29 [a]
S_{30}	55.16 ± 5.30 [b]	12.71 ± 1.40 [b]	15.61 ± 0.57 [c]	183.26 ± 8.19 [c]
S_{60}	37.62 ± 2.70 [d]	10.00 ± 1.07 [d]	21.84 ± 1.06 [b]	127.64 ± 8.39 [d]
S_{90}	35.16 ± 1.90 [d]	9.39 ± 0.69 [d]	25.34 ± 0.92 [a]	94.10 ± 3.83 [e]
SP_0	85.57 ± 4.46 [a]	19.09 ± 0.91 [a]	11.07 ± 0.45 [e]	260.38 ± 8.81 [a]
SP_{30}	64.38 ± 5.01 [b]	14.77 ± 1.08 [b]	11.10 ± 0.95 [e]	230.80 ± 5.48 [b]
SP_{60}	44.76 ± 1.67 [c]	11.23 ± 0.22 [c]	13.84 ± 1.04 [d]	181.60 ± 4.65 [c]
SP_{90}	44.49 ± 2.69 [c]	11.16 ± 0.59 [c]	15.32 ± 1.03 [c]	131.08 ± 5.51 [c]

Significance level at $p \leq 0.05$ was determined for the salt treatment using Duncan's multiple range test. Different letters in each column indicate statistical differences. S: a strawberry grown alone; SP: the strawberry and *P. oleracea* companionship; EL: electrolyte leakage; SC: stomatal conductivity.

EL is considered an important criterion for salt stress parameters. EL was increased with increasing levels of salt. For example, leaf EL was found to be 11.90 and 11.07% at S_0 and SP_0, respectively. Increases of EL were trended from 15.61 to 25.34% with respect to conditions from S_{30} to S_{90}, respectively. When *P. oleracea* was accompanied with strawberry plants in the NaCl conditions, the increase of EC was so minimal that only 11.10 and 15.32% were recorded at SP_{30} and SP_{90}, respectively; see Table 1.

Stomatal conductivity in saline conditions was gradually decreased as the concentration of NaCl increased in plants grown alone in saline conditions (Table 1). However, the cultivation of *P. oleracea* improved the SC of strawberry plants under all NaCl conditions when compared to those grown alone in saline conditions. The improvement of SC was evident in that the increases were from 183.26 to 230.80% from S_{30} to SP_{30} cultivation conditions, respectively. At the higher NaCl concentrations of S_{90} and SP_{90}, the SC was still improved with a lesser efficiency from 94.10 to 131.08%, respectively.

The average fruit weight and yield of strawberry plants under NaCl conditions were reduced in plants grown alone, but the co-cultivation of strawberry plants with *P. oleracea* increased the average and total fruit weight (Table 2).

The employment of *P. oleracea* not only increased the crop yield and physiological parameters but also improved the quality of fruits in terms of lycopene and vitamin C contents. Lycopene and vitamin C contents were gradually decreased as the concentration of NaCl increased. Again, the employment of *P. oleracea* increased the lycopene and vitamin C at all NaCl levels (Table 2). For example, the remarkable effect was more evident at the 90 mmol L^{-1} NaCl conditions, as the both lycopene and vitamin C contents were increased when grown with *P. oleracea* as compared to those of plants grown alone in saline conditions.

Table 2. Yield and some fruit properties of strawberry plants grown alone or in combination with *P. oleracea* at differing NaCl levels.

Treatments	Average Fruit Weight (g Plant^{-1})	Yield (g Plant^{-1})	Lycopene (mg kg^{-1} Fwt)	Vitamin C (mg kg^{-1} Fwt)	TSS (%)
S_0	18.53 ± 0.24 [a]	214.76 ± 25.26 [a]	37.98 ± 1.61 [a]	49.87 ± 2.48 [a]	8.80 ± 0.22 [a]
S_{30}	14.96 ± 0.58 [c]	132.85 ± 19.44 [c]	35.27 ± 1.29 [a]	45.07 ± 2.55 [c]	6.80 ± 0.25 [b]
S_{60}	9.80 ± 1.25 [e]	83.76 ± 15.20 [d]	27.52 ± 1.13 [b]	36.79 ± 1.30 [d]	5.80 ± 0.24 [d]
S_{90}	5.60 ± 0.58 [f]	31.53 ± 5.40 [e]	16.56 ± 1.61 [c]	32.53 ± 0.81 [e]	5.20 ± 0.25 [e]
SP_0	19.09 ± 0.52 [a]	229.40 ± 17.46 [a]	37.20 ± 1.45 [a]	51.88 ± 1.89 [a]	9.00 ± 0.20 [a]
SP_{30}	16.94 ± 0.54 [b]	164.80 ± 23.99 [b]	34.87 ± 1.95 [a]	47.19 ± 1.04 [a]	7.40 ± 0.17 [b]
SP_{60}	15.74 ± 0.53 [c]	147.57 ± 25.67 [c]	33.93 ± 1.82 [a]	41.36 ± 1.70 [c]	5.90 ± 0.20 [c]
SP_{90}	12.65 ± 0.73 [d]	94.00 ± 8.91 [d]	28.55 ± 1.31 [b]	43.12 ± 2.51 [c]	5.60 ± 0.19 [c]

Significance level at $p \leq 0.05$ was determined for the salt treatment using Duncan's multiple range test. Different letters in each column indicate statistical differences. S: a strawberry grown alone; SP: the strawberry and *P. oleracea* companionship; TSS: total soluble solids.

Unlike other parameters, the TSS contents of the fruits in saline conditions were significantly lowered. The co-cultivation of *P. oleracea* did not significantly improve the conditions of strawberry plants (Table 2).

Chl-*a* and Chl-*b* were significantly affected by salinity at the S_{60} and S_{90} mmol L^{-1} NaCl levels ($p \leq 0.05$). For example, the Chl-*a* and Chl-*b* were determined as 0.70 and 0.36 mg g^{-1} Fwt, respectively, at S_{90} mmol L^{-1} NaCl levels in strawberry plants. The positive effects of *P. oleracea* on the Chl-*a* and Chl-*b* contents at SP_{90} mmol L^{-1} NaCl were evident, as the Chl-*a* and Chl-*b* contents were 1.01 and 0.51 mg g^{-1} Fwt, respectively, in strawberry plants (Figure 2A,B).

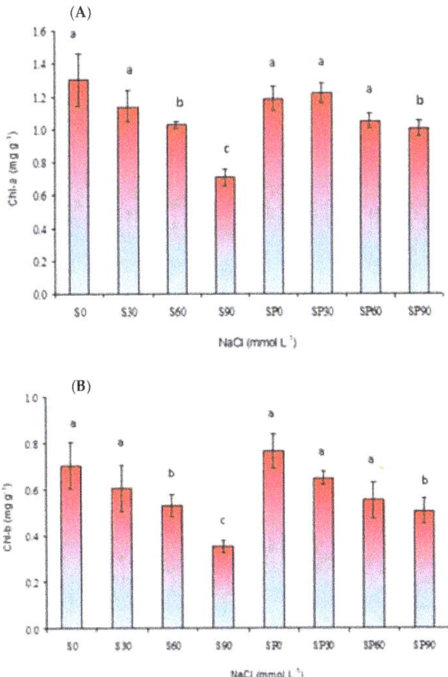

Figure 2. Leaf Chl-*a* (**A**) and Chl-*b* (**B**) contents of strawberry plants grown alone or in combination with *P. oleracea* at differing NaCl levels (0, 30, 60, and 90 mmol L^{-1}). S: a strawberry grown alone; SP: the strawberry and *P. oleracea* companionship; TSS: total soluble solids.

Leaf proline content significantly increased as the concentration of NaCl levels increased as a response to salinity stress. ($p \leq 0.05$); see Figure 3A. The highest proline level was determined as 13.77 μmol g^{-1} Fwt with the S$_{90}$ mmol L^{-1} NaCl treatment, whereas at the SP$_{90}$ condition, the proline level decreased to 4.40 μmol g^{-1}. Therefore, the combination of *P. oleracea* not only improved the physiological and biochemical conditions of strawberry plants but also reduced the stress metabolite levels.

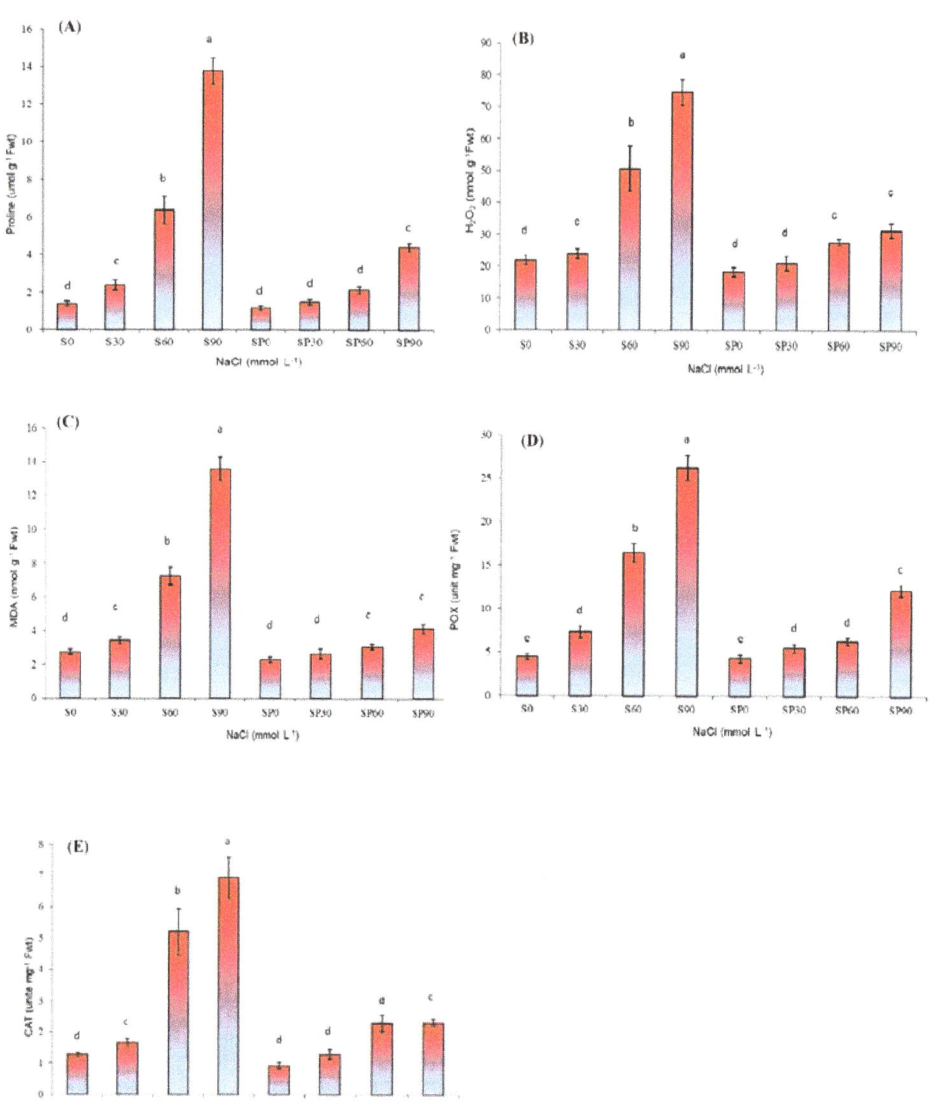

Figure 3. Proline (**A**), H$_2$O$_2$ (**B**), and malondialdehyde (MDA) (**C**) contents; peroxidase enzyme activity (POX) (**D**) and catalase enzyme activity (CAT) (**E**) antioxidant enzyme contents of strawberry plants grown alone or in combination with *P. oleracea* at differing NaCl levels (0, 30, 60, and 90 mmol L^{-1}). S: a strawberry grown alone; SP: the strawberry and *P. oleracea* companionship; TSS: total soluble solids.

The companionship of *P. oleracea* had a remarkable effect to reduce the impact of NaCl stress in strawberry plants. Again, leaf H_2O_2 and MDA contents increased with the increasing levels of salt stress. The highest H_2O_2 and MDA levels were determined as 74.72 and 13.58 nmol g^{-1} Fwt, respectively, at the S_{90} mmol L^{-1} NaCl level. The co-cultivation of *P. oleracea* with strawberry plants reduced the contents of H_2O_2 and MDA levels down to 31.56 and 4.15 nmol g^{-1} Fwt, respectively (Figure 3B,C).

POX and CAT antioxidant enzymes showed parallel patterns to those of previous parameters. The co-cultivation of *P. oleracea* significantly decreased antioxidant enzyme levels at the 60 and 90 mmol L^{-1} NaCl conditions, (Figure 3D,E).

Leaf Mineral Contents

The concentrations of beneficial ions such as those of K^+ and Ca^{2+} decreased with the increases in salinity levels in strawberry plants. The lowest K^+ and Ca^{2+} ions were determined at the S_{90} level. The leaf Mg^{2+} content was not significantly affected upon NaCl stress. *P. oleracea* co-cultivation with strawberry plants enhanced the Mg^{2+} ion level at NaCl treatments; see Table 3. Under saline conditions, gradual increases of Na^+ and Cl^- ions were evident in strawberry plants grown at increasing NaCl salinity, but the employment of *P. oleracea* significantly decreased the Na^+ and Cl^- ion contents; Table 3.

Table 3. Strawberry leaf mineral contents of strawberry plants grown alone or in combination with *P. oleracea* at differing NaCl levels.

Treatments	K^+ (%)	Ca^{2+} (%)	Mg^{2+} (%)	Na^+ (%)	Cl^- (%)
S_0	2.32 ± 0.10 [a]	2.39 ± 0.13 [a]	0.32 ± 0.03 [a]	0.17 ± 0.04 [e]	0.35 ± 0.01 [d]
S_{30}	1.82 ± 0.08 [b]	1.97 ± 0.11 [a]	0.28 ± 0.03 [a]	0.32 ± 0.02 [d]	0.67 ± 0.03 [d]
S_{60}	1.41 ± 0.06 [c]	1.86 ± 0.04 [b]	0.29 ± 0.02 [a]	0.69 ± 0.05 [b]	2.09 ± 0.29 [b]
S_{90}	1.02 ± 0.05 [e]	1.72 ± 0.03 [b]	0.28 ± 0.03 [a]	1.09 ± 0.03 [a]	3.69 ± 0.23 [a]
SP_0	2.42 ± 0.11 [a]	2.25 ± 0.06 [a]	0.35 ± 0.02 [a]	0.09 ± 0.01 [e]	0.25 ± 0.05 [d]
SP_{30}	2.27 ± 0.06 [a]	2.15 ± 0.13 [a]	0.33 ± 0.03 [a]	0.19 ± 0.03 [e]	0.46 ± 0.03 [d]
SP_{60}	1.78 ± 0.15 [b]	2.15 ± 0.25 [a]	0.34 ± 0.03 [a]	0.31 ± 0.05 [d]	1.16 ± 0.10 [c]
SP_{90}	1.66 ± 0.09 [b]	2.14 ± 0.11 [a]	0.31 ± 0.02 [a]	0.49 ± 0.08 [c]	1.29 ± 0.06 [c]

Significance level at $p \leq 0.05$ was determined for the salt treatment using Duncan's multiple range test. Different letters in each column indicate statistical differences. S: a strawberry grown alone; SP: the strawberry and *P. oleracea* companionship.

4. Discussion

Salinity stress is one of the most devastating issues that damages crop plants in terms of quantity and quality. Increased salinity levels not only damage plants during vegetative stages but also negatively affect reproductive stages. Under salt stress, Na^+ is extensively accumulated in the shoots and roots of cultivars and K^+ content is decreased [43]. Quality parameters such as vitamin contents, aromatic substances, and pigments are remarkably reduced. Leaf proline content, as a response to stress, tends to increase. Increasing proline content under salinity conditions indicates the adverse effects of osmotic stress on the plant. Proline and soluble carbohydrates (also known as compatible solutes) are expected to be accumulated under salinity in strawberry [44]. This can be considered to be a criterion for stress tolerance [45]. This study showed that *P. oleracea* in combination with strawberry decreased proline levels under salinity along with the reduction of Na^+ and Cl^- ion levels by reducing the toxic levels of salt ions. *P. oleracea* gave promising results on strawberry plants grown at different NaCl stress levels (0, 30, 60, and 90 mmol L^{-1}). It is important to note that peat has a high bulk density. For example, Nugraha et al. [46] stated that capillary water movement had a very critical role in supplying water to the rooting zones of crop plants or the top parts of the soil. They reported that the rate of capillary water movement progressively corresponded to the increase in bulk density. Farina et al. [47] also stated that NaCl accumulation in peat mulching was much lesser than that of soil. They stated that if the porosity in the surface layers became small enough, irrigation or raindrops could plug macropores in the surface. They either block main avenues for water and roots to move through the soil or they form a cement-like surface layer when the soil dries. The rock-solid upper layer or salt crust then restricts water movement and plant emergence. In our study, evaporation in the greenhouse was not high enough to build up NaCl accumulation in the top part of the soil. Therefore, no salt crust formation, which would have affected the results of our experimental findings, was observed.

Mozafari et al. [48] stated that salinity negatively affected the growth parameters, pigment content, and membrane stability, as well as disturbing the ionic balance in plants. For example, Saied et al. [49] stated that strawberry was considered to be a saline-sensitive plant. Many physiological and biochemical parameters deteriorated. This both directly and indirectly led to diminished productivity in plants [50]. We determined that fresh weight, dry weight, stomatal conductance, fruit average weight, fruit total yield, chlorophyll (Chl-*a* and Chl-*b*), total soluble solids, lycopene content, vitamin C content, and leaf mineral content (K^+ and Ca^{2+}) of strawberry plants significantly decreased with increasing NaCl levels. Strawberry plants grown in companionship with *P. oleracea* improved the condition of plants, and much lesser reductions in terms of total yield and quality were evident. The positive effect on strawberry growth was quite remarkable. The leaf electrolyte, proline, malondialdehyde, H_2O_2 contents, catalase enzyme activities, nd peroxidase enzyme activities, and leaf mineral contents (Na^+ and Cl^-) of strawberry plants increased with an increasing level of salinity. The companion plants helped strawberry plants by reducing toxic ion levels, antioxidant enzyme levels, and stress metabolites. With the improvement of those parameters, electrolyte leakage and stomatal conductance were also improved, and this was reflected in the quality of fruits in terms of lycopene and vitamin C contents. This study proved that mixed planting with *P. oleracea* in saline conditions was an effective phytoremediation technique that might significantly increase the yield production and quality of strawberry. Similar findings were also made for *S. soda* plants by Karakas [51] who suggested that the improvement of tomato plants via companion plants under salt stress (1.3 and 6.5 dS m^{-1}) was achieved with the synthesis of substances used for fruit development instead of building up substances for mechanisms of stress tolerance. It is important to note that synthesizing stress metabolites and antioxidant enzymes is quite costly for plants to cope with abiotic or biotic stress factors [17]. Instead of generating crop plants that can combat stress factors, the strategy that involves removing stress factors would be much appreciated. Any genetic modifications or biochemical approaches that increase the removing capacity of toxic ions or compounds from the soil habitat would be an environmentally friendly approach and a safe strategy. For example, Grafienberg et al. [52] and Karakas et al. [51] stated that reductions in stress metabolites and the uptake of toxic ions enabled tomato plants to use more energy to build up organic components such as lycopene and proteins instead of producing substances for defense mechanisms. In this study, salinity stress resulted in a reduction in vitamin C content and lycopene contents in strawberry. Jamalian et al. [53] showed that salinity reduced the vitamin C content of strawberries, which was in line with the results of the present study. The decrease in the vitamin C content of fruits at high salinity levels can be attributed to the decrease in carbohydrate (sugar) production caused by the decrease in photosynthesis required for vitamin C biosynthesis.

Yaghubi et al. [44] reported that MDA concentration was also high in strawberry plants at salt stress conditions. They reported that reactive oxygen species (ROS) production was muck higher than the scavenging capacity of antioxidant enzymes. The dismutation of O_2^{-2} into H_2O_2 and O_2 was reported to increase H_2O_2 concentration [54]. This was observed by a higher H_2O_2 content in salt-stressed strawberry plants than in control plants. Since H_2O_2 was accompanied by an increase in the key antioxidant enzymes such as CAT, POD, and superoxide dismutase (SOD), a reduction of H_2O_2 was achieved. In our methodology, we achieved the decrease of stress metabolites while suppressing the antioxidant enzymes via the use of *P. oleracea* plants. Though antioxidant enzymes such as CAT, POD, and SOD are known to substantially reduce the levels of O_2^- and H_2O_2 in plants and play a vital role in plant defense against oxidative stress [55], the increase of these enzymes might interfere with the chemical compounds involved in quality parameters such as lycopene and vitamin C. With the use of *P. oleracea*, we were able to reduce stress metabolites and toxic ions, reduced further damages to cell components, and increased the quality-related compounds without increasing defense-related antioxidant compounds. This saved the energy to be used for defense responses, and this saved energy could be used to increase metabolic functions and quality parameters.

5. Conclusions

Strawberry cultivation has become popular recently, and this has led to an increase in cultivated areas. These areas have become saline-polluted, saline-prone, and saline-prevalent. Since strawberry is a salt-sensitive plant, it is easily affected by a mild or moderate level of salinity. A very low level of NaCl could reduce crop yield and reduce the quality of fruits.

In this study, strawberry seedlings were grown alone or in combination with *P. oleracea* under differing NaCl concentrations. Strawberry seedlings under increasing NaCl salinity were negatively affected in terms of physiological, morphological, and biochemical parameters. Defending plants synthesized various stress metabolites such as proline, MDA, H_2O_2, and antioxidant enzymes to ease the negative effects of NaCl toxicity. However, increases of these metabolites were negatively correlated with quality-related metabolites such as vitamin C and lycopene. The cultivation of strawberry plants with *P. oleracea* plants reduced the concentrations of stress metabolites and antioxidant enzyme levels, as well as indirectly contributing to increases of vitamin C and lycopene contents.

We suggest that the employment of *P. oleracea* would remediate the conditions of strawberry parameters by accumulating Na^+ and Cl^- ions, thus causing reductions in the synthesis of stress metabolites. The use of *P. oleracea* is a quite practical and environmentally friendly approach where salinity is prevalent. *P. oleracea* has a high potential that could be used in high saline and in other environmental stress conditions.

Author Contributions: Conceptualization, S.K., M.D. and I.B.; methodology, S.K., I.B. and M.D.; software, S.K. and M.D.; validation, S.K., I.B. and M.D.; formal analysis, S.K. and M.D.; investigation, S.K., I.B. and M.D.; resources, S.K., I.B. and M.D.; data curation, S.K.; writing—original draft preparation, S.K.; writing—review and editing, M.D. and I.B.; visualization, S.K., I.B. and M.D.; supervision, S.K.; project administration, S.K.; funding acquisition, S.K. All authors have read and agreed to the published version of the manuscript.

Funding: This research was funded by Harran University Scientific Research Project (HUBAP), grant number 17247.

Institutional Review Board Statement: Not applicable.

Informed Consent Statement: Not applicable.

Conflicts of Interest: The authors declare no conflict of interest.

References

1. Zhao, G.M.; Han, Y.; Sun, X.; Li, S.H.; Shi, Q.M.; Wang, C.H. Salinity increases secondary metabolites and enzyme activity in safflower. *Ind. Crop. Prod.* **2015**, *64*, 175–181.
2. Zhou, Y.; Tang, N.; Huang, L.; Zhao, Y.; Tang, X.; Wang, K. Effects of Salt Stress on Plant Growth, Antioxidant Capacity, Glandular Trichome Density, and Volatile Exudates of Schizonepeta tenuifolia Briq. *Int. J. Mol. Sci.* **2018**, *19*, 252. [CrossRef]
3. Hasanuzzaman, M.; Nahar, K.; Alam, M.M.; Bhowmik, P.C.; Hossain, M.A.; Rahman, M.M.; Prasad, M.N.V.; Ozturk, M.; Fujita, M. Potential use of halophytes to remediate saline soils. *BioMed Res. Int.* **2014**, *2014*, 589341. [CrossRef] [PubMed]
4. Menason, E.; Betty, T.; Vijayan, K.K.; Anbudurai, P.R. Modification of fatty acid composition in salt adopted Synechocystis 6803 cells. *Ann. Biol. Res.* **2015**, *6*, 4–9.
5. Flowers, T.J.; Colmer, T.D. Salinity tolerance in halophytes. *New Phytol.* **2008**, *179*, 945–963. [CrossRef] [PubMed]
6. Kader, M.A.; Lindberg, S. March Cytosolic calcium and pH signaling in plants under salinity. *Plant. Signal. Behav.* **2010**, *5*, 233–238. [CrossRef]
7. Hossain, M.D.; Inafuku, M.; Iwasaki, H.; Taira, N.; Mostofa, M.G.; Oku, H. Differential enzymatic defense mechanisms in leaves and root of two true mangrove species under long-term salt. *Aquat. Bot.* **2017**, *142*, 32–40. [CrossRef]

8. Munns, R. Comparative physiology of salt and water stress. *Plant Cell Environ.* **2002**, *25*, 239–250. [CrossRef]
9. Acosta-Motos, J.R.; Ortuño, M.F.; Bernal-Vicente, A.; Diaz-Vivancos, P.; Sanchez-Blanco, M.J.; Hernandez, J.A. Plant Responses to Salt Stress: Adaptive Mechanisms. *Agronomy* **2017**, *7*, 18. [CrossRef]
10. Cheeseman, J. The evolution of halophytes, glycophytes and crops, and its implications for food security under saline conditions. *New Phytol.* **2014**, 206. [CrossRef]
11. Meng, X.; Sui, J.Z.N. Mechanisms of salt tolerance in halophytes: Current understanding and recent advances. *Open Life Sci.* **2018**, *13*, 149–154. [CrossRef]
12. Aslam, R.; Bostan, N.; Amen, N.; Maria, M.; Safdar, W. A critical review on halophytes: Salt tolerant plants. *J. Med. Plants Res.* **2011**, *5*, 7108–7118.
13. Flowers, T.J.; Galal, H.K.; Bromham, L. Evolution of halophytes: Multiple originsof salt tolerance in land plants. *Funct. Plant. Biol.* **2010**, *37*, 604–612. [CrossRef]
14. Qadir, M.; Qureshi, R.H.; Ahmad, N. Amelioration of calcareous saline sodic soils through phytoremediation and chemical strategies. *Soil Use Manag.* **2002**, *18*, 381–385. [CrossRef]
15. Dikilitas, M.; Karakas, S. Salt as potential environmental Pollutants, their types, effects on plants, and approaches for their phytoremediation. In *Plant Adaptation and Phytoremediation*; Ashraf, M., Ozturk, M., Ahmad, M.S.A., Eds.; Springer: Berlin/Heidelberg, Germany, 2010; pp. 357–383.
16. Xing, J.C.; Dong, J.; Wang, M.V.; Liu, C.; Zhao, B.Q.; Wen, Z.G.; Zhu, X.M.; Ding, H.R.; Zhao, X.H.; Hong, L.Z. Effects of NaCl stress on growth of *Portulaca oleracea* and underlying mechanisms. *Braz. J. Bot.* **2019**, *42*, 217–226. [CrossRef]
17. Karakas, S.; Dikilitas, M.; Tıpırdamaz, R. Phytoremediation of Salt-Affected Soils Using Halophytes. In: Grigore MN. (eds) Handbook of Halophytes. *Springercham* **2020**, 1–18. [CrossRef]
18. De Lacerda, L.P.; Lange, L.C.; Costa França, M.G.; Diniz Leão, M.M. Growth and differential salinity reduction between *Portulaca oleracea* and *Eichhornia crassipes* in experimental hydroponic units. *Env. Technol.* **2018**, *22*, 1–9. [CrossRef]
19. Grieve, C.M.; Suarez, D.L. Purslane (*Portulaca oleracea* L.): A halophytic crop for drainage water reuse systems. *Plant. Soil* **1997**, *192*, 277–283. [CrossRef]
20. Karakas, S.; Çullu, M.A.; Dikilitas, M. In Vitro kosullarda halofit bitkilerden *Salsola soda* ve *Portulaca oleracea'* nın NaCl stresine karşı çimlenme ve gelisim durumları. *Harran Tarım Ve Gıda Bilimleri Derg.* **2015**, *19*, 66–74.
21. Folta, K.M.; Davis, T.M. Strawberry genes and genomics. *Crit. Rev. Plant. Sci.* **2006**, *25*, 399–415. [CrossRef]
22. Shulaev, V.; Korban, S.S.; Sosinski, B.; Abbott, A.G.; Aldwinckle, H.S.; Folta, K.M.; Iezzoni, A.; Main, D.; Arús, P.; Dandekar, A.M.; et al. Multiple models for rosaceae genomics. *Plant. Physiol.* **2008**, *147*, 985–1003. [CrossRef] [PubMed]
23. Neocleous, D.; Vasilakakis, M. Effects of NaCl stress on red raspberry (*Rubus idaeus* L. 'Autumn Bliss'). *Sci. Hortic.* **2007**, *112*, 282–289. [CrossRef]
24. Keutgen, A.; Pawelzik, E. Quality and nutritional value of strawberry fruit under long term salt stress. *Food Chem.* **2008**, *107*, 1413–1420. [CrossRef]
25. Jamalian, S.; Gholami, M.; Esna-Ashari, M. Abscisic acid-mediated leaf phenolic compounds, plant growth and yield is strawberry under different salt stress regimes. *Theor. Exp. Plant. Physiol.* **2013**, *25*, 291–299.
26. Garriga, M.; Muñoz, C.A.; Caligari, P.D.; Retamales, J.B. Effect of salt stress on genotypes of commercial (*Fragaria x ananassa*) and Chilean strawberry (*F. chiloensis*). *Sci. Hortic.* **2015**, *195*, 37–47. [CrossRef]
27. Bohlin, C.; Holmberg, P. Peat dominating growing medium in Swedish horticulture. *Acta Hortic.* **2004**, *644*, 177–181. [CrossRef]
28. Bin Mohamad, H.; Zainorabidin, A.; Razali, S.; Zolkefle, S. Assessment for applicability of microwave oven in rapid determination of moisture content in peat soil. *J. Eng. Sci. Technol.* **2020**, *15*, 2110–2118.
29. Catania, P.; Comparetti, A.; De Pasquale, C.; Morello, G.; Vallone, M. Effects of the Extraction Technology on Pomegranate Juice Quality. *Agronomy* **2020**, *10*, 1483. [CrossRef]
30. Barrett, D.M.; Anthon, G. Lycopene content of california-grown tomato varieties. *Acta Hortic.* **2001**, *542*, 165–174. [CrossRef]
31. Karakas, S. Development of Tomato Growing in Soil Differing in Salt Levels and Effects of Companion Plants on Same Physiological Parameters and Soil Remediation. Ph.D. Thesis, University of Harran, Sanlıurfa, Turkey, 2013.
32. Oz, A.T. Effects of two differrent temperatures on l-ascorbic acid content (Vitamın C), lenght of storage time and fruit quality. *Bahce* **2002**, *31*, 51–57.
33. Lutts, S.; Kinet, J.M.; Bouharmont, J. NaCl-induced senescence in leaves of rice (*Oryza sativa* L.) cultivars differing in salinity resistance. *Ann. Bot.* **1996**, *78*, 389–398. [CrossRef]
34. Karlidag, H.; Yildirim, E.; Turan, M. Role of 24-epibrassinolide in mitigating the adverse effects of salt stress on stomatal conductance, membrane permeability, and leaf water content, ionic composition in salt stressed strawberry (Fragaria × ananassa). *Sci. Hortic.* **2011**, *130*, 133–140. [CrossRef]
35. Arnon, D.L. A copper enzyme is isolated chloroplast polyphenol oxidase in Beta vulgaris. *Plant. Physiol.* **1949**, *24*, 1–15. [CrossRef] [PubMed]
36. Bates, L.S.; Waldren, R.P.; Teare, I.D. Rapid determination of free proline for water-stress studies. *Plant. Soil* **1973**, *39*, 205–207. [CrossRef]
37. Velikova, V.; Yordanov, I.; Edreva, A. Oxidative stress and some antioxidant systems in acid raintreated bean plants: Protective role of exogenous polyamines. *Plant. Sci.* **2000**, *151*, 59–66. [CrossRef]

38. Karakas, S.; Dikilitas, M.; Tıpırdamaz, R. Biochemical and molecular tolerance of *Carpobrotus acinaciformis* L. halophyte plants exposed to high level of NaCl stress. *Harran J. Agric. Food Sci.* **2019**, *23*, 99–107.
39. Sairam, R.K.; Sexena, D. Oxidative stress and antioxidants in wheat genotypes: Possible mechanism of water stress tolerance. *J. Agron. Crop. Sci.* **2000**, *184*, 55–61. [CrossRef]
40. Milosevic, N.; Slusarenko, A.J. Active Oxygen Metabolism and Lignifications in The Hypersensitive Response in Bean. *Physiol. Mol. Plant. Pathol.* **1996**, *49*, 143–158. [CrossRef]
41. Cvikrova, M.; Hrubcova, M.; Vagner, M.; Machackova, I.; Eder, J. Phenolic acids and peroxidase activity in Alfalfa (*Medicago sativa*) embryogenic cultures after ethephon treatment. *Plant. Physiol.* **1994**, *91*, 226–233. [CrossRef]
42. Chapman, H.D.; Pratt, P.F. *Methods of Analysis for Soils, Plants, and Waters*; University of California, Division of Agricultural Sciences: Riverside, CA, USA, 1961.
43. Saidimoradi, D.; Ghaderi, N.; Javadi, T. Salinity stress mitigation by humic acid application in strawberry (*Fragaria x ananassa* Duch.). *Sci. Hortic.* **2019**, *256*, 594. [CrossRef]
44. Yaghubi, K.; Ghaderi, N.; Vafaee, Y.; Javadi, T. Potassium silicate alleviates deleterious effects of salinity on two strawberry cultivars grown under soilless pot culture. *Sci. Hortic.* **2016**, *213*, 87–95. [CrossRef]
45. Munns, R.; Tester, M. Mechanisms of salinity tolerance. *Annu. Rev. Plant. Biol.* **2008**, *59*, 651–681. [CrossRef] [PubMed]
46. Nugraha, M.I.; Annisa, W.; Syaufina, L.; Anwar, S. Capillary water rise in peat soil as affected by various groundwater levels. *Indones. J. Agric. Sci.* **2016**, *17*, 75–83. [CrossRef]
47. Farina, E.; Allera, C.; Paterniani, T.; Palagi, M. Mulching as a technique to reduce salt accumulation in soilless culture. *Acta Hortic.* **2003**, *609*, 459–466. [CrossRef]
48. Mozafari, A.A.; Dedejani, S.; Ghaderi, N. Positive responses of strawberry (*Fragaria×ananassa Duch.*) explants to salicylic an iron nanoparticle application undersalinity conditions. *Plant. Celltissue Organ. Cult.* **2018**, *134*, 267–275. [CrossRef]
49. Saied, A.S.; Keutgen, A.J.; Noga, G. The influence of NaCl salinity on growth, yield and fruit quality of strawberry cvs. 'Elsanta'and 'Korona'. *Sci. Hortic.* **2005**, *103*, 289–303. [CrossRef]
50. Joseph, B.; Jini, D.; Sujatha, S. Insight into the role of exogenous salicylic acid on plants grown under salt environment. *Asian J. Crop. Sci.* **2010**, *2*, 226–235. [CrossRef]
51. Karakas, S.; Cullu, M.A.; Kaya, C.; Dikilitas, M. Halophytic companion plants improve growth and physiological parameters of tomato plants grown under salinity. *Pak. J. Bot.* **2016**, *48*, 21–28.
52. Grafienberg, A.; Botrini, L.; Giustiniani, L.; Filippi, F.; Curadi, M. Tomato growing in saline conditions with biodesalinating plants: *Salsola soda* and *Portulaca oleracea*. *Acta Hortic.* **2003**, *609*, 301–305. [CrossRef]
53. Jamalian, S.; Tehranifar, A.; Tafazoli, E.; Eshghi, S.; Davarynejad, G.H. Paclobutrazol application ameliorates the negative effect of salt stress on reproductive growth, yield, and fruit quality of strawberry plants. *Hortic. Environ. Biotechnol.* **2008**, *49*, 1–6.
54. Singh, R.; Flowers, T. Physiology and molecular biology of the effects of salinity on rice. *Handb. Plant. Crop. Stress* **2010**, 901–942.
55. Ashraf, M. Biotechnological approach of improving plant salt tolerance using antioxidants as markers. *Biotechnol. Adv.* **2009**, *27*, 84. [CrossRef] [PubMed]

 *horticulturae*

Communication

Effects of Short-Term Exposure to Low Temperatures on Proline, Pigments, and Phytochemicals Level in Kale (*Brassica oleracea* var. *acephala*)

Valentina Ljubej [1], Erna Karalija [2], Branka Salopek-Sondi [1] and Dunja Šamec [1,3,*]

[1] Department of Molecular Biology, Ruđer Bošković Institute, Bijenička Cesta 54, 10000 Zagreb, Croatia; kruk.valentina@gmail.com (V.L.); salopek@irb.hr (B.S.-S.)

[2] Department for Biology, Faculty of Science, University of Sarajevo, Zmaja od Bosne 33-35, 71000 Sarajevo, Bosnia and Herzegovina; erna.karalija@gmail.com

[3] Department of Food Technology, University North, Trg dr. Žarka Dolinara 1, 48000 Koprivnica, Croatia

* Correspondence: dsamec@unin.hr

Citation: Ljubej, V.; Karalija, E.; Salopek-Sondi, B.; Šamec, D. Effects of Short-Term Exposure to Low Temperatures on Proline, Pigments, and Phytochemicals Level in Kale (*Brassica oleracea* var. *acephala*). *Horticulturae* **2021**, *7*, 341. https://doi.org/10.3390/horticulturae7100341

Academic Editors: Agnieszka Hanaka, Jolanta Jaroszuk-Ściseł, Małgorzata Majewska and Alessandra Francini

Received: 6 September 2021
Accepted: 22 September 2021
Published: 24 September 2021

Publisher's Note: MDPI stays neutral with regard to jurisdictional claims in published maps and institutional affiliations.

Abstract: Kale (*Brassica oleracea* var. *acephala*) is known as a vegetable with good tolerance of environmental stress and numerous beneficial properties for human health, which are attributed to different phytochemicals. In the present study, investigation of how low temperatures affect proline, pigments and specialized metabolites content was performed using 8-weeks old kale plants subjected to chilling (at 8 °C, for 24 h) followed by short freezing (at −8 °C, for 1 h after previous acclimation at 8 °C, for 23 h). Plants growing at 21 °C served as a control. In both groups of plants (exposed to low temperatures and exposed to short freezing) a significant increase in proline content (14% and 49%, respectively) was recorded. Low temperatures (8 °C) induced an increase of pigments (total chlorophylls 7%) and phytochemicals (phenolic acids 3%; flavonoids 5%; carotenoids 15%; glucosinolates 21%) content, while exposure to freezing showed a different trend dependent upon observed parameter. After freezing, the content of chlorophylls, carotenoids, and total phenolic acids retained similar levels as in control plants and amounted to 14.65 ± 0.36 mg dw g^{-1}, 2.58 ± 0.05 mg dw g^{-1} and 13.75 ± 0.07 mg dw CEA g^{-1}, respectively. At the freezing temperature, total polyphenol content increased 13% and total flavonoids and glucosinolates content decreased 21% and 54%, respectively. Our results suggest that acclimatization (23 h at 8 °C) of kale plants can be beneficial for the accumulation of pigments and phytochemicals, while freezing temperatures affect differently specialized metabolite synthesis. The study suggests that growing temperature during kale cultivation must be considered as an important parameter for producers that are orientated towards production of crops with an increasing content of health-related compounds.

Keywords: *Brassica oleracea* var. *acephala*; short-term cold stress; phytochemicals; pigments

1. Introduction

Climate changes that generally cause global warming can cause sudden, low temperature episodes. These types of events can significantly reduce crop production and quality. According to the review article by Ritonga and Chen [1], low temperature is one of the most harmful environmental stresses that higher plants face, along with drought stress. Low temperature stress significantly limits the geographical distribution of plants, but with climate changes and climate disturbances, crops can experience periods of extremely low temperatures uncommon for the region [2]. Crops that experience low-temperature stress suffer impaired growth, which affects and delays essential processes in their development such as flowering, development and fruiting. That may cause loss in crop yields and quality, and in extreme cases, the plant freezes, wilts and dies [2]. Plants differ in their tolerance to chilling (0–15 °C) and freezing (<0 °C) temperatures, and resistance is species- or cultivar-dependent. Chilling tolerant plants often grow in temperate climatic regions, and they can increase their freezing tolerance by being exposed to chilling, non-freezing temperatures, a process known as cold acclimation [3]. Plant tolerance to low temperature may involve mechanisms related to carbohydrates, specialized metabolites and energy metabolism, transcription, signal transduction, protein transport and degradation [3–5].

The family Brassicaceae, often called Cruciferae or mustard family, represents a monophyletic group distributed worldwide, except for Antarctica, comprising of approximately 341 genera and 3977 species. The number of genera and species changes frequently in parallel with development of new methods that dissect the genetic diversity of the family in a greater detail [6]. One of the important members is the genus *Brassica*, which includes significant crop plants such as rapeseed and some commonly used vegetables with a long history of agricultural use around the world [7]. One of the key species is *Brassica oleracea* that includes vegetables with significant morphological diversity of plant organs and good adaptation to adverse environmental conditions [8]. According to the morphology and other characteristics *B. oleracea* includes several cultivar groups (Acephala, Botrytis, Capitata group, etc.). Among these groups, plants from Acephala group include leafy, non-heading cabbage varieties with common names of kale and collards. Kale (*B.oleracea* var. *acephala*) is known for its good tolerance of environmental stress such as salt, drought and frost hardiness and resistance [8–14]. Kale is easy and cheap for cultivation and production significantly increased from 3994 to 6256 harvested acres in US, in the time from 2007 to 2012, respectively [15]. In recent years, kale has been marked as a superfood [8] due to its beneficial effects on human health, which are related to the presence of specialized metabolites from the group of polyphenols, carotenoids and glucosinolates, but also important minerals, vitamins and dietary fibres [8,11,13,14]. Kale and collards have higher content of Ca, folate, riboflavin, vitamin C, K and A content than other cruciferous vegetables while their phytochemical content is comparable with other *Brassica* vegetables [8]. Kale possesses biological activities such as antioxidant and anticarcinogens activity, and protective properties in cases of cardiovascular and gastrointestinal system diseases (reviewed by Šamec et al. [8]). Kale tolerates both high summer temperatures and low winter temperatures, even freezing temperatures for a short period of time. Improved understanding of mechanisms related to kale low temperature tolerance can introduce improved cold resistance varieties, better equipped to deal with unusually low temperatures that might arise during the growth season. Low temperatures during kale cultivation have been reported to affect polyphenolic compounds [16,17], soluble sugars and glucosinolates [18], but most of these studies were conducted in the field where it is difficult to control other environmental parameters.

Tolerance of low temperatures in kale is associated with the content of specialized metabolites that can serve not only as a protective mechanism against environmental stressors for the plant, but also as a source of benefit compounds for human health. We hypothesize that phytochemicals and pigments may play a significant role in kale stress adaptation capability and that short term low temperature exposure affects specialized metabolite content. In the present study, we investigated cold induced changes in specialized metabolite content: polyphenols, glucosinolates and carotenoids, under short term chilling temperatures (8 °C and −8 °C) compared with the control (21 °C). In addition, we determined the effect of short cold stress on basic growth physiological parameters (chlorophyll *a*, chlorophyll *b*, total chlorophylls) and stress markers (proline content).

2. Materials and Methods

2.1. Plant Growing and Stress Experiments

Seeds were purchased from the family farm "Srđan Franić", Vrgorac, Croatia. Seeds were sterilized and germinated as we have previously reported [13]. Approximately 20 seeds per plate were transferred to Petri dishes containing 1% agar (*w/v*). Plates were first stored in dark at 4 °C for 48 h, then transferred to the growth chamber and maintained at 21 °C for three days with a 16/8 h photoperiod. Seedlings were then placed in pots (diameter 55–73 mm; height 105 mm) containing Stender A240 substrate (Stender GmbH,

Schermbeck, Germany) and maintained in a growth chamber at 16/8 h photoperiod (light/dark) at 21 °C. All plants were watered regularly with the same amount of water and transplanted after 4 weeks into larger pots (diameter 130–145 mm; height 110 mm) with fresh substrate to ensure sufficient plant nutrients throughout the experimental period. After 8 weeks of growth, 30 healthy representative plants were selected for cold stress experiments.

The experimental design is shown in Figure 1. Plants from all three groups (control, chilling and freezing group) were harvested at exactly the same time, quickly frozen in liquid nitrogen, and stored at −80 °C followed by freeze-drying (Lyovac GT 2; Steris Deutschland GmbH, Köln, Germany) for further extraction and analysis.

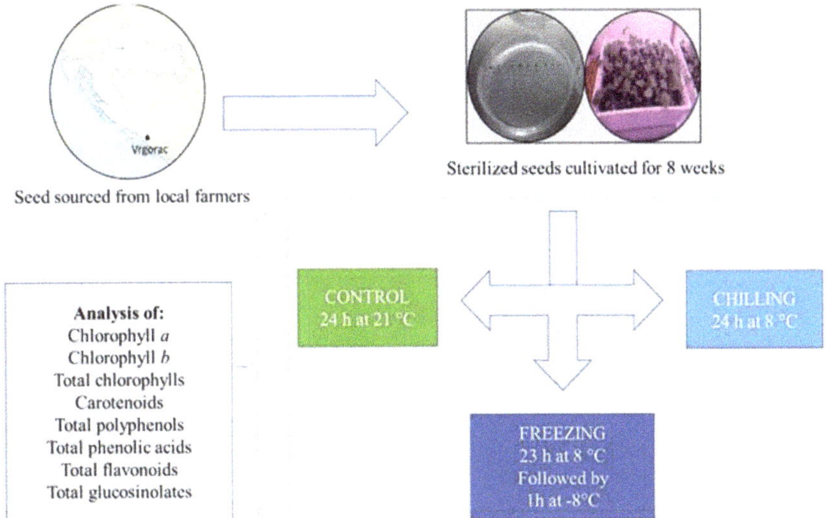

Figure 1. Schematic diagram of the short-term cold experiments.

2.2. Determination of Proline Content

For proline determination, we extracted 30 mg of freeze-dried tissue with 70% ethanol, as previously reported [10]. For the reaction, a mixture of 100 µL of extract and 1 mL of the reaction mixture (containing 1% ninhydrin, 60% acetic acid, and 20% ethanol) was heated at 95 °C for 20 min, followed by cooling on ice. Absorbance was measured at 520 nm against control using a UV–VIS spectrophotometer (BioSpec-1601, Shimadzu, Kyoto, Japan). Quantification of proline was done according to calibration curve of proline (0–1.6 mM) and expressed as μmol g^{-1} dw^{-1}.

2.3. Determination of Chlorophylls and Carotenoids Content

To determine concentration of chlorophyll *a*, chlorophyll *b*, total chlorophylls, and carotenoids, 10 mg of freeze-dried tissue was extracted using 80% acetone until tissue discoloration. Samples were centrifuged and supernatants were collected for quantification by spectrophotometric absorbance reading at 663.2 nm for chlorophyll *a*, 646.8 nm for chlorophyll *b*, and 470 nm for carotenoids [19] and expressed as mg g^{-1} dw^{-1}.

2.4. Determination of Polyphenolic Compounds

For the determination of the content of polyphenolic compounds, extracts were prepared by mixing 60 mg of dried material with 2 mL of 80% methanol and the extraction was performed as previously reported [14].

For the determination of total polyphenols, total phenolic acids and total flavonoids, we used well-established methods suitable for small-scale volumes and for Brassicacea plants [20]. For the determination of total phenols content, Folin–Ciocâlteu method was used and the calibration curve was constructed using gallic acid as the standard. Results are expressed as milligrams of gallic acid equivalents per gram of dry weight (mg GAE $g^{-1}dw^{-1}$). For the determination of phenolic acids, a mixture of sodium nitrite and sodium molybdate was used and caffeic acid was used to construct the calibration curve. Results are expressed as caffeic acid equivalents per gram of dry weight (mg CAE $g^{-1}dw^{-1}$). Flavonoids were determined using Al_2O_3 method, catechin was used as the standard to construct the calibration curve, and results are expresses as catechin equivalents per gram of dry weight (mg CE g^{-1} dw^{-1}).

2.5. Determination of Total Glucosinolates

For the determination of total glucosinolates, an established method based on the reaction with sodium tetrachloropalladate II (Na_2PdCl_4) [21] was adapted as previously reported [14]. Sinigrin standard was used to construct the calibration curve and the results are expressed as sinigrin equivalents per gram of dry weight (mg g^{-1} dw^{-1}).

2.6. Statistical Analysis

All analyses were performed in at least three replicates and results are expressed as mean $\pm$ standard deviation (SD). All statistical analyses were performed using the free software PAST [22]. One-way ANOVA and post hoc multiple mean comparison (Tukey's HSD test) were performed and differences between measurements were considered significant at $p < 0.05$.

3. Results

3.1. Effect of Cold Stress on Proline Content in Kale

Increase in proline content was noticed after exposure to cold stress was strongly related to the drop of the temperature (Figure 2). After 24 h at chilling temperatures (8 °C), proline content was 14% higher than in the control plants, with additional increase to 49% in kale exposed to −8 °C for 1 h after 23 h of acclimation. The rise of proline, as one of cold stress markers, is a clear indication that kale plants perceive the administered temperatures as cold stress.

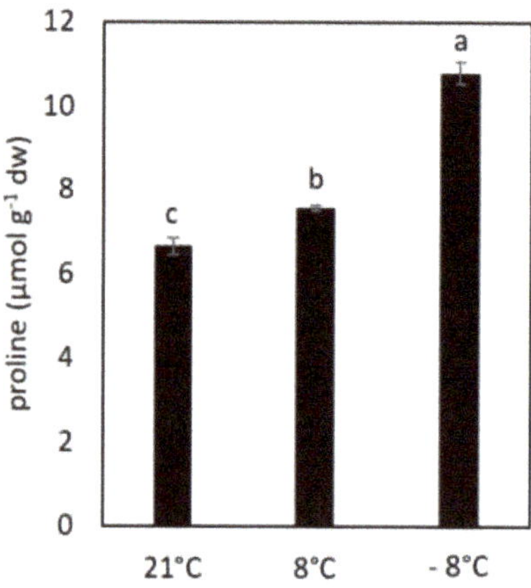

Figure 2. The proline content in kale plants exposed to cold stress. Value marked with different letters are significantly different at $p < 0.05$.

3.2. Effect of Cold Stress on Content of Chlorophylls and Carotenoids in Kale

Concentration of chlorophylls in chilled plants showed a significant increase in content of chlorophyll a, b and total chlorophylls (Figure 3), indicating that cold stressed plants increased the rate of synthesis of chlorophylls. In case of previous acclimatization of plants to chilling temperatures followed by exposure for freezing temperatures, the plant's photosynthetic apparatus was less affected and the content of photosynthetic pigments remained similar to the levels recorded for stress free control plants.

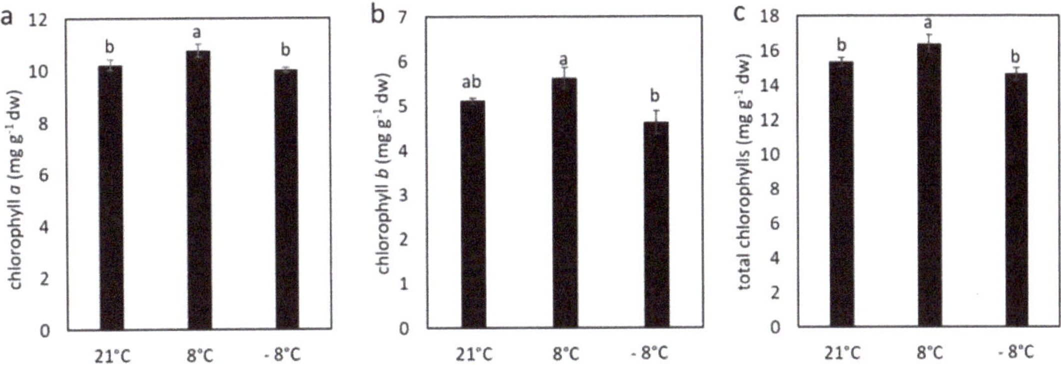

Figure 3. The content of chlorophyl a (**a**), chlorophyl b (**b**) and total chlorophylls (**c**) in kale exposed to low temperatures. Value marked with different letters are significantly different at $p < 0.05$.

The content of carotenoids in kale under low temperatures is shown in Figure 4. In the control, the content of carotenoids was 2.71 ± 0.06 mg g^{-1} and their amount followed the same trend as that of chlorophylls—under 8 °C increase, but their content in plants subjected to freezing temperature was comparable with those in the control plants.

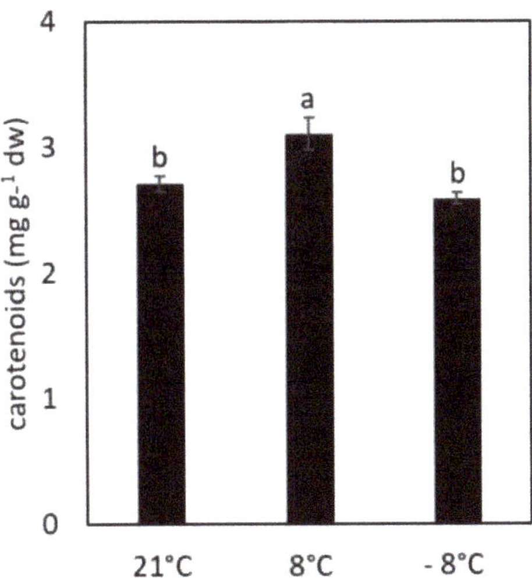

Figure 4. The content of carotenoids in kale exposed to low temperatures. Value marked with different letters are significantly different at $p < 0.05$.

The ratio of chlorophyll a and chlorophyll b, as well as the ratio of total chlorophylls and total carotenoids could provide useful information about adaptation of plants to stress conditions [19]. Our values for those parameters are shown at Table 1. The ratio of chlorophyll a/chlorophyll b remained stable for under chilling temperatures while significant increase was recorded for plants exposed to freezing temperature. The ratio of total chlorophylls/total carotenoids remained stable under cold stress.

Table 1. The ratio of chlorophyll a/chlorophyll b and total chlorophylls/total carotenoids in kale exposed to low temperatures. Value marked with different letters are significantly different at $p < 0.05$.

	Control (21 °C)	Chilling 24 h (8 °C)	Freezing 1 h after Acclimatization 23 h (−8 °C)
Chlorophyll *a*/ chlorophyll *b*	2.00 ± 0.02 [b]	1.91 ± 0.04 [b]	2.17 ± 0.10 [a]
Total chlorophylls/ total carotenoids	5.67 ± 0.07 [a]	5.29 ± 0.14 [a]	5.67 ± 0.24 [a]

3.3. Effect of Cold Stress on Content of Polyphenolic Compounds in Kale

Significant changes in polyphenolic compounds was recorded in plants exposed to freezing temperatures and results are shown at Figure 5. For total phenolic acids, the trend was different and only plants at chilling temperature showed significantly higher content than in the control. For flavonoids, a slight increase was observed at chilling temperature, while total flavonoid content decreased at freezing temperature compared with the control and plants at chilling temperature, suggesting that polyphenolic compounds are main metabolites synthesized during acclimation and counteracting cold stress induced by freezing temperatures.

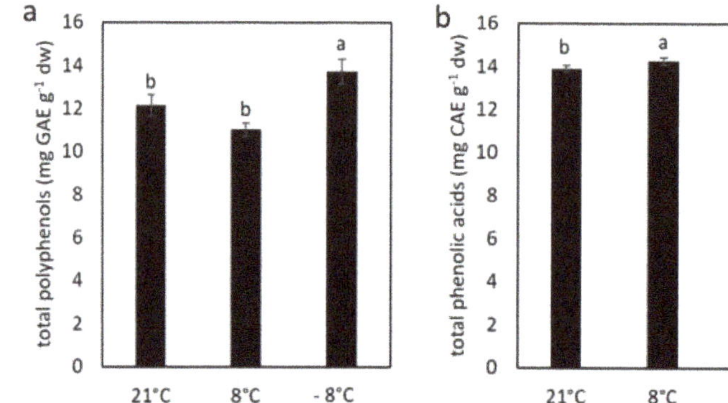

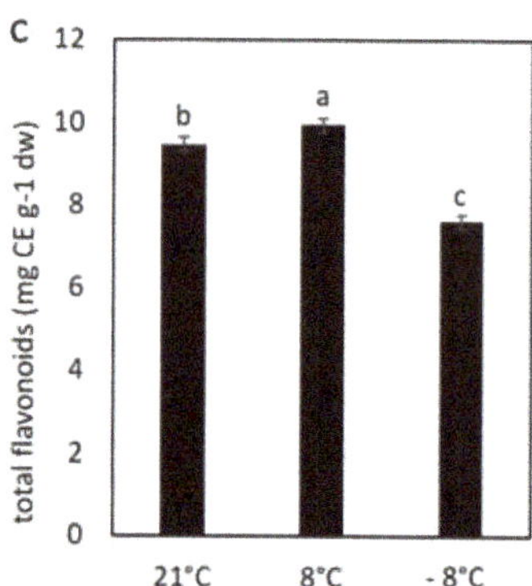

Figure 5. The content of total polyphenols (**a**), total phenolic acids (**b**) and total flavonoids (**c**) in kale exposed to low temperatures. Value marked with different letters are significantly different at $p < 0.05$.

3.4. Content of Glucosinolates under Low Temperature

The level of glucosinolates was majorly affected by cold stress as is shown at Figure 6. Exposure to chilling temperatures induced an increase in glucosinolates synthesis while a significant drop in glucosinolates content in plants exposed to freezing temperatures was recorded.

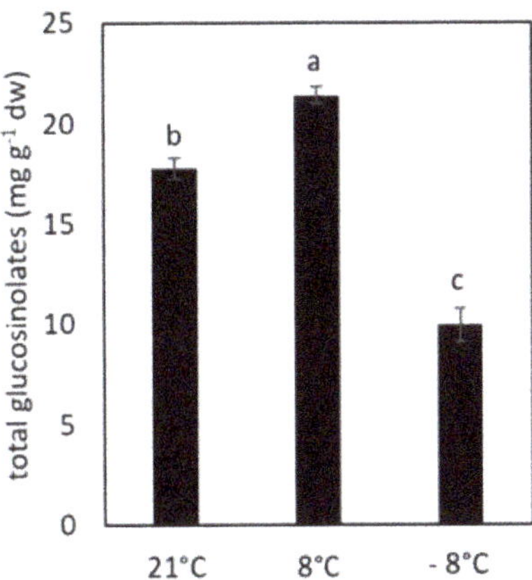

Figure 6. The content of total glucosinolates in kale exposed to low temperatures. Value marked with different letters are significantly different at $p < 0.05$.

4. Discussion

Proline is an amino acid that plays an important role in plant metabolism and development. Its accumulation is associated with plant stress response where it acts as an osmolyte, metal chelator, antioxidant defence molecule and signalling molecule (reviewed by Hayat et al. [23]. In the presented study, proline also acted as a protective molecule against cold stress with recorded increase in content correlating to the temperature drop (Figure 2). Similar results were reported for frost-resistant mutants of cauliflower (*Brassica oleracea* var. *botrytis*) by Hady et al. [24] and Fuller et al. [25] Proline accumulation is reported to be associated with cold stress tolerance in many other crops such as beans [26], sorghum [27], and rice as well [28]. Stressful environment cause changes in a variety of physiological, biochemical, and molecular processes in plants, which is evident from the proline levels mentioned above, but it can also affect photosynthesis, the most fundamental and complicated physiological process in all green plants [29]. Photosynthesis is an essential process for plants, and it is very sensitive to changes in environmental conditions [30]. Chlorophylls play important role in photosynthesis in harvesting and converting light energy in the antenna systems and charge separation and electron transport in the reaction centres [31]. Therefore, chlorophyll content is an important parameter to evaluate photosynthetic capacity. The level of chlorophyll *a*, *b*, and total chlorophylls and carotenoids was decreased after exposure to chilling temperature, while staying stable under freezing temperature in similar concentration as in control plants. Freezing temperatures normally cause a decrease in the ratio of total chlorophyll to carotenoids [12], but the acclimatization period in chilling period was much longer, so decrease in chlorophylls can also be a response to long exposure to chilling rather than freezing itself.

The ratio of total chlorophylls and carotenoids is an indicator of the plant greenness [19] and in our experiment chilling temperatures did not change significantly the ratio. The ratio of chlorophyll *a*/chlorophyll *b* may be an indicator of the functional pigment endowment and light adaptation of the photosynthetic apparatus [19]. Under stress conditions in plants, both chlorophyll *a* and chlorophyll *b* tend to decrease while, the chlorophyll *a*/*b* ratio tends to increase due to greater reduction in chlorophyll *b* compared to chlorophyll *a* [29]. Accordingly, our results indicate notable stress under freezing temperatures.

Kale is also considered a good source of carotenoids [32]. Carotenoids are essential pigments in photosynthetic organs along with chlorophylls where they act as photoprotectors, antioxidants, colour attractants, and precursors of plant hormones in non-photosynthetic organs of plants [33]. In addition, they play important roles in humans such as precursors of vitamin A, photoprotectants, antioxidants, enhancers of immunity, and others [33]. Under our experimental conditions, in control plants, total carotenoid content was 2.71 ± 0.06 µg g^{-1}, increased to 3.10 ± 0.13 µg g^{-1} under 8 °C and decreased to 2.58 ± 0.05 µg g^{-1} under −8 °C. Similarly, Hwang et al. [34] reported a gradual increase in total carotenoid content of kale during cold acclimatizationon for 3 days. They noted an increase in zeaxanthin content under cold stress. Mageney et al. [32] also reported high zeaxanthin contents in kale under traditional harvest conditions consistent with low temperatures. Consistent with reported papers, our results show that cold acclimatization may be beneficial for carotenoid content accumulation. According to our previously published work comparing the carotenoid content in kale with that of white cabbage and Chinese cabbage, kale tends to accumulate a higher amount of carotenoids, which may be related to its better tolerance to low temperatures [10,11].

Polyphenol compounds are considered an important player in the response of plants to abiotic stress, including low temperature stress [35]. The content of total polyphenols, phenolic acids and flavonoids in our experiments is shown in Figure 6. Polyphenol content differed between chilling and freezing where decrease and significant increase was recorded, respectively. Jurkow et al. [36] also reported increased levels of polyphenols in kale after moderate (−5 °C) and severe (−15 °C) frost compared to the pre-freezing (>0 °C) period. Similarly, Lee and Oh [16] reported increased levels of polyphenolic compounds in 3 weeks old kale exposed to 4 °C for 3 days. They also reported an increase in the content of phenolic acids, caffeic acid and ferulic acid, and flavonoid kaempferol after low temperature treatment at 4 °C. The same authors in another study [17] observed no differences in total flavonoid content in kale grown at 10 °C compared to the control. Our results for total flavonoids content are consistent with these findings; we found no significant changes in total flavonoids and phenolic acids at 8 °C, but a decrease in flavonoids at freezing (−8 °C) temperature was observed. The duration of exposure to low temperature and prior acclimatization may have an influence on the level of phenolic compounds under low temperature stress [18], and probably this is the reason for different trends published in different papers. In addition, individual phenolic compounds, or their conformation (e.g., conjugates, etc.) may change at low temperatures, which cannot always be observed by measuring total phenolic compounds content [37]. Low temperatures induce changes in cellular ultrastructure and localization of phenolic compounds, as has been reported for rapeseed (*B. napus* var. *oleifera*) leaves [38] and this may also affect polyphenolic compounds extraction and their detection.

Glucosinolates are well-known specialized metabolites characteristic for Brassicacea plants. Their contents in plants exposed to low temperatures compared to the control in our experiments are shown in Figure 6, where we observed an interesting trend. Under chilling temperature their content increased, while under additional freezing temperature their content decreased significantly. This trend is different from the one we reported previously, where the content of glucosinolates increases under chilling and freezing temperatures [12]. This different trend may be due to different experimental conditions and age of plants as reported in our previous work [12], where we exposed plants to freezing temperature for 1 h after 1 week of acclimation. Differences in glucosinolates trend reported in different papers are probably due to different experimental conditions, e.g., temperatures, duration of acclimation, etc., which may influence accumulation of different phytochemicals or be related to the kale variety.

5. Conclusions

In a present study, 8 weeks old kale plants were exposed to short-term low temperature stress (chilling and freezing) to study changes in proline, phytochemicals and pigments content under low temperature. Increases in proline content indicate that stress protective mechanisms were activated in plants at low temperatures and confirmed proline as a reliable stress marker for low temperature stress. Compared to the control, plants exposed to 8 °C (chilling temperatures) for 24 h accumulated significantly higher content of chlorophyll *a*, *b*, total chlorophylls, and carotenoid as well as content of total polyphenols, flavonoids and glucosinolates. This may indicate that short term (24 h) chilling temperature is beneficial for the accumulation of phytochemicals in kale. However, our results also suggest that freezing temperatures (-8 °C) may cause significant stress and decrease in pigments and phytochemical levels. Changes in measured phytochemicals indicate that they play significant role in chilling and freezing tolerance in used kale variant, but future more advanced metabolomics and transcriptomic studies will explain the mechanisms in more detail.

Author Contributions: Conceptualization, D.Š.; methodology, D.Š.; formal analysis, D.Š.; V.L.; investigation, V.L., D.Š., B.S.-S.; writing—original draft preparation, D.Š.; writing—review and editing, D.Š.; B.S.-S., B.S.-S., E.K.; visualization, D.Š.; supervision, D.Š.; project administration, D.Š.; B.S.-S.; funding acquisition, D.Š.; B.S.-S. All authors have read and agreed to the published version of the manuscript.

Funding: This research was funded by Unity through Knowledge Fund, grant number 12/17, and by the Operational Programme Competitiveness and Cohesion 2014–2020 and the Croatian European regional fund under a specific scheme to strengthen applied research in proposing actions for climate change adaptation (Project No. KK.05.1.1.02.0005).

Institutional Review Board Statement: Not Applicable.

Informed Consent Statement: Not Applicable.

Data Availability Statement: Additional data are available upon request.

Acknowledgments: We thank Srđan Franić for providing the seeds used in the experiment.

Conflicts of Interest: The authors declare no conflict of interest.

References

1. Ritonga, F.N.; Chen, S. Physiological and Molecular Mechanism Involved in Cold Stress Tolerance in Plants. *Plants* **2020**, *9*, 560. [CrossRef]
2. Hussain, H.A.; Hussain, S.; Khaliq, A.; Ashraf, U.; Anjum, S.A.; Men, S.; Wang, L. Chilling and Drought Stresses in Crop Plants: Implications, Cross Talk, and Potential Management Opportunities. *Front. Plant Sci.* **2018**, *9*, 393. [CrossRef]
3. Sanghera, G.S.; Wani, S.H.; Hussain, W.; Singh, N.B. Engineering Cold Stress Tolerance in Crop Plants. *Curr. Genom.* **2011**, *12*, 30–43. [CrossRef]
4. Carvajal, F.; Rosales, R.; Palma, F.; Manzano, S.; Cañizares, J.; Jamilena, M.; Garrido, D. Transcriptomic changes in Cucurbita pepo fruit after cold storage: Differential response between two cultivars contrasting in chilling sensitivity. *BMC Genom.* **2018**, *19*, 125. [CrossRef] [PubMed]
5. Yang, L.; Wen, K.-S.; Ruan, X.; Zhao, Y.-X.; Wei, F.; Wang, Q. Response of Plant Secondary Metabolites to Environmental Factors. *Molecules* **2018**, *23*, 762. [CrossRef] [PubMed]
6. Franzke, A.; Lysak, M.A.; Al-Shehbaz, I.A.; Koch, M.A.; Mummenhoff, K. Cabbage family affairs: The evolutionary history of *Brassicaceae*. *Trend. Plant Sci.* **2011**, *16*, 108–116. [CrossRef]
7. Šamec, D.; Salopek-Sondi, B. Cruciferous (*Brassicaceae*) vegetables. In *Nonvitamin and Nonmineral Nutritional Supplements*; Nabavi, S.M., Sanches Silva, T., Eds.; Academic Press: Cambridge, MA, USA, 2019; pp. 195–202.
8. Šamec, D.; Urlić, B.; Salopek-Sondi, B. Kale (*Brassica oleracea* var. *acephala*) as a superfood: Review of the scientific evidence behind the statement. *Crit. Rev. Food Sci. Nutr.* **2019**, *59*, 2411–2422. [CrossRef] [PubMed]
9. Pavlović, I.; Petrík, I.; Tarkowska, D.; Lepeduš, H.; Vujčić Bok, V.; Radić Brkanac, S.; Novák, O.; Salopek Sondi, B. Correlations between Phytohormones and Drought Tolerance in Selected Brassica Crops: Chinese Cabbage, White Cabbage and Kale. *Int. J. Mol. Sci.* **2018**, *19*, 2866. [CrossRef]
10. Linić, I.; Šamec, D.; Grúz, J.; Vujčić Bok, V.; Strnad, M.; Salopek Sondi, B. Involvement of Phenolic Acids in Short-Term Adaptation to Salinity Stress is Species-Specific among Brassicaceae. *Plants* **2019**, *8*, 155. [CrossRef]
11. Šamec, D.; Linić, I.; Salopek-Sondi, B. Salinity Stress as an Elicitor for Phytochemicals and Minerals Accumulation in Selected Leafy Vegetables of Brassicaceae. *Agronomy* **2021**, *11*, 361. [CrossRef]
12. Ljubej, V.; Radojčić Redovniković, I.; Salopek-Sondi, B.; Smolko, A.; Roje, S.; Šamec, D. Chilling and Freezing Temperature Stress Differently Influence Glucosinolates Content in *Brassica oleracea* var. *acephala*. *Plants* **2021**, *10*, 1305. [CrossRef]
13. Šamec, D.; Pavlović, I.; Radojčić Redovniković, I.; Salopek-Sondi, B. Comparative analysis of phytochemicals and activity of endogenous enzymes associated with their stability, bioavailability and food quality in five *Brassicaceae* sprouts. *Food Chem.* **2018**, *269*, 96–102. [CrossRef]
14. Šamec, D.; Kruk, V.; Ivanišević, P. Influence of Seed Origin on Morphological Characteristics and Phytochemicals Levels in *Brassica oleracea* var. *acephala*. *Agronomy* **2019**, *9*, 502. [CrossRef]

15. USDA, National Agricultural Statistics Service. 2012; Census of Agriculture. Available online: https://www.agcensus.usda.gov/Publications/2012/Full_Report/Volume_1,_Chapter_1_US/st99_1_038_038.pdf (accessed on 18 September 2021).
16. Lee, J.-H.; Oh, M.-M. Short-term Low Temperature Increases Phenolic Antioxidant Levels in Kale. *Hortic. Environ. Biotechnol.* **2015**, *56*, 588–596. [CrossRef]
17. Lee, J.-H.; Kwon, M.C.; Jung, E.S.; Lee, C.H.; Oh, M.-M. Physiological and Metabolomic Responses of Kale to Combined Chilling and UV-A Treatment. *Int. J. Mol. Sci.* **2019**, *20*, 4950. [CrossRef] [PubMed]
18. Steindal, A.L.H.; Rødven, R.; Hansen, E.; Mølmann, J. Effects of photoperiod, growth temperature and cold acclimatisation on glucosinolates, sugars and fatty acids in kale. *Food Chem.* **2015**, *174*, 44–51. [CrossRef] [PubMed]
19. Lichtenthaler, H.K.; Buschmann, C. Chlorophylls and carotenoids: Measurement and characterization by UV–VIS spectroscopy. In *Current Protocols in Food Analytical Chemistry*; John Wiley & Sons, Inc.: Hoboken, NJ, USA, 2001; pp. F4.3.1–F4.3.8.
20. Šamec, D.; Bogović, M.; Vincek, D.; Martinčić, J.; Salopek-Sondi, B. Assessing the authenticity of the white cabbage (*Brassica oleracea* var. *capitata* f. *alba*) cv. 'Varaždinski' by molecular and phytochemical markers. *Food Res. Int.* **2014**, *60*, 266–272. [CrossRef]
21. Aghajanzadeh, T.; Hawkesford, M.J.; Kok, L.J. The significance of glucosinolates for sulfur storage in Brassicaceae seedlings. *Front. Plant Sci.* **2014**, *5*, 704. [CrossRef] [PubMed]
22. Hammer, O.; Harper, A.A.T.; David, A.T.; Ryan, P.D. PAST: Paleontological Statistics Software Package for Education and Data Analysis. *Palaeontol. Electron* **2001**, *4*, 1–9.
23. Hayat, S.; Hayat, Q.; Alyemeni, M.N.; Wani, A.S.; Pichtel, J.; Ahmad, A. Role of proline under changing environments. *Plant Signal. Behav.* **2012**, *7*, 1456–1466. [CrossRef]
24. Hadi, F.; Gilpin, M.; Fuller, M.P. Identification and expression analysis of CBF/DREB1 and COR15 genes in mutants of Brassica oleracea var. botrytis with enhanced proline production and frost resistance. *Plant Physiol. Biochem.* **2011**, *49*, 1323–1332. [CrossRef] [PubMed]
25. Fuller, M.P.; Metwali, E.M.R.; Eed, M.H.; Jellings, A.J. Evaluation of Abiotic Stress Resistance in Mutated Populations of Cauliflower (*Brassica oleracea* var. *botrytis*). *Plant Cell Tissue Organ. Cult.* **2006**, *86*, 239. [CrossRef]
26. Neto, N.B.M.; Custódio, C.C.; Gatti, A.B.; Priolli, M.R.; Cardoso, V.J.M. Proline: Use as an indicator of temperature stress in bean seeds. *Crop Breed. Appl. Biotechnol.* **2004**, *4*, 330–337. [CrossRef]
27. Vera-Hernández, P.; Ramírez, M.A.O.; Núñez, M.M.; Ruiz-Rivas, M.; Rosas-Cárdenas, F.F. Proline as a probable biomarker of cold stress tolerance in Sorghum (*Sorghum bicolor*). *Mex. J. Biotechnol.* **2018**, *3*, 77–86. [CrossRef]
28. de Freitas, G.M.; Thomas, J.; Liyanage, R.; Lay, J.O.; Basu, S.; Ramegowda, V.; do Amaral, M.N.; Benitez, L.C.; Braga, E.J.B.; Pereira, A. Cold tolerance response mechanisms revealed through comparative analysis of gene and protein expression in multiple rice genotypes. *PLoS ONE* **2019**, *14*, e0218019. [CrossRef]
29. Ashraf, M.; Harris, P.J.C. Photosynthesis under stressful environments: An overview. *Photosynthetica* **2013**, *51*, 163–190. [CrossRef]
30. Ensminger, I.; Busch, F.; Huner, N.P.A. Photostasis and cold acclimation: Sensing low temperature through photosynthesis. *Phisiol. Plant.* **2006**, *126*, 28–44. [CrossRef]
31. Zhao, Y.; Han, Q.; Ding, C.; Huang, Y.; Liao, J.; Tao Chen, T.; Feng, S.; Zhou, L.; Zhang, Z.; Chen, Y.; et al. Effect of Low Temperature on Chlorophyll Biosynthesis and Chloroplast Biogenesis of Rice Seedlings during Greening. *Int. J. Mol. Sci.* **2020**, *21*, 1390. [CrossRef]
32. Mageney, V.; Baldermann, S.; Albach, D.C. Intraspecific Variation in Carotenoids of *Brassica oleracea* var. *sabellica*. *J. Agric. Food Chem.* **2016**, *64*, 3251–3257. [CrossRef] [PubMed]
33. Maoka, T. Carotenoids as natural functional pigments. *J. Nat. Med.* **2020**, *74*, 1–16. [CrossRef] [PubMed]
34. Hwang, S.-J.; Chun, J.-H.; Kim, S.-J. Effect of Cold Stress on Carotenoids in Kale Leaves (*Brassica oleracea*). *Korean J. Environ. Agric.* **2017**, *36*, 106–112. [CrossRef]
35. Šamec, D.; Karalija, E.; Šola, I.; Vujčić-Bok, V.; Salopek-Sondi, B. The Role of Polyphenols in Abiotic Stress Response: The Influence of Molecular Structure. *Plants* **2021**, *10*, 118. [CrossRef] [PubMed]
36. Jurkow, R.; Wurst, A.; Kalisz, A.; Sękara, A.; Cebula, S. Cold stress modifies bioactive compounds of kale cultivars during fall–winter harvests. *Acta Agrobot.* **2019**, *72*, 1761. [CrossRef]
37. Schmidt, S.; Zietz, M.; Schreiner, M.; Rohn, S.; Kroh, L.W.; Krumbein, A. Genotypic and climatic influences on the concentration and composition of flavonoids in kale (*Brassica oleracea* var. *sabellica*). *Food. Chem.* **2010**, *119*, 1293–1299. [CrossRef]
38. Stefanowska, M.; Kuras, M.; Kacperska, A. Low Temperature-induced Modifications in Cell Ultrastructure and Localization of Phenolics in Winter Oilseed Rape (*Brassica napus* L. var. *oleifera* L.) Leaves. *Ann Bot.* **2002**, *90*, 637–645. [CrossRef]

horticulturae

Article

Effect of Cold Stress on Growth, Physiological Characteristics, and Calvin-Cycle-Related Gene Expression of Grafted Watermelon Seedlings of Different Gourd Rootstocks

Kaixing Lu [1], Jiutong Sun [1], Qiuping Li [1], Xueqin Li [2],* and Songheng Jin [2],*

[1] College of Science & Technology, Ningbo University, Ningbo 315211, China; lukaitong@nbu.edu.cn (K.L.); sunjiutong@126.com (J.S.); liqiuping@nbu.edu.cn (Q.L.)
[2] Jiyang College, Zhejiang A&F University, Zhuji 311800, China
* Correspondence: lxqin@zafu.edu.cn (X.L.); shjin@zafu.edu.cn (S.J.)

Abstract: Recently, grafting has been used to improve abiotic stress resistance in crops. Here, using watermelon 'Zaojia 8424' (*Citrullus lanatus*) as scions, three different gourds (*Lagenaria siceraria*, 0526, 2505, and 1226) as rootstocks, and non-grafted plants as controls (different plants were abbreviated as 0526, 2505, 1226, and 8424), the effect of cold stress on various physiological and molecular parameters was investigated. The results demonstrate that the improved cold tolerance of gourd-grafted watermelon was associated with higher chlorophyll and proline content, and lower malondialdehyde (MDA) content, compared to 8424 under cold stress. Furthermore, grafted watermelons accumulated fewer reactive oxygen species (ROS), accompanied by enhanced antioxidant activity and a higher expression of enzymes related to the Calvin cycle. In conclusion, watermelons with 2505 and 0526 rootstocks were more resilient compared to 1226 and 8424. These results confirm that using tolerant rootstocks may be an efficient adaptation strategy for improving abiotic stress tolerance in watermelon.

Keywords: watermelon; rootstock; cold stress; antioxidant enzymes; gene expression

Citation: Lu, K.; Sun, J.; Li, Q.; Li, X.; Jin, S. Effect of Cold Stress on Growth, Physiological Characteristics, and Calvin-Cycle-Related Gene Expression of Grafted Watermelon Seedlings of Different Gourd Rootstocks. *Horticulturae* **2021**, *7*, 391. https://doi.org/10.3390/horticulturae7100391

Academic Editors: Agnieszka Hanaka, Jolanta Jaroszuk-Ściseł and Małgorzata Majewska

Received: 23 August 2021
Accepted: 30 September 2021
Published: 11 October 2021

1. Introduction

Cold stress is one of the key environmental factors that severely affect plant growth and development, especially for watermelon (*Citrullus lanatus*). Watermelon is an important global fruit due to its economic importance and nutritional qualities, it contains a great quantity of antioxidants and may mitigate oxidative damage in tissues. Originating from Africa and tropical regions in Asia, watermelon is sensitive to low temperatures. The best growing conditions for watermelon are between 21 and 29 °C. When the temperature drops to 10 °C, watermelon stops growing and it even dies at temperatures below 1 °C [1]. Extreme weather frequently occurs in China, especially during the seedling stage, such as winter, late autumn, and early spring, in which watermelon usually suffers cold stress. Thus, it is hard to grow the fruit and obtain high yields. Therefore, the study of how to improve cold resistance in watermelon has become a vital part of watermelon breeding projects. One way to improve cold tolerance is to graft plants onto rootstocks with higher cold tolerance [2–7].

Grafting originated in Japan in the late 1920s and was initially used to improve resistance to soil pathogens [8]. Since then, the use of grafting has spread around the world. In Korea and Japan, approximately 95% and 92% of the land area, respectively, is cultivated with grafted watermelon [9]. Recently, grafting was proposed as a promising approach to improve tolerance to abiotic stress and can represent an efficient technique for reducing or eliminating losses in production. Over the years, many studies have demonstrated the ability of grafting to enhance salt tolerance in different plants, such as melon (*Cucumis melo* L.) [10], pumpkin (*Cucurbita moschata* D.) [4], cucumber (*Cucumis sativus* L.) [11], grapevines

(*Vitis vinifera* L.) [12], and watermelon [13]. Other studies have demonstrated that grafting induces resistance to high temperatures [14] and water stress [15], improves the yield and quality of watermelon under low potassium supply [16], and enhances nutrient uptake [17], among other things. However, to date, there are few studies concerning the results of grafting with appropriate rootstocks for cold tolerance in watermelon.

Accumulating evidence has demonstrated that certain stresses lead to oxidative stress by overproducing reactive oxygen species (ROS), such as superoxide anions (O_2^-) and hydrogen peroxide (H_2O_2). The improved performance of grafted plants, especially under abiotic stress, is usually associated with the higher accumulation of osmolytes, such as proline, as well as higher antioxidant enzyme activity, such as that of superoxide dismutase (SOD), peroxidase (POD), and catalase (CAT), which protect the cellular systems from the cytotoxic damage by ROS [18,19]. Moreover, some studies have shown that grafted plants have enhanced tolerance to stress via the induced the expression of some key genes, especially those related to photosynthesis. Li et al. [20] identified novel photosynthetic proteins in grafted cucumber seedlings. Similarly, Yang et al. [21] found that the amelioration of photosynthetic capacity in grafted watermelon seedlings under salt stress might be due to enzymes of the Calvin cycle. They further proved that rootstock grafting watermelon seedlings enhanced the gene expression of enzymes related to ribulose-1,5-bisphosphate (RuBP) regeneration under salt stress [22]. Xu et al. [7], who analyzed the transcriptomic results of grafted watermelon under cold stress, found 702 genes that were differentially expressed, among which 180 genes associated with photosynthesis were downregulated. These studies demonstrated that rootstocks could regulate gene expression patterns, especially those related to photosynthesis, in scions under stress.

Recently, different authors have demonstrated that the effects of grafting on plant growth and stress tolerance depend on different rootstocks [12,13,23,24]. Therefore, selecting suitable rootstocks with higher compatibility and resistance is a promising strategy. However, studies on rootstock screening to improve watermelon's cold tolerance are lacking and urgently required. Due to their vigorous root systems, bottle gourd (*L. siceraria* Standl) and pumpkin have been used as rootstocks by watermelon growers to improve fruit quality and sensory parameters [4]. Recently, some local bottle gourd genotypes that may have potential for use as a rootstock against stress tolerance were selected and bred by the Ningbo Academy of Agricultural Sciences, but no studies have been performed on them. Information about the molecular responses of grafted watermelon to cold stress is also limited. Therefore, this investigation aimed to understand the mechanisms by which grafted watermelon can develop improved cold resistance. The 'ZaoJia 8424' watermelon was chosen for grafting onto three gourd rootstocks to evaluate ROS and osmolyte accumulation, membrane stability, antioxidant defense system, and the expression level of photosynthesis-related genes. The performed study will help to explain cold-tolerant mechanisms in grafted horticultural crops, which may be useful when exploring potential rootstocks to improve watermelon cold tolerance.

2. Materials and Methods

2.1. Plant Material and Treatments

The watermelon (*Citrullus lanatus* (Thunb.) Matsum. and Nakai) cv. ZaoJia 8424 was used as a scion, and two cold-tolerant gourds (*L. siceraria* Standl. cv. LS0525, and LS0526) and a cold-sensitive gourd (*L. siceraria* Standl. cv. LS1226) were selected as rootstocks. The rootstock-grafted watermelon plants have been abbreviated as 2505, 0526, and 1226, respectively. The nongrafted watermelon plants (abbreviated as 8424) were used as controls. The watermelon and gourds used here were selected and bred by the Ningbo Academy of Agricultural Sciences of China.

The seeds of the gourds (rootstocks) and watermelons (scion and nongrafted plants) were sown on May 5 and 8 in a sand/soil/peat (1:1:1 by volume) mixture. The scion plants were grafted onto the rootstock six days after the watermelon scion seeds were sown using the method of Lee [25]. To improve graft formation, transparent plastic film was used to cover the seedlings. All grafted seedlings were kept in the shade for 72 h. To ensure the stability of relative humidity, the plastic film was opened every day and completely removed 7 days after grafting. Then, all different seedlings were grown under the following conditions in the greenhouse: approximately $28/23 \pm 1\,°C$ (D/N) with 60% relative humidity, a 12 h photoperiod, and a photosynthetic photon flux density (PPFD) of $300\ \mu mol·m^{-2}·s^{-}$. After the third true leaves had fully unfolded, 12 plants of uniform size from each group were transplanted into different pots (one seedling per pot) and cultivated in an illuminating incubator under the controlled environment described above. After another two days of acclimation, the plants were randomly distributed to two identical illuminating incubators; one was a control group, and the other was a cold stress group. The control group remained in the same conditions, while the experimental group was subjected to cultivation for two days with a temperature of $20/15\,°C$ (D/N) and then for five days under $10/6\,°C$ (D/N). Samples were collected at the same time, i.e., after one week from both control and cold-stressed plants (2-day acclimation period and 5 days under cold exposure). The third leaves from the tops of different groups were harvested and used for H_2O_2 and O_2^- detection, each replicated three times. The remaining leaves were also sampled immediately and frozen in liquid N_2 before being kept at $-80\,°C$. Leaves from the same position were used to measure the same parameter.

2.2. Chlorophyll Content and Plant Growth Measurements

Chlorophyll content was measured spectrophotometrically according to Porra et al. [26]. The height and stem diameter of seedlings from each treatment were measured using a ruler and vernier caliper, respectively. The dry weights were determined after drying at $105\,°C$ for 10 min and then at $70\,°C$ for 72 h. The total number of leaves was counted on each plant.

2.3. Measurements of ROS, MDA, and Proline Content

Histochemical staining of O_2^- was performed using nitroblue tetrazolium (NBT) [27]. Leaf stalks were immediately combined with 0.1% NBT (w/v) in 25 mmol L^{-1} K-HEPES buffer (pH7.8) and kept at $25\,°C$ in the dark for 4 h. After the leaves were boiled in 95% ethanol, they were photographed. The different staining colors indicated the different degrees of lipid peroxidation.

Detection of H_2O_2 in the leaves and measurements of O_2^- generation, H_2O_2, and MDA content were all carried out according to our previous methods [28]. Proline content was analyzed according to the method of Bates et al. [29].

2.4. Antioxidant Enzyme Analysis

The extraction of antioxidant enzymes and the determination of those enzyme activities (SOD, CAT, and POD) were all carried out according to our previous methods [18], while the activity of GPX was measured following Huang et al. [30]. Electrophoretic separation of POD and CAT bands was performed as described by Lu et al. [18] and Woodbury et al. [31].

2.5. Gene Expression Analysis

Leaves of different watermelons were used to extract total RNA using the TRI Reagent (Takara Bio Inc. Dalian, China) according to the manufacturer's protocol. Reverse transcription reactions were performed using a PrimeScript RT Reagent Kit with genomic DNA Eraser (Takara Bio Inc. Dalian, China). Suitable primers were designed based on the National Center for Biotechnology Information (NCBI) and the watermelon genome sequence in the cucurbit genomics database. All primers used in this study are provided in Table 1. Quantitative gene expression analysis was performed using a Light Cycler 480 II, Roche Real Time PCR System (Roche Diagnostics Ltd., Basel, Switzerland). qPCR was carried out in a 20 μL reaction mixture containing 1 μL diluted first-strand cDNA, 125 nM of each primer, and 10 μL Lightcycler 480 SYBR Green I Master (Roche Diagnostics GmbH,

Mannhein, Germany). The $2^{-\Delta\Delta Ct}$ method was used to determine the relative change in gene expression [32], and melt curve analysis (from 55 to 94 °C) was used to determine the specificity of PCR amplification. All measurements were performed in triplicate, and the data are presented as the means and their standard errors.

Table 1. Gene-specific primers designed for qRT-PCR.

Gene	Forward Primer	Reverse Primer
rbcL	TCTTGGCAGCATTCCGAGTAA	TCGCAAATCCTCCAGACGTAG
TPI	GAAATTCTTCGTCGGTGGC	GAACCCAACAATTCCGTGCTG
FBPA	GTTGGTCCCTATTGTGGAGCC	CCTTGTAAACCGCAGC
FBPase	TCACAGCCCTCGAATTTA	CTTCGGAAACAAGGATACAAG
SBPase	TCGAGGCCTTGAGATACTCAC	GCCATCGCTGCTGTAACC
PRK	GGGCTGAGAAGATTACC	GAAGGATCTACAATCTCATGG

Note: *rbcL*: Rubisco large subunit (RBCL) gene; *TPI*: oftriose-3-phosphate isomerase gene; *FBPA*: fructose-1,6-bisphosphate aldolase gene; *FBPase*: fructose-1,6-bisphosphatase gene; *SBPase*: sedoheptulose-1,7-bisphosphatase gene; *PRK*: ribulose-5-phosphate kinase gene.

2.6. Statistical Methods

All data were statistically analyzed with SPSS 13.0 software (SPSS Chicago, IL, USA). Two-way analyses of variance (ANOVA) were used to evaluate the effects of rootstock and cold treatment. Tukey's honestly significant difference (HSD, $p \leq 0.05$) post hoc test was performed to test the existence of statistical differences between different rootstocks under normal and cold-stressed conditions.

3. Results

3.1. Growth Parameters

Under control conditions, all the grafted seedlings exhibited higher growth rates than non-grafted 8424, among which 2505 and 0526 exhibited the best performance trend, followed by 1226 (Table 2). Cold stress caused remarkable changes, and the growth of 8424 was significantly inhibited, whereas the growth inhibition of grafted plants was clearly alleviated, and the level significantly varied depending on rootstock genotypes, among which 2505 and 0526 were better than 1226. All growth parameters under cold stress decreased significantly compared with the control, with 2505 and 0526 having a smaller decrease than 1226. These results indicate that grafting increased the tolerance of watermelon seedlings to cold stress, and the degree of resistance depended on the rootstock. These growth parameters were significantly influenced by both different rootstocks and cold stress, and a significant interaction of rootstock and cold was observed (Table 2).

Table 2. Effects of graft combinations and cold stress on watermelon seedlings.

Treatment	Seedlings	Plant Height (cm)	Stem Diameter (mm)	Dry Weight (g)	Number of Leaves
Normal	2505	16.69 ± 0.87 ^{c}	3.32 ± 0.35 ^{c}	1.04 ± 0.11 ^{c}	9.11 ± 0.51 ^{c}
	0526	17.08 ± 0.28 ^{c}	2.90 ± 0.27 ^{ab}	1.15 ± 0.08 ^{c}	8.53 ± 0.60 ^{c}
	1226	15.20 ± 1.03 ^{b}	2.67 ± 0.16 ^{b}	0.92 ± 0.07 ^{b}	7.18 ± 0.57 ^{b}
	8424	8.47 ± 0.64 ^{a}	2.13 ± 0.12 ^{a}	0.51 ± 0.05 ^{a}	4.07 ± 0.65 ^{a}
Cold	2505	15.06 ± 0.99 ^{c}	3.04 ± 0.27 ^{b}	0.92 ± 0.06 ^{c}	7.62 ± 0.43 ^{c}
	0526	15.82 ± 0.43 ^{d}	2.47 ± 0.19 ^{b}	0.81 ± 0.05 ^{d}	7.02 ± 0.68 ^{b}
	1226	13.03 ± 0.82 ^{b}	2.18 ± 0.14 ^{b}	1.06 ± 0.10 ^{b}	5.57 ± 0.48 ^{b}
	8424	7.08 ± 0.38 ^{a}	1.81 ± 0.08 ^{a}	0.48 ± 0.04 ^{a}	3.82 ± 0.31 ^{a}
Two way ANOVA					
Fs		10.05 ***	5.33 **	6.27 **	3.97 **
Fc		6.85 ***	17.19 *	6.43 **	10.58 **
Fs×c		3.27 ***	6.84 *	9.51 *	2.20 **

Values are means of three replicates ± standard error (SE). Small case superscript letters in the same column show statistically significant differences among different rootstocks for the same parameter under normal and cold stress at $p \leq 0.05$ according to Tukey's tests. F_S: rootstock effect, F_C: cold effect, $F_{S \times C}$: rootstock × cold interaction effect. *, **, and ***: significant at $p \leq 0.05$, 0.01, and 0.001, respectively.

3.2. Chlorophyll Content

Photosynthetic pigments play a significant role in plant photosynthesis. Under control conditions, the chlorophyll content of grafted watermelon was significantly higher than that of non-grafted 8424 (Table 3). Under cold stress conditions, chlorophyll content decreased significantly, and the reduction was greater in 8424 (37%) and 1226 (34%) than in 2505 (26%) and 0526 (25%) compared with normal conditions (Table 3). No significant changes were found in the ratios of Chl a/b under normal conditions, while the ratio of Chl a/b in grafted seedlings was significantly higher than that in non-grafted 8424 under cold stress.

Table 3. Variability of total chlorophyll content and ratio of Chl a/b between different graft combinations after cold stress.

Treatment	Seedlings	Chlorophyll (mg g^{-1} FW)	Chl a/b
Normal	2505	1.84 ± 0.04 [b]	3.46 ± 0.10 [a]
	0526	1.86 ± 0.02 [b]	3.41 ± 0.03 [a]
	1226	1.71 ± 0.07 [b]	3.39 ± 0.08 [a]
	8424	1.46 ± 0.13 [a]	3.29 ± 0.10 [a]
Cold	2505	1.36 ± 0.03 [c]	2.44 ± 0.03 [c]
	0526	1.39 ± 0.03 [b]	2.35 ± 0.02 [c]
	1226	1.13 ± 0.07 [ab]	2.28 ± 0.04 [b]
	8424	0.92 ± 0.09 [a]	2.03 ± 0.02 [a]
Two-way ANOVA			
F_S		10.15 **	8.14 **
F_C		13.24 **	6.82 **
$F_{S×C}$		12.49 *	9.37 *

Values are means of three replicates ± standard error (SE). Small case superscript letters in the same column show statistically significant differences among different rootstocks for the same parameter under normal and cold stress at $p \leq 0.05$ according to Tukey's tests. F_S: rootstock effect, F_C: cold effect, $F_{S×C}$: rootstock × cold interaction effect. *, **, and ***: significant at $p \leq 0.05$, 0.01, and 0.001, respectively.

3.3. Oxidative Stress Evaluation

The accumulation of H_2O_2 in leaves was reflected by necrotic lesions, which were easily detected by chlorophyll bleaching. Obviously, under control conditions (Figure 1A), the accumulation of H_2O_2 in grafted 1226 and non-grafted 8424 was higher than in grafted 2505 and 0526. Necrotic lesions were mainly located on the leaf base and main veins in 2505 and 0526 leaves, while in 1226 and 8424 leaves, even the small veins had brown spots. Furthermore, significant brown spot accumulation was observed in 1226 and 8424 leaves under cold stress conditions, while this phenomenon was slight in 2505 and 0526 leaves (Figure 1A). The measurement of H_2O_2 content (Figure 1C) matched the effects of H_2O_2 accumulation in leaves. Under control conditions, no significant difference was found in H_2O_2 content between grafted and non-grafted watermelons, whereas cold stress noticeably increased the H_2O_2 content in 8424 and 1226, which, compared with the controls, increased by 56.3% and 21.1%, respectively.

Figure 1B shows the results of O_2^- generation and accumulation in the inoculated leaves by nitroblue tetrazolium staining. The trend in O_2^- concentration was similar to that of H_2O_2. Under control conditions, more dark blue staining was detected in 8424, followed by 1226, whereas very little dark blue staining was found in 2505 and 0526. Cold stress improved the accumulation of O_2^-, which was reflected in greater NBT staining spots, especially in 8424, where the blue color was sinificantly deepened. The influence of cold stress on the O_2^- production rates of different seedlings was also measured (Figure 1D). Under control conditions, no significant difference was observed between grafted seedlings in terms of O_2^- production rate, whereas in 8424, the O_2^- production rate was significantly higher. Furthermore, cold stress induced a sharp increase in the O_2^- production rate in 8424 (34.0%), while a lower increase was observed in 2505, 0526, and 1226, the rates for which were increased by 26.8%, 24.3%, and 14.9%, respectively. The interaction of rootstocks × cold stress significantly affected the H_2O_2 content and O_2^- production rate.

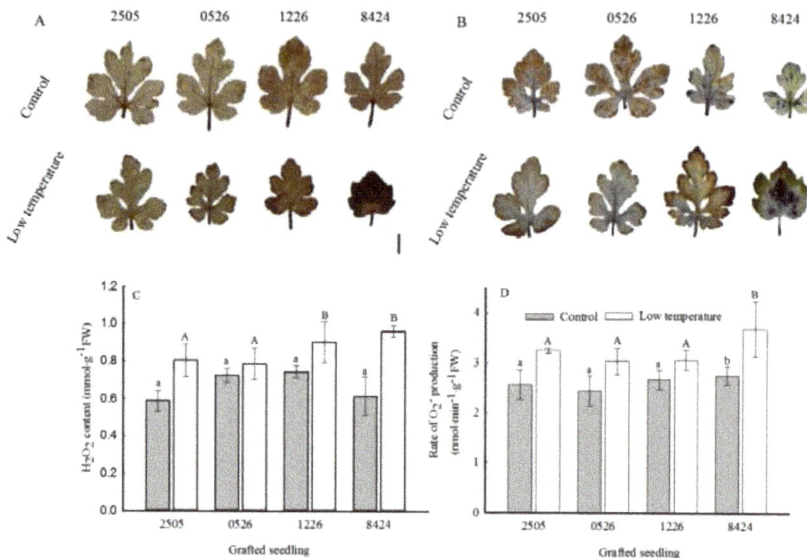

Figure 1. Effects of cold stress on the accumulation of H_2O_2 (**A**) and O_2^- (**B**), as well as the content of H_2O_2 (**C**) and O_2^- producing rate (**D**) in leaves of different watermelon seedlings. The numbers 0526, 2505, and 1226 represent watermelon grafted onto three different gourds, while 8424 means non-grafted watermelon. Scale bars correspond to 2 cm. Data are the mean ± SE. Bars with different letters indicate a significant difference ($p < 0.05$).

3.4. Lipid Peroxidation and Proline Accumulation

Lipid peroxidation, reflected by MDA content, usually accompanies ROS accumulation. Under control conditions, the MDA content was significantly higher in 8424 than in grafted seedlings (Figure 2A). A sharp increase was observed when seedlings were exposed to cold stress. The MDA content of 8424 underwent the greatest increase under cold stress (50.0%), and 0526 and 1226 showed intermediate increases (39.5% and 34.1%, respectively). A minimal increase was displayed by 2505 (18.6%). Under control conditions, no significant differences in the level of proline content were observed between grafted and non-grafted seedlings (Figure 2B). Although cold stress promoted leaf proline content, a significantly greater increase was observed in grafted seedlings than in non-grafted 8424. When compared with the corresponding controls, proline accumulation in 2505 and 1226 was increased by 79.2% and 66.9%, respectively, while in 0526 and 8424, the increase was only 57.4% and 49.1%, respectively. Moreover, the interaction of rootstocks × cold stress significantly affected the MDA and proline contents.

3.5. Antioxidant Enzyme Activity

The effect of cold stress on the activities of SOD, CAT, POD, and GPX in leaves of different watermelons is shown in Figure 3A–D. SOD is the first enzyme in the enzymatic antioxidative pathway. Under control conditions, SOD enzymatic activity was significantly higher in grafted 2505 and 0526 than in grafted 1226 and non-grafted 8424 (Figure 3A). Cold stress decreased the SOD activity of both grafted and self-rooted watermelons compared with the control, while the reduction was less notable in 2505 and 0526 (25% and 28%, respectively) than in 1226 and 8424 (31% and 32%, respectively). In contrast, cold stress increased the CAT activity of both grafted and self-rooted seedlings (Figure 3B). However, the values in grafted watermelons were significantly higher than those in non-grafted 8424 under both control and cold stress conditions. Conversely, the activity of POD was severely inhibited in both grafted and non-grafted seedlings when they were exposed to cold stress (Figure 3C). A lesser decrease was observed in plants grafted onto 0526 (17.7%), 1226 (23.4%), and 2505 (24.5%) than in non-grafted 8424 (30.3%). The GPX activity was reduced by cold stress in non-grafted 8424 (25.2%) (Figure 3D), while in grafted seedlings, cold stress led to a considerable increase in GPX activity (5.5%, 16.8%, and 21.3% for 0526, 2505, and 1226, respectively). Moreover, the activities of those enzymes were all significantly affected by the interaction of rootstocks × cold.

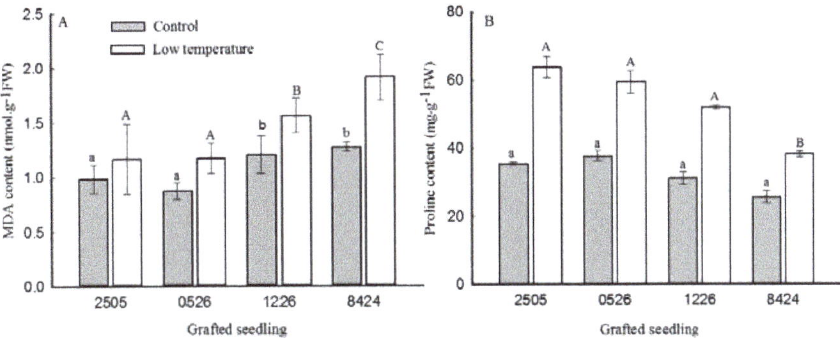

Figure 2. Effects of low temperature on MDA (**A**) and proline (**B**) content in leaves of different grafted seedlings. Data are the mean ± SE. Bars with different letters indicate a significant difference ($p < 0.05$).

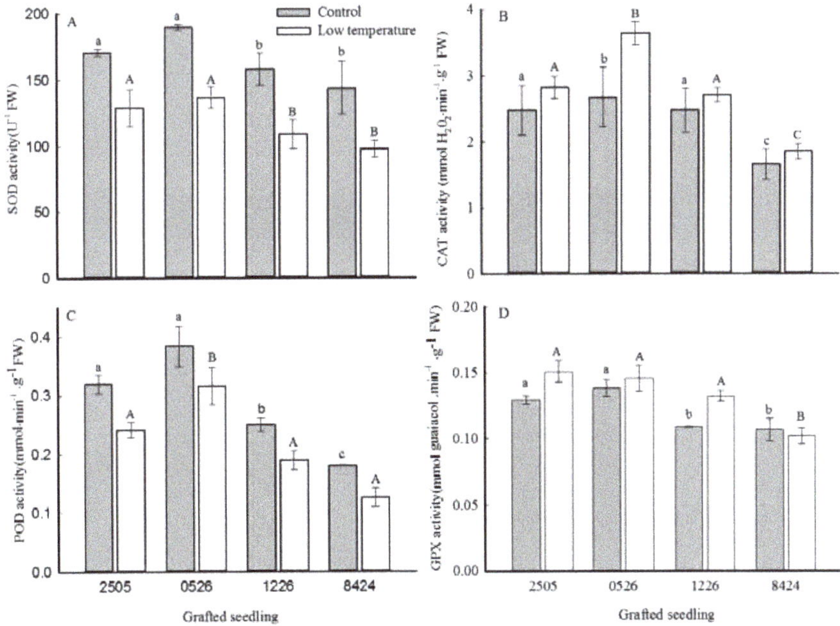

Figure 3. Effect of low temperature on activities of SOD, CAT, POD, and GPX (**A–D**) in leaves of different grafted seedlings. Data are the mean ± SE. Bars with different letters indicate a significant difference ($p < 0.05$).

To evaluate the enzyme activity as stimulated by cold stress, polyacrylamide gel electrophoresis analysis of enzyme isoform patterns was performed. No significant differences were found in POD and GPX isoforms (data not shown), but there was a significant difference in SOD and CAT isoforms (Figures 4 and 5). At least five isoforms were detected in the leaves of different watermelon seedlings under both control and cold stress conditions, while CAT 6 was only found under cold stress conditions. The cold-induced increase in CAT enzyme activity coincided with the increased expression of CAT isoforms. Under both control and stress conditions, the expression of CAT isoforms in grafted watermelons was much higher than that in non-grafted 8424 (Figure 4). Conversely, the expression of SOD isoforms was downregulated by cold stress (Figure 5). Under control conditions, at least nine isoforms were detected in different seedlings, with SOD1–3 and 6–8 being the major ones, while under cold stress conditions, the expression of these isoforms was greatly diminished. Similar to CAT isoforms, the expression of SOD isoforms in grafted watermelons was also much stronger than in non-grafted 8424, among which 0526 displayed the highest expression, followed by 2505 and 1226. The differences observed in the SOD isoform patterns match the changes in total SOD activity described above.

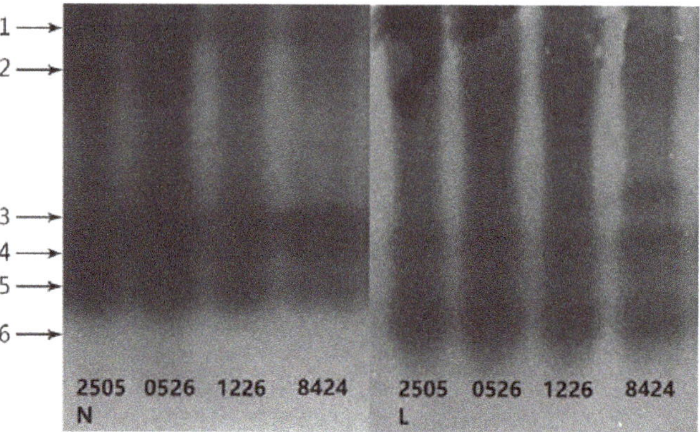

Figure 4. CAT isozymes in leaves of different grafted seedlings (different numbers represent different bands).

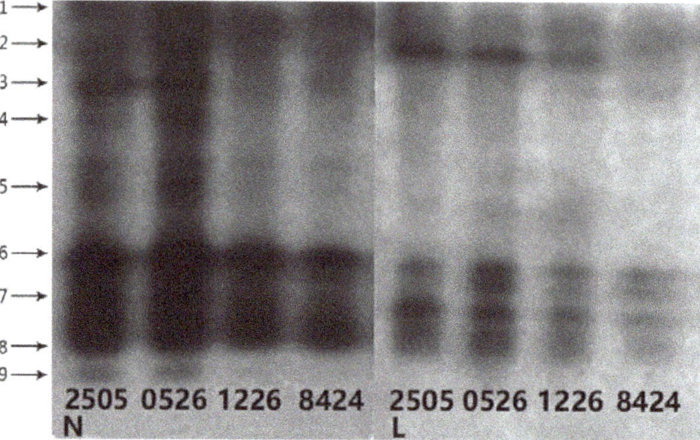

Figure 5. SOD isozymes in leaves of different grafted seedlings (different numbers represent different bands).

3.6. Gene Expression

Quantitative expression analyses were used to evaluate the association of the selected genes with cold resistance, as well as their differential expression under stress conditions in different watermelon seedlings, and the results are shown in Figure 6. Grafting increased the expressions of *RbcL*, *TPI*, *FBPA*, *SBPase,* and *PRK*, whereas the expression of *FBPase* was only slightly upregulated in 0526, while it decreased in 2505 and 1226. The expression of *SBPase* was significantly upregulated by cold stress in both grafted and non-grafted seedlings, while the expressions of *TPI*, *FBPA*, and *SBPase* were only enhanced by cold stress in grafted watermelons and decreased in non-grafted 8424. The expressions of *RbcL* and *PRK* were reduced in all four seedlings by cold stress. The expression of these six genes in grafted watermelons was higher than those in non-grafted 8424 under cold stress. Moreover, the interaction of rootstocks × cold significantly affected the expressions of *RbcL*, *TPI*, *FBPA,* and *SBPase.*

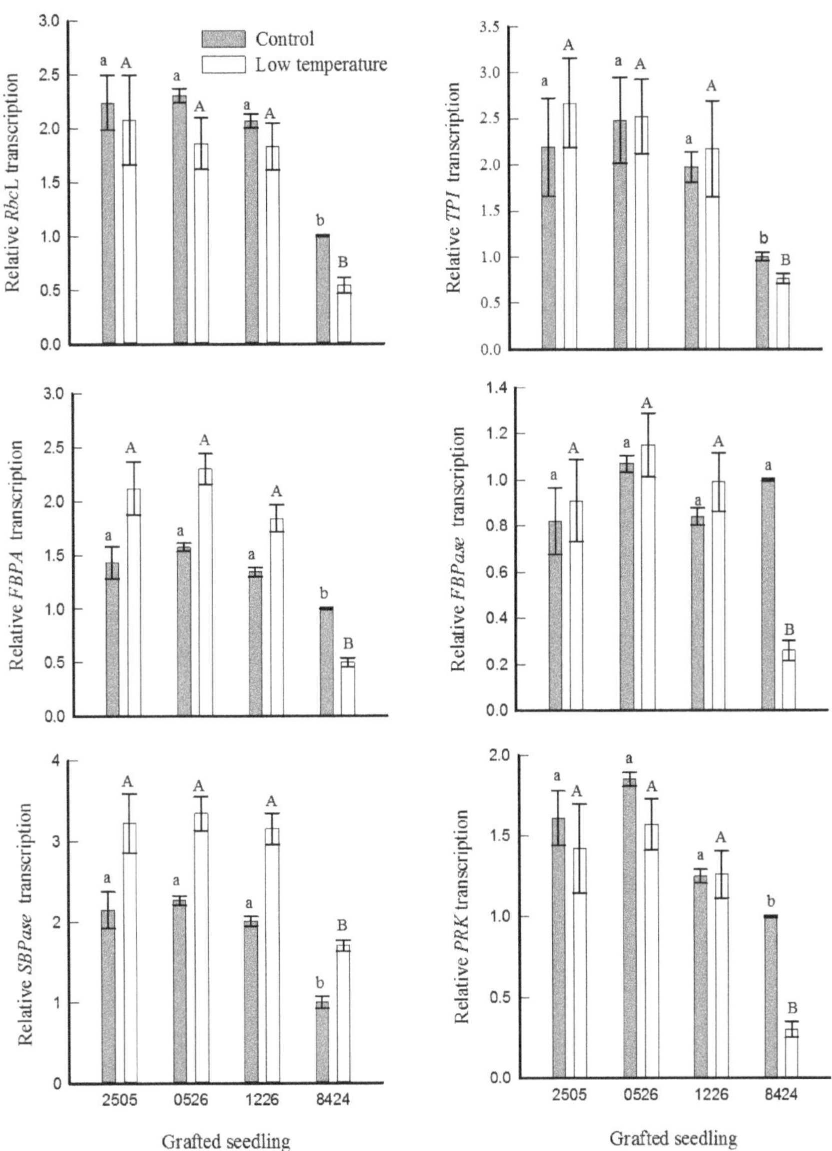

Figure 6. qRT-PCR analysis of the studied genes. Total RNA was extracted from scion leaves, converted to cDNA, and subjected to comparative real-time RT-PCR quantification. Relative transcript levels from qRT-PCR are the mean ± SE of three replicates. Bars with different letters indicate a significant difference ($p < 0.05$).

4. Discussion

Grafting is a widely used technique in plants for improving tolerance to abiotic stress [10–13]. The results of the present study showed that grafted plant growth was significantly better than that of non-grafted 8424, especially under cold stress (Table 1). Furthermore, plants grafted on 2505 and 0526 rootstocks showed better tolerance to cold stress than the plants grafted onto 1226. The results demonstrate that grafting improved watermelon tolerance to cold stress, but the inhibition level varied depending on rootstock genotypes. Similarly, Ramón et al. [17] reported that sweet pepper (*Capsicum annuum* L.), when grafted with appropriate rootstocks, could overcome the negative effects of heat stress conditions. Yan et al. [13] also stated that the use of 'Kaijia No.1' rather than 'Jingxin No.2' or 'Quanneng Tiejia' improved watermelon salt tolerance. In the present study, the superior performances of 2505 and 0526 under cold stress conditions were attributed to the use of an appropriate rootstock, which can absorb more water and nutrients than 8424 and 1226. Our results indicate that rootstock genotypes might play a crucial role in resistance to abiotic stress.

The chlorophyll content reflects the physiological status of plants and is closely related to photosynthetic potential and primary production. The present study shows that the chlorophyll content increased significantly in grafted plants under normal conditions (Table 2), while no significant difference was found in the ratio of *Chl* a/b. Under cold stress conditions, a reduction in chlorophyll content was observed in 8424 and grafted watermelons, whereas these values were higher in the latter, indicating that the grafted watermelon alleviated the photosynthetic inhibition induced by cold stress. Similarly, Yan et al. [13] showed that the chlorophyll content increased in grafted watermelons under salt stress conditions, and Tao et al. [33] also indicated that cucumber plants grafted onto bitter melon (*Momordica charantia* L.) had higher chlorophyll contents than self-grafted plants. The reduced chlorophyll content in stressed plants could be attributed to decreased chlorophyll synthesis, faster chlorophyll degradation, or both.

Environmental stress leads to the accumulation of ROS, which are highly toxic to plants, causing necrotic symptoms [34]. The ROS concentration is a good indicator of oxidative stress. Our results show the greater accumulation of H_2O_2 and O_2^- in non-grafted 8424 (Figure 1), whereas in gourd-grafted plants, this accumulation was relatively lower, indicating that gourd grafting alleviated the inhibition of oxidative stress. This difference was even clearer under cold stress conditions. Cold stress induced more severe oxidative stress in non-grafted 8424 plants than in grafted plants, as manifested by their high ROS accumulation. Among the three rootstocks, 2505 and 0526 were considered more cold-resistant than 1226, which suggests that the different degrees of oxidative stress induced by cold stress were closely related to the rootstock genotype. Similar observations for cold tolerance in grafted grapevines (*Vitis vinifera* L.) [35] and salt tolerance in grafted watermelons [13] have been reported, whereby plants using tolerant rootstocks displayed lower ROS accumulation than those using sensitive rootstocks.

An increase in ROS accumulation under abiotic stress parallels increased lipid peroxidation. MDA, as the final product of peroxidation, is responsible for cell membrane damage [36]. Our results show that, under control conditions, the degree of lipid peroxidation in 8424 and 1226 was significantly higher than that in 0526 and 2505, while upon exposure to cold stress, the MDA content in 8424 increased more pronouncedly than that in grafted watermelons (Figure 2A), indicating that the damage caused by cold injury in 8424 was more serious than that in grafted watermelons. Among the tested rootstocks, 2505 showed the lowest increase. The results confirm that rootstock grafting enhanced the cold tolerance of watermelon, but this was genotype dependent. These findings validate the results of previous investigations, wherin chilling injury increased MDA accumulation [19].

As a compatible solute, proline plays an important role in protecting enzymes from denaturation in order to stabilize the machinery of protein synthesis [37]. Proline accumulation during stress conditions was also thought to be correlated with tolerance. Our results also showed that cold stress promoted the accumulation of proline (Figure 2B), especially in grafted seedlings, indicating that grafting improvs the cold tolerance of watermelon by promoting proline accumulation. These results are also similar to those of Krasensky and Jonak [38], who found that proline was accumulated under abiotic stress conditions in many plant species.

To alleviate oxidative damage under stress conditions, plants have developed a series of effective detoxification mechanisms, among which antioxidant enzymes play an important role [15,35]. In the present research, the activities of these antioxidant enzymes (SOD, CAT, POD, and GPX) were higher in grafted seedlings than in the non-grafted 8424 (Figure 3), which indicates that grafted seedlings counteracted oxidative stress by elevation the antioxidant enzyme activity so as to scavenge ROS and protected the membrane from damage. Compared to normal conditions, the CAT and GPX activities increased when plants were exposed to low temperatures, while the POD and SOD activities decreased. Among the three different rootstocks, 2505 and 0526 showed higher increments and lower reductions. This suggests that CAT and GPX might play a crucial role in antioxidant defense. The cold-induced increase in CAT enzyme activity was likely due to the enhanced expression of CAT isoforms (Figure 4), while the decrease in SOD enzyme activity was also correlated with the decreased expression of SOD isoforms (Figure 5). Similar results were also obtained in other studies, which reported that CAT played an important role in alleviating the oxidative injuries induced by low temperatures [39,40]. Other studies found that SOD activity [39] and POD activity [41] increased under cold stress conditions. Recently, Yan et al. [13] reported that SOD activity, as well as CAT, POD, and APX activity, were increased in watermelon grafted onto P2, which suggests this grafted seedling had a higher H_2O_2 scavenging capacity than those using other rootstocks. These contrasting reports indicate that different plants might employ different pathways to cope with oxidative stress. Here, CAT and GPX might play a more important role in impeding the accumulation of ROS under cold stress.

Abiotic stress is known to influence gene expression. As the last rate-limiting step in carbon fixation, the carboxylated activity of Rubisco is closely related to the Rubisco large subunit (RBCL) and RuBP [42]. In this study, the transcript levels of the Rubisco large subunit (*RbcL*) and enzymes involved in RuBP regeneration were analyzed (Figure 6). As shown in Figure 6, the higher transcript level of *RbcL* in grafted plants, especially under stress conditions, implies the existence of a correlation between cold tolerance and *RbcL* expression, while the latter ameliorates the photosynthetic capacity. As the main genes encoding enzymes required for RuBP regeneration, triose-3-phosphate isomerase (TPI) and fructose-1,6-bisphosphate aldolase (FBPA) catalyze two triose-3-phosphates to fructose-1,6-bisphosphate (FBP). Fructose-1,6-bisphosphatase (FBPase) catalyzes the hydrolysis of FBP to fructose-6-bisphosphatase (Fru6P), while sedoheptulose-1,7-bisphosphatase (SBPase) catalyzes the conversion of sedoheptulose-1,7-bisphosphate (SBP) to sedoheptulose 7-phosphate (Sed7P). Ribulose phosphate epimerase (RuPE) and ribulose-5-phosphate kinase (PRK) are the last rate-limiting enzymes in RuBP synthesis. Except for *RuPE*, which was almost unchanged between different gourd-grafted and non-grafted plants (data not shown), and *FBPase*, in which there was no significant difference between plants under control conditions, the other genes of gourd-grafted plants were expressed much more highly than those of non-grafted plants, especially under cold stress. These results imply that gourd-grafted watermelons could sense cold-induced changes early, and reponse with quick regulations at the transcript level so as to guarantee greater RuBP regeneration, which would activate the carboxylated activity of Rubisco and result in higher carboxylation efficiency.

5. Conclusions

The selection of stress-tolerant rootstocks might be a promising approach to alleviate the detrimental effects of abiotic stress on watermelon productivity. In the present study, the results show that the cold-induced inhibition of growth was significantly ameliorated in gourd-grafted watermelons, as manifested by physiological indices, such as much better growth parameters; much higher chlorophyll and proline contents; lower levels of ROS and lipid peroxidation; higher antioxidant enzyme activities, especially CAT and GPX; and higher expression levels of enzymes related to the Calvin cycle. Overall, as evidenced by the presented data, watermelon grafted with 2505 and 0526 rootstocks showed better resilience than those grafted with 1226 and 8424.

Author Contributions: K.L. performed the experiments. K.L. and J.S. wrote the manuscript. Q.L. participated in the low temperature stress treatment and the determination of physiological characteristics. X.L. participated in organization and discussion regarding the manuscript. X.L. and S.J. designed the research. All authors have read and agreed to the published version of the manuscript.

Funding: This research was funded by the National Key Research and Development Program of China, grant number 2019YFE0118900, the Zhejiang Provincial Natural Science Foundation of China, grant number LY18C06000, and Ningbo Natural Science Foundation, grant number 2017A610293.

Institutional Review Board Statement: Not applicable.

Informed Consent Statement: Not applicable.

Data Availability Statement: All datasets generated for this study are included in the article.

Acknowledgments: We thank the National Key Research and Development Program of China (2019YFE0118900), the Zhejiang Provincial Natural Science Foundation of China (LY18C06000) and Ningbo Natural Science Foundation (2017A610293) for supporting this work.

Conflicts of Interest: The authors declare no conflict of interest.

References

1. Noh, J.; Kim, J.M.; Sheikh, S.; Lee, S.G.; Lim, J.H.; Seong, M.H.; Jung, G.T. Effect of heat treatment around the fruit set region on growth and yield of watermelon [*Citrullus lanatus* (Thunb.) Matsum. and Nakai]. *Physiol. Mol. Biol. Plants* **2013**, *19*, 509–514. [CrossRef]
2. Hao, J.; Gu, F.; Zhu, J.; Lu, S.; Liu, Y.; Li, Y.; Chen, W.; Wang, L.; Fan, S.; Xian, C.J. Low night temperature affects the phloem ultrastructure of lateral branches and Raffinose Family Oligosaccharide (RFO) accumulation in RFO-Transporting plant Melon (*Cucumismelo*, L.) during fruit expansion. *PLoS ONE* **2016**, *11*, e0160909. [CrossRef]
3. Machado, D.F.S.P.; Ribeiro, R.V.; da Silveira, J.A.G.; Filho, J.R.M.; Machado, E.C. Rootstocks induce contrasting photosyn-thetic responses of orange plants to low night temperature without affecting the antioxidant metabolism. *Theor. Exp. Plant Phys.* **2013**, *25*, 26–35. [CrossRef]
4. Niu, M.; Xie, J.; Sun, J.; Huang, Y.; Kong, Q.; Nawaz, M.A.; Bie, Z. A shoot based Na+ tolerance mechanism observed in pumpkin—An important consideration for screening salt tolerant rootstocks. *Sci. Hortic.* **2017**, *218*, 38–47. [CrossRef]
5. Ntatsi, G.; Savvas, D.; Kläring, H.-P.; Schwarz, D. Growth, Yield, and metabolic responses of temperature-stressed tomato to grafting onto rootstocks differing in cold tolerance. *J. Am. Soc. Hortic. Sci.* **2014**, *139*, 230–243. [CrossRef]
6. Wang, Z.M.; Chai, F.M.; Zhu, Z.F.; Elias, G.K.; Xin, H.P.; Liang, Z.C.; Li, S.H. The inheritance of cold tolerance in seven in-terspecifific grape populations. *Sci. Hortic.* **2020**, *266*, 109260. [CrossRef]
7. Xu, J.H.; Zhang, M.; Liu, G.; Yang, X.P.; Hou, X.L. Comparative transcriptome profifiling of chilling stress responsiveness in grafted watermelon seedlings. *Plant Physiol. Biochem.* **2016**, *109*, 561e570. [CrossRef]
8. Kawaide, T. Utilization of rootstocks in cucurbits production in Japan. *Jpn. Agric. Res. Q.* **1985**, *18*, 284–289.
9. Lee, J.-M.; Kubota, C.; Tsao, S.J.; Bie, Z.; Echevarria, P.H.; Morra, L.; Oda, M. Current status of vegetable grafting: Diffusion, grafting techniques, automation. *Sci. Hortic.* **2010**, *127*, 93–105. [CrossRef]
10. Yarşi, G.; Altuntas, O.; Sivaci, A.; Dasgan, H.Y. Effects of salinity stress on plant growth and mineral composition of grafted and un-grafted Galia C8 melon cultivar. *Pak. J. Bot.* **2017**, *49*, 819–822.
11. Xu, Y.; Guo, S.-R.; Li, H.; Sun, H.-Z.; Na, L.; Shu, S.; Sun, J. Resistance of cucumber grafting rootstock pumpkin cultivars to chilling and salinity stresses. *Hort. Sci. Technol.* **2017**, *35*, 220–231. [CrossRef]
12. Lo'ay, A.A.; Abo EL-Ezz, S.F. Performance of 'Flame seedless' grapevines grown on different rootstocks in response to soil salinity stress. *Sci. Hortic.* **2021**, *275*, 109704. [CrossRef]
13. Yan, Y.; Wang, S.; Wei, M.; Gong, B.; Shi, Q. Effect of different rootstocks on the salt stress tolerance in watermelon seedlings. *Hortic. Plant J.* **2018**, *4*, 239–249. [CrossRef]
14. Li, H.; Liu, S.S.; Yi, C.Y.; Wang, F.; Zhou, J.; Xia, X.J.; Shi, K.; Zhou, Y.H.; Yu, J.Q. Hydrogen peroxide mediates abscisic acid-induced HSP 70 accumulation and heat tolerance in grafted cucumber plants. *Plant Cell Environ.* **2014**, *37*, 2768–2780. [CrossRef]
15. Sánchez-Rodríguez, E.; Romero, L.; Ruiz, J. Accumulation of free polyamines enhances the antioxidant response in fruits of grafted tomato plants under water stress. *J. Plant Physiol.* **2016**, *190*, 72–78. [CrossRef] [PubMed]

16. Zhong, Y.Q.; Chen, C.; Nawaz, M.A.; Jiao, Y.Y.; Zheng, Z.H.; Shi, X.F.; Xie, W.T.; Yu, Y.G.; Guo, J.; Zhu, S.H.; et al. Using rootstock to increase watermelon fruit yield and quality at low potassium supply: A comprehensive analysis from agronomic, physiological and transcriptional perspective. *Sci. Hortic.* **2018**, *241*, 144–151. [CrossRef]

17. Ramón, G.-M.; Yaiza, G.P.; Mary-Rus, M.-C.; Salvador, L.-G.; Ángeles, C. Suitable rootstocks can alleviate the effects of heat stress on pepper plants. *Sci. Hortic.* **2021**, *290*, 110529.

18. Chen, S.; Yi, L.; Korpelainen, H.; Yu, F.; Liu, M. Roots play a key role in drought-tolerance of poplars as suggested by reciprocal grafting between male and female clones. *Plant Physiol. Biochem.* **2020**, *153*, 81–91. [CrossRef]

19. Kumar, P.; Rouphael, Y.; Cardarelli, M.; Colla, G. Vegetable grafting as a tool to improve drought resistance and water use efficiency. *Front. Plant Sci.* **2017**, *8*, 1–4. [CrossRef]

20. Li, Y.J.; Liang, G.Y.; Liu, X.J.; Liu, D.C.; Fang, C. Proteomic study on grafted and non-grafted cucumber (*Cucumis sativus* L.). *Acta. Hortic. Sin.* **2009**, *36*, 1147–1152.

21. Yang, Y.; Wang, L.; Tian, J.; Li, J.; Sun, J.; He, L.; Guo, S.; Tezuka, T. Proteomic study participating the enhancement of growth and salt tolerance of bottle gourd rootstock-grafted watermelon seedlings. *Plant Physiol. Biochem.* **2012**, *58*, 54–65. [CrossRef]

22. Yang, Y.; Yu, L.; Wang, L.; Guo, S. Bottle gourd rootstock-grafting promotes photosynthesis by regulating the stomata and non-stomata performances in leaves of watermelon seedlings under NaCl stress. *J. Plant Physiol.* **2015**, *186*, 50–58. [CrossRef]

23. Han, Q.; Guo, Q.; Korpelainen, H.; Niinemets, Ü.; Li, C. Rootstock determines the drought resistance of poplar grafting com-binations. *Tree Physiol.* **2019**, *39*, 1855–1866. [CrossRef]

24. Usanmaz, S.; Abak, K. Plant growth and yield of cucumber plants grafted on different commercial and local rootstocks grown under salinity stress. *Saudi J. Biol. Sci.* **2019**, *26*, 1134–1139. [CrossRef] [PubMed]

25. Lee, J.-M. Cultivation of grafted vegetables, I. current status, grafting methods, and benefits. *HortScience* **1994**, *29*, 235–239. [CrossRef]

26. Porra, R.J.; Thompson, W.A.; Kriedemann, P.E. Determination of accurate extinction coefficients and simultaneous equations for assaying chlorophylls a and b extracted with four different solvents: Verification of the concentration of chlorophyll standards by atomic absorption spectroscopy. *Biochim. Biophys. Acta Bioenerg.* **1989**, *975*, 384–394. [CrossRef]

27. Airaki, M.; Leterrier, M.; Mateos, R.M.; Valderrama, R.; Chaki, M.; Barroso, J.B.; Delrío, L.A.; Palma, J.M.; Corpas, F.J. Metabolism of reactive oxygen species and reactive nitrogen species in pepper (*Capsicum annuum* L.) plants under low temperature stress. *Plant Cell Environ.* **2012**, *35*, 281–295. [CrossRef] [PubMed]

28. Lu, K.; Ding, W.; Zhu, S.; Jiang, D. Salt-induced difference between Glycine cyrtoloba and G. max in anti-oxidative ability and K+ vs. Na+ selective accumulation. *Crop. J.* **2016**, *4*, 129–138. [CrossRef]

29. Bates, L.S.; Waldren, R.P.; Teare, I.D. Rapid determination of free proline for water-stress studies. *Plant Soil* **1973**, *39*, 205–207. [CrossRef]

30. Huang, Z.A.; Jiang, D.A.; Yang, Y.; Sun, J.W.; Jin, S.H. Effects of nitrogen deficiency on gas exchange, chlorophyll fluorescence, and antioxidant enzymes in leaves of rice plants. *Photosynthetica* **2004**, *42*, 357–364. [CrossRef]

31. Woodbury, W.; Spencer, A.; Stahmann, M. An improved procedure using ferricyanide for detecting catalase isozymes. *Anal. Biochem.* **1971**, *44*, 301–305. [CrossRef]

32. Livak, K.J.; Schmittgen, T.D. Analysis of relative gene expression data using real-time quantitative PCR and the $2^{-\Delta\Delta CT}$ method. *Methods* **2001**, *25*, 402–408. [CrossRef] [PubMed]

33. Tao, M.-Q.; Jahan, M.S.; Hou, K.; Shu, S.; Wang, Y.; Sun, J.; Guo, S.-R. Bitter melon (*Momordica charantia* L.) rootstock improves the heat tolerance of cucumber by regulating photosynthetic and antioxidant defense pathways. *Plants* **2020**, *9*, 692. [CrossRef]

34. Soares, C.; Carvalho, M.; Azevedo, R.A.; Fidalgo, F. Plants facing oxidative challenges—A little help from the antioxidant networks. *Environ. Exp. Bot.* **2019**, *161*, 4–25. [CrossRef]

35. Lo'ay, A.A.; Doaa, M.H. The potential of vine rootstocks impacts on 'Flame Seedless' bunches behavior under cold storage and antioxidant enzyme activity performance. *Sci. Hortic.* **2020**, *260*, 108844. [CrossRef]

36. Wakeel, A.; Xu, M.; Gan, Y. Chromium-induced reactive oxygen species accumulation by altering the enzymatic antioxidant system and associated cytotoxic, genotoxic, ultrastructural, and photosynthetic changes in plants. *Int. J. Mol. Sci.* **2020**, *21*, 728. [CrossRef] [PubMed]

37. Szabados, L.; Savouré, A. Proline: A multifunctional amino acid. *Trends Plant Sci.* **2010**, *15*, 89–97. [CrossRef]

38. Krasensky, J.; Jonak, C. Drought, salt, and temperature stress-induced metabolic rearrangements and regulatory networks. *J. Exp. Bot.* **2012**, *63*, 1593–1608. [CrossRef]

39. Petrić, M.; Jevremović, S.; Trifunović, M.; Tadić, V.; Milošević, S.; Dragićević, M.; Subotić, A. The effect of low temperature and GA3 treatments on dormancy breaking and activity of antioxidant enzymes in *Fritillaria meleagris* bulblets cultured in vitro. *Acta Physiol. Plant.* **2013**, *35*, 3223–3236. [CrossRef]

40. Xu, S.-C.; Li, Y.-P.; Hu, J.; Guan, Y.-J.; Ma, W.-G.; Zheng, Y.-Y.; Zhu, S.-J. Responses of Antioxidant enzymes to chilling stress in tobacco seedlings. *Agric. Sci. China* **2010**, *9*, 1594–1601. [CrossRef]

41. Song, G.; Hou, W.; Wang, Q.; Wang, J.; Jin, X. Effect of low temperature on eutrophicated waterbody restoration by *Spirodela polyrhiza*. *Bioresour. Technol.* **2006**, *97*, 1865–1869. [CrossRef] [PubMed]

42. Whitney, S.M.; Houtz, R.L.; Alonso, H. Advancing our understanding and capacity to engineer nature's CO_2-sequestering enzyme, Rubisco. *Plant Physiol.* **2011**, *155*, 27–35. [CrossRef] [PubMed]

horticulturae

Article

Exogenous EBR Ameliorates Endogenous Hormone Contents in Tomato Species under Low-Temperature Stress

Parviz Heidari [1,*], Mahdi Entazari [1], Amin Ebrahimi [1], Mostafa Ahmadizadeh [2], Alessandro Vannozzi [3], Fabio Palumbo [3] and Gianni Barcaccia [3]

[1] Faculty of Agriculture, Shahrood University of Technology, Shahrood 3619995161, Iran; mhdentezari@gmail.com (M.E.); aminebrahimii@shahroodut.ac.ir (A.E.)
[2] Minab Higher Education Center, University of Hormozgan, Bandar Abbas 7981634314, Iran; ahmadizadeh.mostafa@gmail.com
[3] Laboratory of Genomics for Breeding, DAFNAE, Campus of Agripolis, University of Padova, 35030 Legnaro, Italy; alessandro.vannozzi@unipd.it (A.V.); fabio.palumbo@unipd.it (F.P.); gianni.barcaccia@unipd.it (G.B.)
*   Correspondence: heidarip@shahroodut.ac.ir; Tel.: +98-912-0734-034

Citation: Heidari, P.; Entazari, M.; Ebrahimi, A.; Ahmadizadeh, M.; Vannozzi, A.; Palumbo, F.; Barcaccia, G. Exogenous EBR Ameliorates Endogenous Hormone Contents in Tomato Species under Low-Temperature Stress. *Horticulturae* **2021**, *7*, 84. https://doi.org/10.3390/horticulturae7040084

Academic Editors: Agnieszka Hanaka, Jolanta Jaroszuk-Ściseł and Małgorzata Majewska

Received: 9 March 2021
Accepted: 12 April 2021
Published: 17 April 2021

Publisher's Note: MDPI stays neutral with regard to jurisdictional claims in published maps and institutional affiliations.

Abstract: Low-temperature stress is a type of abiotic stress that limits plant growth and production in both subtropical and tropical climate conditions. In the current study, the effects of 24-epi-brassinolide (EBR) as analogs of brassinosteroids (BRs) were investigated, in terms of hormone content, antioxidant enzyme activity, and transcription of several cold-responsive genes, under low-temperature stress (9 °C) in two different tomato species (cold-sensitive and cold-tolerant species). Results indicated that the treatment with exogenous EBR increases the content of gibberellic acid (GA3) and indole-3-acetic acid (IAA), whose accumulation is reduced by low temperatures in cold-sensitive species. Furthermore, the combination or contribution of BR and abscisic acid (ABA) as a synergetic interaction was recognized between BR and ABA in response to low temperatures. The content of malondialdehyde (MDA) and proline was significantly increased in both species, in response to low-temperature stress; however, EBR treatment did not affect the MDA and proline content. Moreover, in the present study, the effect of EBR application was different in the tomato species under low-temperature stress, which increased the catalase (CAT) activity in the cold-tolerant species and increased the glutathione peroxidase (GPX) activity in the cold-sensitive species. Furthermore, expression levels of cold-responsive genes were influenced by low-temperature stress and EBR treatment. Overall, our findings revealed that a low temperature causes oxidative stress while EBR treatment may decrease the reactive oxygen species (ROS) damage into increasing antioxidant enzymes, and improve the growth rate of the tomato by affecting auxin and gibberellin content. This study provides insight into the mechanism by which BRs regulate stress-dependent processes in tomatoes, and provides a theoretical basis for promoting cold resistance of the tomato.

Keywords: cold stress; cold-responsive genes; anti-oxidants; proline; malondialdehyde; hormone profiling

1. Introduction

Low-temperature stress in plants, categorized as freezing stress or chilling stress, is one of the main environmental stresses that adversely affects plant production across the world, especially in subtropical and tropical climate conditions. This environmental extreme is escalating due to global climate change and is, therefore, threatening sustainable crop production [1,2]. Cold stress impacts the photosynthetic system, impairing the cycle of carbon reduction, the thylakoid electron transport, and the stomatal control of CO_2, providing enhanced accumulation of sugars, lipids peroxidation, and water balance disturbance [3–6]. Moreover, a low temperature negatively impacted plants, especially in regards to macromolecules activity, altering the fluidity of the membrane, and reducing osmotic potential of the cell [7]. Regarding the metabolic processes and pathways—cold stress affects antioxidant enzyme activities, membrane fatty acid compositions, and adjusting of the redox state and gene expression [8].

Once plants are exposed to stresses, such as cold stress, different kinds of reactive oxygen species (ROS) are generated, which can undertake a series of oxidation–reduction reactions. Plants defend themselves by enzymatic and non-enzymatic antioxidants [9]. An alternative defense strategy could be supplementing hormones, which could enhance antioxidant and detoxification ability in order to cope and tolerate stressful conditions [10,11]. Previous studies identified numerous hormones and signaling molecules associated with plant responses to particular stress. For instance: ethylene engaged in red light-induced anthocyanin biosynthesis in cabbage [12], the antioxidant system, and ABA in brassinosteroid-induced water stress tolerance of grapevines [13]. Coordination of signaling molecules and hormones positively influences the plant's responses to stress and ultimately its preservation in unfavorable conditions [14–16]. Moreover, brassinosteroids (BRs) as steroidal hormones are involved in an array of physiological and developmental processes via their active engagement in processes, such as antioxidant metabolism [4,17,18], photosynthesis [4,19], nitrogen metabolism [20,21], plant–water relations [22], and osmolyte accumulation [23] in various conditions [1,24–26]. Treatment with 24-epi-brassinolide (EBR) regulates the ascorbate–glutathione (AsA–GSH) component cycle in low-temperature stress on a temporal basis, leading to increased low-temperature tolerance in grapevines at the seedling stage [7]. Transcriptome analysis revealed that treatment with EBR in cold conditions raises the transcript levels of genes related to photosynthesis and chlorophyll biosynthesis, including those encoding for photosystem II (PSII) oxygen-evolving enhancer protein, photosystem I (PSI) subunit, light-harvesting chlorophyll protein complexes I and II, and ferredoxin [27].

Furthermore, BRs illustrated engagement in the regulation of ROS metabolism through the expression of many antioxidant genes that enhance the activity of antioxidant enzymes, such as catalase (CAT), superoxide dismutase (SOD), and peroxidase (POX) [17]. Under low-temperature conditions, plants, by active BR signaling and accumulation of the activate Brassinazole resistant 1 (BZR1) (a BR signaling positive regulator protein), elevate the respiratory burst oxidase homolog 1 *(RBOH1)* transcript levels and the apoplastic H_2O_2 production [28]. The *RBOH1* encodes NADPH oxidase that is involved in ROS in the apoplast, mainly for signaling purposes [25] Moreover, crosstalk between the alternative oxidase (AOX) pathway and BR plays a pivotal role in ameliorating plant tolerance to cold stress, and it has been shown that BR-induced AOX synthesis protects photosystems by bounding ROS synthesis exposed to low-temperature stress [29]. In young grapevine seedlings, foliar application of 24-epi-brassinolide adjusted proteins, free proline contents, and soluble sugars activates the antioxidant machinery to increase chilling stress tolerance [30].

Tomato is a popular garden fruit worldwide because of its edible fruits, rich in antioxidants, and capable of fighting against ROS. Overexpression of DWARF or exogenous EBR application enhances low-temperature tolerance by diminishing oxidative damage in tomato plants [31]. It is worth noting that ROS may also act as a signal in mediating BR-adjusted responses in low-temperature tolerance [32]. A previous study showed that, to protect the plants from oxidative damage, glutaredoxin (GRX), 2-cysteine peroxiredoxin (2-Cys Prx), and RBOH1 participate in a signaling cascade to mediate BR-induced low-temperature tolerance in tomatoes [31]. It was shown that BRs can interact with auxin, salicylic acid, cytokinin, abscisic acid, jasmonic acid, gibberellin, and ethylene, in controlling several morpho-physiological processes in plants [33]. The objective of the current study was to evaluate the effects of 24-epi-brassinolide (EBR) treatment on hormone content, antioxidant activity, the content of malondialdehyde (MDA) and proline, and gene expression of cold-responsive genes on the tomato species under low-temperature stress. This study provides insight into the mechanisms by which BRs regulate stress-dependent processes in tomatoes and provides a theoretical basis for promoting cold resistance in tomato.

2. Materials and Methods

2.1. Plant Materials and Growth Condition

In this study, seeds of the two contrasting tomato species, cold-sensitive (*Solanum lycopersicum* cv. 'Moneymaker') and cold-tolerant (*S. habrochaites*, Accession 'LA1777') [34], were selected to investigate the effects of BR on tomato seedlings under low-temperature stress. Firstly, seeds were sterilized using 2% sodium hypochlorite solution for 12 min and then washed with double distilled water and dried. The three sterilized seeds of each species were sown in plates containing 50% vermicompost and 50% perlite. In this study, 30 plates were used for each tomato species. The plates were maintained in a growth chamber at $23 \pm 2\ ^\circ C$ with the 16 h light/8 h dark cycling. After 40 days, tomato plates of each species were divided two groups; half of them were sprayed with 5 mg/L of 24-epi-brassinolide (EBR) and repeated after 6 h. After 3 h from the last spraying, each treatment was divided into two groups; the first group was transferred to a growth chamber at $9 \pm 1\ ^\circ C$ and the second group was maintained at $23 \pm 1\ ^\circ C$. After three days, the leaves of each sample were cut and stored in liquid nitrogen and transferred to $-80\ ^\circ C$ for the next analysis. In the present study, control plants were cultivated at $23\ ^\circ C$ and without spraying EBR.

2.2. Hormone Profiling

To analyze the free forms of the hormones, including abscisic acid (ABA), indole-3-acetic acid (IAA), and gibberellin (GA3), the young leaves (2.0 g) of each treatment were well-powdered using liquid nitrogen and then samples were crushed by cold methanol. The extract was achieved using 30 mL of 80% cold aqueous methanol in darkness at $4\ ^\circ C$. To determine the hormone content, 10 μL of the extract was injected. The concentration of each hormone was determined using HPLC (Unicam, Cambridge, UK) with a C18 reverse-phase column (4.6×250 mm Diamonsic C18, 5 μm, PerkinElmer, Ohio, USA) and column temperature was $35\ ^\circ C$, gradient elution, mobile phase in methanol, and 1 mL/min flow rate at a wavelength of 254 nm. The peak area of the standard was considered to determine the sample concentration. Moreover, the standards of ABA, IAA, and GA3 were received from Sigma–Aldrich (Steinheim, Germany). The content of IAA and GA3 was measured based on the method defined by Tang et al. [35]. The content of ABA was determined according to the method characterized by Li et al. [36].

2.3. Lipid Peroxidation Assay and Proline Content

The malondialdehyde (MDA) content has been identified as a marker of lipid peroxidation rate associated with oxidative stress. In the current study, 200 mg of fresh leaves were homogenized using 1% TCA (w/v). The MDA content was measured according to the method defined by Campos et al. [37]. Moreover, 0.5 g of leaves were homogenized by 10 mL of 3% sulfosalicylic acid to determine the proline content in each sample. In the current study, the free proline content was analyzed using a method described by Zhang and Huang [38].

2.4. Enzyme Activity

The tomato leaves (300 mg) were ground to a powder in liquid nitrogen and mixed in 3 mL of 0.1 M extraction phosphate buffer (pH 7.5) and the mixed sample was shortly vortexed. The homogenized samples were centrifuged at 13,000 rpm for 15 min at $4\ ^\circ C$. The supernatant of each sample was transferred to determine the enzyme activities. The glutathione peroxidase (GPX; EC 1.11.1.9) activity was distinguished using a method described by Mittova et al. [39], and catalase (CAT; EC 1.11.1.6) activity was measured as described by Aebi [40].

2.5. RNA Extraction and Real Time PCR

The leaves of tomato seedlings were well powdered in liquid nitrogen and the total RNA of each sample was extracted using RNX ^{TM}-Plus (SinaClon, Tehran, Iran), based on the manufacturer's protocols. The extracted RNA samples were then treated by RNase-free DNase I (Thermo Fisher Scientific, Wilmington, MA, USA). The Nano Photometer (Implen N50, Munich, Germany) and a 1% (*W/V*) agarose gel were used to check the quality and quantity of extracted RNA samples. The cDNA was synthesized using 1 µg total RNA and M-MULV reverse transcriptase (Thermo Fisher Scientific, Wilmington, MA, USA) based on the instructions of manufacture. The real-time PCR reactions were run using RealQ Plus 2x Master Mix Green High ROXTM (Ampliqon, Odense, Denmark) on an ABI StepOne system. In this study, four genes belonging to the apetala2/ethylene responsive factor (AP2/ERF) gene family, involved in response to low-temperature stress [41] and the inducer of CBF expression 1 (ICE1), as a key transcription factor gene involved in cold stress tolerance [42], were selected to study the expression patterns by real-time PCR. The *elongation factor 1α* (*EF-1α*; *Solyc06g005060*) gene, as an internal control gene, was used to calculate the relative expression of target genes. The specific real-time PCR primers of genes were designed and evaluated by the online Primer3 Plus tool (Table 1). Finally, the relative expression levels of selected genes were calculated using the $2^{-\Delta\Delta Ct}$ method [43].

Table 1. List of used primers in real-time PCR reactions.

Gene Name	Gene ID	Primer (5′-3′)	Product Size (bp)
ERF.B13	Solyc08g078190	F: GGTGAAGAAGTATGAATGGATCGA R: TCACAGGAACCGAAACAATCG	71
ERF2	Solyc01g090310	F: CTTATGACCAAGCCGCATTC R: ACCCGAGCCGATTAAATGAG	74
ERF52	Solyc03g117130	F: CATTGGGGATCTTGGGTTTC R: TTAGTGCGTGCTGTTGAACC	143
ERF13	Solyc04g054910	F: TCAAGTATGGCCTCCTGCAA R: GAGCAACCTTCACTATTACATGAC	88
ICE1	Solyc03g118310	F: ATGGAGGAACTGGTTCTTGG R: TCCACACCTCCATCATCAAC	139
EF-1α	Solyc06g005060	F: GGAACTTGAGAAGGAGCCTAAG R: CAACACCAACAGCAACAGTCT	158

2.6. Statistical Analyses

All experiments were run in triplicate with three technical replicates, and the effect of the low temperature and EBR treatments on analyzed variables within each species was analyzed by one-way ANOVA and Tukey test using Minitab software (version 17). The final graphs were created using Prism 6 software (GraphPad Software Inc., San Diego, CA, USA) based on the average of each treatment and the standard deviation (SD).

3. Results

3.1. Effects of EBR on Endogenous Hormones in Tomato Leaves Exposed to a Low Temperature

After three days of exposure to low-temperature stress, the content of both GA3 and IAA hormones significantly decreased in cold-sensitive species, but not in cold-tolerant species, although a slight decrease of IAA content under stress was still observed (Figure 1). Exogenous EBR treatment could significantly increase the content of GA3 and IAA in cold-sensitive species in comparison to the control under low-temperature stress. Moreover, the ABA content in both tomato species significantly increased in response to low-temperature stress. A sharp increase in the ABA content was observed in cold-sensitive species that received EBR treatment compared with the control under the low-temperature stress. The treatment with EBR had different effects in cold-sensitive and cold-tolerant species. In fact, although the trend of accumulation in response to a low temperature was similar to that observed in an untreated plant (-EBR), in the cold sensitive species, the treatment with EBR did not significantly affect the GA3, ABA, and IAA content in unstressed conditions, with respect to untreated plants (-EBR). However, it led to a higher accumulation of hormones under stress with respect to untreated plants. The opposite was true for cold-tolerant species: in all cases (IAA, ABA, GA3), treatment with EBR led to a higher accumulation of hormones in unstressed plants with respect to untreated ones, whereas it did not affect the hormone concentration in stressed plants.

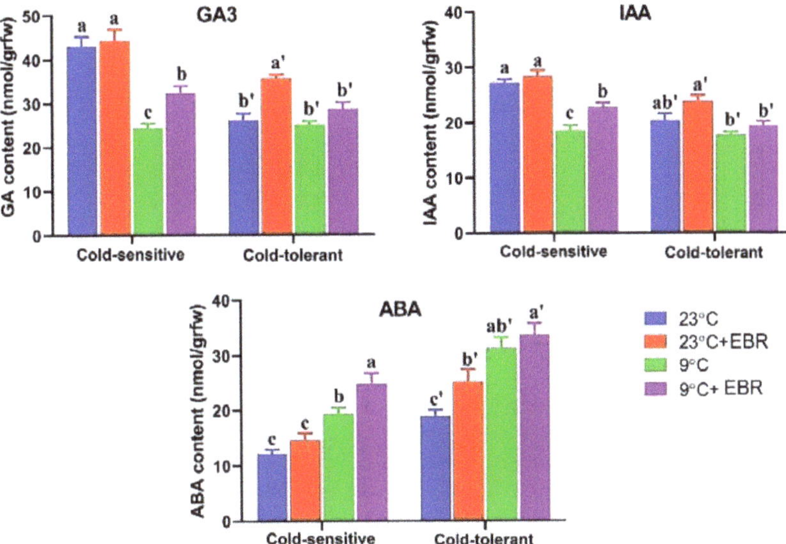

Figure 1. Profile of hormone content (in nanomoles per gram fresh weight (nmol/grfw)) of tomato leaves in response to temperature change and EBR application. Different letters above a bar show significant difference according to the Tukey's range test at $p < 0.05$.

The ABA/GA3 and ABA/IAA ratio increased in both species after three days of exposure to low-temperature stress (Figure 2). Interestingly, the ABA/GA3 and ABA/IAA ratio in cold-tolerant species were higher than the cold-sensitive species. However, the results of the current study revealed that EBR could not affect the ABA/GA3 and ABA/IAA ratio.

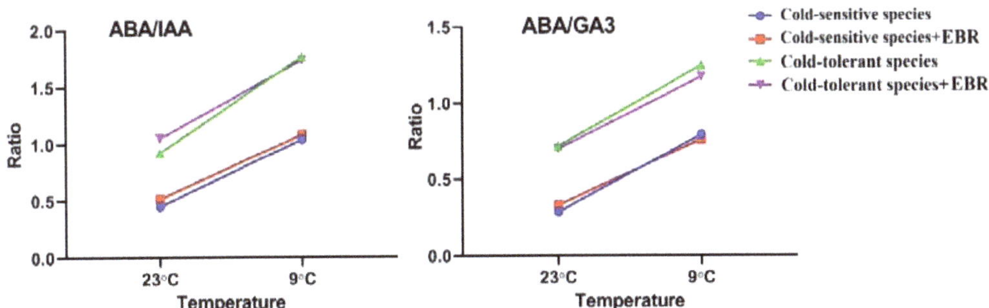

Figure 2. The ration of ABA/IAA and ABA/GA3 under low-temperature stress and EBR application.

3.2. MDA and Proline Are Increased By a Low Temperature

The content of MDA and proline significantly increased in both species, especially in the cold-tolerant species in response to low-temperature stress (Figure 3). Interestingly, EBR treatment could not affect the content of MDA and proline when compared with the control under normal temperature and low-temperature stress. However, in general, the EBR treatment slightly reduced the content of MDA and proline.

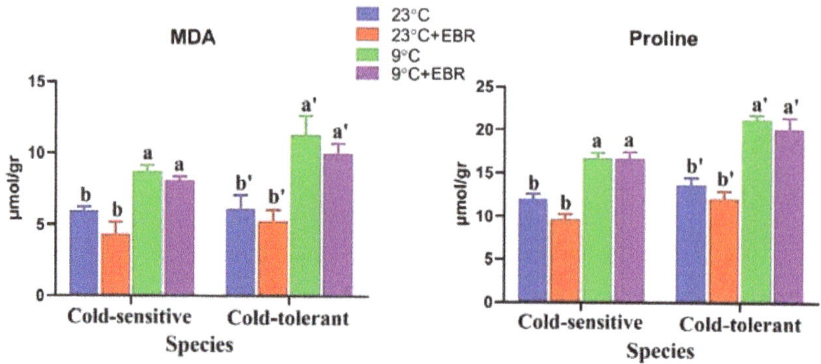

Figure 3. The profile of MDA and proline content under low-temperature stress and EBR treatment. Different letters above a bar show significant difference according to the Tukey's range test at $p < 0.05$.

3.3. Effects of EBR on Activity of Antioxidant Enzymes

According to the results of antioxidant activity, EBR treatment could affect the GPX activity in both tomato species (Figure 4). In normal conditions (23 °C), the EBR treatment increased the GPX activity compared to the untreated plants in both species, whereas in low-temperature stress, the treatment significantly enhanced GPX activity only in the cold-sensitive species. EBR treatment showed different effects in the tomato species under low-temperature stress (Figure 4). The cold-sensitive species, in plants not treated with EBR, showed an increase in CAT activity under cold stress, whereas the treatment with EBR seemed to impair the CAT response to cold. Interestingly, EBR treatment significantly increased the CAT activity in the cold-tolerant species.

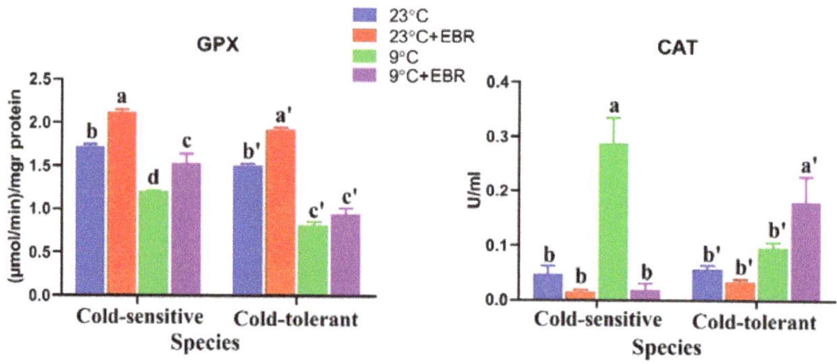

Figure 4. The profile of GPX and CAT activity under low-temperature stress and EBR treatment. Different letters above a bar show significant difference according to the Tukey's range test at $p < 0.05$.

3.4. Effect of EBR on Expression Pattern of ERF Genes

The expression pattern of four members of the ERF multigenic family, together with the MYC-like bHLH transcriptional activator ICE1, was investigated. The expression levels of ERF genes, as well as the ICE1 gene, were significantly affected by low temperatures compared to normal temperatures (Figure 5). Most studied genes were significantly upregulated in both species in response to low-temperature stress while the expression pattern of ERF13 was sharply downregulated in the cold-tolerant species. Furthermore, EBR treatment increased the expression levels of the ERF2, ERF13, ERF.B13, and ICE1 genes in the cold-sensitive genotype comparing to the control, while they were not affected by EBR treatment in the cold-tolerant species. Our results revealed that studied genes are involved in response to low-temperature stress and BR may associate with cold tolerance.

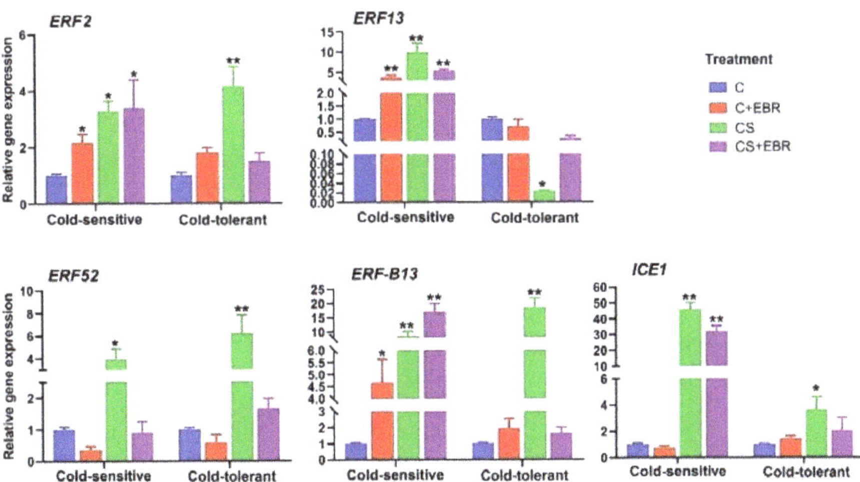

Figure 5. Expression patterns of studied genes under a low temperature (CS) and EBR treatment application; * and ** above a bar shows a significant difference between the applied treatments and normal temperature treatments as control (C) at *p*-value < 0.05 and *p*-value < 0.01, respectively (according to Student's *t*-test).

4. Discussion

4.1. EBR Improves Cold Tolerance by Affecting ABA Content

Various interactions between plant hormones induce a heterogeneous network of plant responses that make it challenging to predict plant performance in response to adverse conditions [44,45]. Moreover, BR can regulate stress responses by cross-talking with other phytohormones [33,46]. In this study, the synergetic interaction was observed between BR and ABA in response to low-temperature stress, where endogenous ABA content significantly increased in the cold-sensitive species under low-temperature stress and EBR treatment. However, the ABA/GA3 and ABA/IAA ratios were not influenced by EBR application. Abscisic acid (ABA) is known as a stress hormone that is influenced by stress and raises plant durability during abiotic stresses, such as drought and cold stress [47]. Moreover, ABA can decrease the damage of dehydration by closing the stomatal pore and maintaining the cellular water [48,49]. However, several antagonistic effects have been observed between signaling components of BRs and ABA under different stress conditions [47,50]. One well-known case of crosstalk occurs at the GSK3-like kinase BIN2 (BRASSINOSTEROID-INSENSITIVE 2), which inhibits the signaling components of the BR pathway, but can be activated by ABA [51]. Moreover, it was stated that ABA negatively controls the BR signaling pathway via phosphorylation of BES1 (bri1-EMS-SUPPRESSOR 1) as a BR signaling positive regulator [52]. Furthermore, Divi et al. found that BR effects are masked by ABA in *Arabidopsis* responses to heat stress, and only in the ABA-deficient *aba1-1* mutant, BR application could make the positive effect [53]. On the other hand, Bajguz stated that BR can enhance the ABA content in *Chlorella vulgaris* under stress conditions [54]. Overall, it seems that the interaction between ABA and BR plays important role in increasing stress tolerance through controlling the synthesizing antioxidants, photosynthesis, and expression of stress response genes [55]. Our results indicated that application of BR is involved in low-temperature stress tolerance by directly/indirectly affecting the ABA content of the tomato species.

4.2. EBR Application Affects the Auxin and GA Content under Low-Temperature Stress

The synergetic interactions are stated between BR and auxin in regulating the cellular processes related to growth, such as cell proliferation and cell expansion [53,56,57]. Furthermore, it was defined that BR and GA are involved in several common cellular processes and BR can regulate cell elongation from GA metabolism [58]. In the current study, the content of IAA and GA was significantly decreased in cold-sensitive species under low-temperature stress. The decrease in the content of GA and IAA through cold stress limits the plant growth and lets it withstand adverse environmental conditions, such as cold, salt, and osmotic stress [59,60]. Moreover, previous studies stated that the IAA content is decreased in response to abiotic stresses, including salinity and cold stress [61,62]. Cold stress can inhibit the activity of acropetal auxin transport by controlling the PIN2 as an auxin efflux carrier [62]. It seems that the transport of auxin from root to shoot is reduced in tomato seedlings under cold stress. Furthermore, the ABA/GA3 and ABA/IAA ratio were increased under low-temperature stress that revealed an antagonist interaction between ABA with auxin and GA in response to low-temperature stress.

The content of GA3 and IAA was significantly increased by EBR application in the cold-sensitive species, compared with the control, under low-temperature stress. Various studies on the role of plant hormones in response to adverse conditions have been performed, but the exact interaction between BR with auxin and GA has not yet been determined, based on molecular information. The expression of many target genes that are involved in growth processes and stress response are commonly controlled by both BR and auxin [57,63,64]. Furthermore, BR and auxin may be involved in the induction of the phosphoinositide and calcium–calmodulin signaling as a second messenger in cellular signal transduction [57]. In the present study, the IAA content increased under low-temperature stress. It seems that BR might affect the polar transporter of auxin [65], and under low-temperature stress, the auxin transfer may increase from root to shoot. Moreover, BR can induce the expression of genes involved in GA biosynthesis, such as *GA3ox-2* [58]. Furthermore, BR can interact with DELLAs, as the GA suppressors, from BZR1, a BR signaling positive regulator [66–68]. In general, it seems that the application of EBR may affect the GA biosynthesis and increase the GA content in the tomato seedlings under low-temperature stress. Overall, the use of EBR treatment as a stimulant may induce some cellular signaling pathways associated with stress tolerance and reduce the adverse effects of stress on growth by increasing the content of growth-regulating hormones, such as GA and auxin.

4.3. MDA and Proline Are Not Affected by EBR Treatment

Abiotic stresses, such as low-temperature stress, hurt the cell membrane through enforced lipid peroxidation and membrane oxidation [11,69]. Antioxidant enzymes activity and the proline content were enhanced by the 28-homobrassinolide treatment in the *Brassica juncea* under cadmium stress. Moreover, the content of proline in roots was higher than in the leaves [70]. The content of MDA under salinity stress in rice seedlings was reduced by EBL treatment [71]. In this study, we discovered that MDA content significantly enhanced in the cold-sensitive species in low-temperature conditions, showing that the plasma membrane was affected and lipid peroxidation increased. In the same line, the increased activities of the antioxidant systems, as a result of BR applications, remarkably defeat the chilling injury of the tomato species by minimizing membrane lipid peroxidation in stress conditions. Moreover, proline content increased in response to low-temperature stress. During stress, proline, as an osmolyte, plays a critical role in controlling cell turgor and stability of membranes [72]. Furthermore, proline can reduce lipid peroxidation and acts as an antioxidant to overcome the oxidative stress created by cold stress [72,73]. Application of brassinosteroid in peppermint (*Mentha piperita* L.) under salinity hampered the death of the plant even at severe stress (150 mM) and prevented negative impact of salinity stress through elevating the activities of antioxidant enzymes and reducing the lipid peroxidation [74]. Moreover, our results revealed that the content of MDA and proline were not influenced by EBR treatment. Overall, it seems that BRs work from a proline-independent pathway to increase endurance to low-temperature stress, although more detailed studies are needed.

4.4. Effects of EBR on Activity of Antioxidant Enzymes

To enhance uncontrolled free radicals, plants respond by non-enzymatic and enzymatic antioxidants to regulate cellular homeostasis and mitigate oxidants [75,76]. Moreover, maintenance and regulation of redox homeostasis seem to be crucial to elevate chilling tolerance in tomato plants [77,78]. Thus, adjustment of the antioxidant system is remarked as a significant mechanism for increasing tomato chilling tolerance. It was demonstrated that using BRs induces antioxidant enzyme activity as well as non-enzymatic antioxidants. For instance, maize seedlings treated by brassinolide (BL) increased the activities of SOD, CAT, APX, carotenoid content, and ascorbic acid [10]. Antioxidative enzymes activity and mRNA expression of Cat A, Mn-SOD, Cat B, Cu/Zn-SOD, GR, and APX remarkably enhanced with EBL treatment under heavy metal stress in *Oryza sativa* L. [79]. In the present study, the effect of EBR application on the antioxidant activity, CAT, and GPX, in the tomato species, was different under low-temperature stress. Different behaviors of CAT were observed in both cold sensitive and cold tolerant species, or, in other words, the cold stress led to an increase in CAT activity in both species. However, the enhancement was much higher in cold sensitive compared to the cold tolerant species (Figure 4). There were substantial differences based on the genotype considered in BR treated plants; the CAT activity in the cold-tolerant species was increased by EBR treatment compared with untreated plants under low-temperature stress. In fact, cold tolerant plants treated with EBR showed an increase in CAT activity, which was higher than untreated plants. In cold sensitive genotypes the EBR treatment seem to impair the CAT activity, whereas in cold-sensitive species, the GPX activity was more influenced by EBR application (Figure 4). It seems that the effect of BR on the activity of antioxidant enzymes depends on the plant species, likely depending on the amount of stress received, the tomato species uses different mechanisms to reduce the induced oxidants. Previous studies on the effect of BR on elevated tolerance of resistant and susceptible tomato species in low temperatures indicated that EBR treatment enhanced the activities of the enzymes in pepper [80], cucumber [81], and eggplant [82] in low temperatures. It seems that oxidative stress is induced by low-temperature stress in the tomato species. Antioxidant enzymes, CAT, and GPX are induced to reduce the oxidants and keep cellular redox. From these results, it could be concluded that BR treatment could play a significant role in the alleviation of ROS damage by increasing antioxidant the defense system, resulting in elevating the tolerance of the tomato species to chilling stress.

4.5. Effects of Low-Temperature Stress on Cold-Related Genes

Ethylene responsive factor (ERF) genes belong to the *AP2/ERF* gene family, known as a large gene family of transcription factors [83,84]. The ERF gene family, as a key regulator, plays an important role in response to adverse conditions, such as cold stress in plant species [85,86]. In this study, the expression level of cold-responsive genes, *ERF* genes, and *ICE1*, selected based on previous studies [41,42], was significantly induced by low-temperature stress. *ICE1* as an upstream transcription factor can regulate cold-responsive genes, such as *CBF* genes [42]. Interestingly under EBR application in a cold-tolerant species, expression patterns of *ERF* genes and *ICE1* reversed to normal temperature conditions. However, Kagale et al. indicated that the EBL application could induce the expression of cold-related genes [87]. Extensive studies were performed on the role of *AP2/ERF* gene family in response to abiotic and biotic stresses as well as hormone treatments, but the effect of BR on this gene family has not yet been investigated. Recently, Xie et al. stated that TINY, an AP2/ERF transcription factor, may negatively affect the expression of BR-responsive genes while it positively controls drought-responsive genes in *Arabidopsis* [88]. In addition, previously, it has been revealed that EBR treatment increases the basic thermo-tolerance of *Brassica napus* [89]. The merit of EBR to grant tolerance in plants to different stresses was corroborated via expression analysis of a subset of cold and drought stress marker genes [87]. Brassinosteroid induced auxin-related genes and cell wall-modifying in soybeans, contrarily, transcriptome analysis demonstrated the twisted BR roles in various biological processes by suppressing some WRKY genes engaged in senescence and stress responses [90]. Overall, our results disclosed that exogenous EBR application might interact with endogenous hormones and reduce the negative effects of low temperatures that induce the expression of cold-responsive genes, *ERFs* and *ICE1*, to return to a state similar to that without stress.

5. Conclusions

In the current study, the effect of the EBR application was investigated in two tomato species under low-temperature stress. The results depicted that low-temperature stress can create oxidative stress and reduce the content of growth-regulatory hormones, IAA and GA3. Moreover, the EBR application increased the content of endogenous ABA, and a synergetic interaction was observed between BR and ABA in response to low-temperature stress. Furthermore, our findings indicated that ABA/GA3 and ABA/IAA ratios are not affected by EBR treatment. In the current study, we found that EBR treatment could not affect the content of MDA and proline under low-temperature stress, but could increase the content of antioxidant enzymes to reduce ROS induced by low-temperature stress. Overall, we concluded that EBR diminishes the adverse effect of low-temperature stress by increasing the content of endogenous phytohormones, increasing the content of antioxidant enzymes, and controlling the gene expression. Furthermore, it seems that BR effects are dependent on the tomato species.

Author Contributions: P.H., M.E. and A.E. designed the study. P.H. and M.E. conducted the experiments and analyzed the data. P.H. and M.A. investigated. All authors contributed to writing the manuscript. A.V., F.P. and G.B. contributed to the interpretation, presentation, and discussion of the data. All authors contributed to writing and editing the manuscript. All authors have read and agreed to the published version of the manuscript.

Funding: This research received no external funding.

Institutional Review Board Statement: Not applicable.

Informed Consent Statement: Not applicable.

Data Availability Statement: All datasets generated or analyzed during this study are included in the manuscript.

Conflicts of Interest: The authors declare no conflict of interest.

References

1. Ahammed, G.J.; Xia, X.-J.; Li, X.; Shi, K.; Yu, J.-Q.; Zhou, Y.-H. Role of brassinosteroid in plant adaptation to abiotic stresses and its interplay with other hormones. *Curr. Protein Pept. Sci.* **2015**, *16*, 462–473. [CrossRef]
2. Heidari, P. Comparative analysis of C-repeat binding factors (CBFs) in tomato and arabidopsis. *Braz. Arch. Biol. Technol.* **2019**, *62*. [CrossRef]
3. Jiang, Y.; Huang, L.; Cheng, F.; Zhou, Y.; Xia, X.; Mao, W.; Shi, K.; Yu, J. Brassinosteroids accelerate recovery of photosynthetic apparatus from cold stress by balancing the electron partitioning, carboxylation and redox homeostasis in cucumber. *Physiol. Plant* **2013**, *148*, 133–145. [CrossRef]
4. Fariduddin, Q.; Yusuf, M.; Ahmad, I.; Ahmad, A. Brassinosteroids and their role in response of plants to abiotic stresses. *Biol. Plant* **2014**, *58*, 9–17. [CrossRef]
5. Ahammed, G.J.; Li, X.; Liu, A.; Chen, S. Brassinosteroids in Plant Tolerance to Abiotic Stress. *J. Plant Growth Regul.* **2020**, 1–14. [CrossRef]
6. Faraji, S.; Ahmadizadeh, M.; Heidari, P. Genome-wide comparative analysis of Mg transporter gene family between Triticum turgidum and Camelina sativa. *BioMetals* **2021**, *4*, 1–22.
7. Chen, Z.-Y.; Wang, Y.-T.; Pan, X.-B.; Xi, Z.-M. Amelioration of cold-induced oxidative stress by exogenous 24-epibrassinolide treatment in grapevine seedlings: Toward regulating the ascorbate–glutathione cycle. *Sci. Hortic.* **2019**, *244*, 379–387. [CrossRef]
8. Khan, T.A.; Yusuf, M.; Ahmad, A.; Bashir, Z.; Saeed, T.; Fariduddin, Q.; Hayat, S.; Mock, H.-P.; Wu, T. Proteomic and physiological assessment of stress sensitive and tolerant variety of tomato treated with brassinosteroids and hydrogen peroxide under low-temperature stress. *Food Chem.* **2019**, *289*, 500–511. [CrossRef] [PubMed]
9. Bajguz, A.; Hayat, S. Effects of brassinosteroids on the plant responses to environmental stresses. *Plant Physiol. Biochem.* **2009**, *47*, 1–8. [CrossRef]
10. Sun, Y.; He, Y.; Irfan, A.R.; Liu, X.; Yu, Q.; Zhang, Q.; Yang, D. Exogenous Brassinolide Enhances the Growth and Cold Resistance of Maize (Zea mays L.) Seedlings under Chilling Stress. *Agronomy* **2020**, *10*, 488. [CrossRef]

11. Huang, X.; Chen, M.-H.; Yang, L.-T.; Li, Y.-R.; Wu, J.-M. Effects of exogenous abscisic acid on cell membrane and endogenous hormone contents in leaves of sugarcane seedlings under cold stress. *Sugar Tech.* **2015**, *17*, 59–64. [CrossRef]
12. Wang, F.; Ahammed, G.J.; Li, G.; Bai, P.; Jiang, Y.; Wang, S.; Chen, S. Ethylene is involved in red light-induced anthocyanin biosynthesis in cabbage (Brassica oleracea). *Int. J. Agric. Biol.* **2019**, *21*, 955–963.
13. Wang, Y.-T.; Chen, Z.-Y.; Jiang, Y.; Duan, B.-B.; Xi, Z.-M. Involvement of ABA and antioxidant system in brassinosteroid-induced water stress tolerance of grapevine (Vitis vinifera L.). *Sci. Hortic.* **2019**, *256*, 108596. [CrossRef]
14. Ahmadizadeh, M.; Chen, J.-T.; Hasanzadeh, S.; Ahmar, S.; Heidari, P. Insights into the genes involved in the ethylene biosynthesis pathway in Arabidopsis thaliana and Oryza sativa. *J. Genet. Eng. Biotechnol.* **2020**, *18*, 1–20. [CrossRef]
15. Rezaee, S.; Ahmadizadeh, M.; Heidari, P. Genome-wide characterization, expression profiling, and post- transcriptional study of GASA gene family. *Gene Rep.* **2020**, *20*, 100795. [CrossRef]
16. Heidari, P.; Mazloomi, F.; Nussbaumer, T.; Barcaccia, G. Insights into the SAM synthetase gene family and its roles in tomato seedlings under abiotic stresses and hormone treatments. *Plants* **2020**, *9*, 586. [CrossRef]
17. Ogweno, J.O.; Song, X.S.; Shi, K.; Hu, W.H.; Mao, W.H.; Zhou, Y.H.; Yu, J.Q.; Nogués, S. Brassinosteroids alleviate heat-induced inhibition of photosynthesis by increasing carboxylation efficiency and enhancing antioxidant systems in Lycopersicon esculentum. *J. Plant Growth Regul.* **2008**, *27*, 49–57. [CrossRef]
18. Hayat, S.; Hasan, S.A.; Yusuf, M.; Hayat, Q.; Ahmad, A. Effect of 28-homobrassinolide on photosynthesis, fluorescence and antioxidant system in the presence or absence of salinity and temperature in Vigna radiata. *Environ. Exp. Bot.* **2010**, *69*, 105–112. [CrossRef]
19. Yu, J.Q.; Huang, L.F.; Hu, W.H.; Zhou, Y.H.; Mao, W.H.; Ye, S.F.; Nogués, S. A role for brassinosteroids in the regulation of photosynthesis in Cucumis sativus. *J. Exp. Bot.* **2004**, *55*, 1135–1143. [CrossRef]
20. Yuan, L.; Yuan, Y.; Du, J.; Sun, J.; Guo, S. Effects of 24-epibrassinolide on nitrogen metabolism in cucumber seedlings under Ca (NO3) 2 stress. *Plant Physiol. Biochem.* **2012**, *61*, 29–35. [CrossRef]
21. Yusuf, M.; Fariduddin, Q.; Ahmad, I.; Ahmad, A. Brassinosteroid-mediated evaluation of antioxidant system and nitrogen metabolism in two contrasting cultivars of Vigna radiata under different levels of nickel. *Physiol. Mol. Biol. Plants* **2014**, *20*, 449–460. [CrossRef]
22. Yuan, G.-F.; Jia, C.-G.; Li, Z.; Sun, B.; Zhang, L.-P.; Liu, N.; Wang, Q.-M. Effect of brassinosteroids on drought resistance and abscisic acid concentration in tomato under water stress. *Sci. Hortic.* **2010**, *126*, 103–108. [CrossRef]
23. Yusuf, M.; Fariduddin, Q.; Ahmad, A. 24-Epibrassinolide modulates growth, nodulation, antioxidant system, and osmolyte in tolerant and sensitive varieties of Vigna radiata under different levels of nickel: A shotgun approach. *Plant Physiol. Biochem.* **2012**, *57*, 143–153. [CrossRef] [PubMed]
24. Ahanger, M.A.; Ashraf, M.; Bajguz, A.; Ahmad, P. Brassinosteroids regulate growth in plants under stressful environments and crosstalk with other potential phytohormones. *J. Plant Growth Regul.* **2018**, *37*, 1007–1024. [CrossRef]
25. Choudhary, S.P.; Yu, J.-Q.; Yamaguchi-Shinozaki, K.; Shinozaki, K.; Tran, L.-S.P. Benefits of brassinosteroid crosstalk. *Trends Plant Sci.* **2012**, *17*, 594–605. [CrossRef]
26. Heidari, P.; Ahmadizadeh, M.; Izanlo, F.; Nussbaumer, T. In silico study of the CESA and CSL gene family in Arabidopsis thaliana and Oryza sativa: Focus on post-translation modifications. *Plant Gene* **2019**, *19*, 100189. [CrossRef]
27. Zhao, M.; Yuan, L.; Wang, J.; Xie, S.; Zheng, Y.; Nie, L.; Zhu, S.; Hou, J.; Chen, G.; Wang, C. Transcriptome analysis reveals a positive effect of brassinosteroids on the photosynthetic capacity of wucai under low temperature. *BMC Genom.* **2019**, *20*, 1–19. [CrossRef] [PubMed]
28. Fang, P.; Yan, M.; Chi, C.; Wang, M.; Zhou, Y.; Zhou, J.; Shi, K.; Xia, X.; Foyer, C.H.; Yu, J. Brassinosteroids act as a positive regulator of photoprotection in response to chilling stress. *Plant Physiol.* **2019**, *180*, 2061–2076. [CrossRef]
29. Deng, X.-G.; Zhu, T.; Zhang, D.-W.; Lin, H.-H. The alternative respiratory pathway is involved in brassinosteroid-induced environmental stress tolerance in Nicotiana benthamiana. *J. Exp. Bot.* **2015**, *66*, 6219–6232. [CrossRef]
30. Xi, Z.; Wang, Z.; Fang, Y.; Hu, Z.; Hu, Y.; Deng, M.; Zhang, Z. Effects of 24-epibrassinolide on antioxidation defense and osmoregulation systems of young grapevines (V. vinifera L.) under chilling stress. *Plant Growth Regul.* **2013**, *71*, 57–65. [CrossRef]
31. Xia, X.; Fang, P.; Guo, X.; Qian, X.; Zhou, J.; Shi, K.; Zhou, Y.; Yu, J. Brassinosteroid-mediated apoplastic H2O2-glutaredoxin 12/14 cascade regulates antioxidant capacity in response to chilling in tomato. *Plant Cell Environ.* **2018**, *41*, 1052–1064. [CrossRef]
32. Cui, J.; Zhou, Y.; Ding, J.; Xia, X.; Shi, K.A.I.; Chen, S.; Asami, T.; Chen, Z.; Yu, J. Role of nitric oxide in hydrogen peroxide-dependent induction of abiotic stress tolerance by brassinosteroids in cucumber. *Plant Cell Environ.* **2011**, *34*, 347–358. [CrossRef]
33. Ahmad, F.; Singh, A.; Kamal, A. Crosstalk of brassinosteroids with other phytohormones under various abiotic stresses. *J. Appl. Biol. Biotech.* **2018**, *6*, 56–62.
34. Ntatsi, G.; Savvas, D.; Ntatsi, G.; Kläring, H.P.; Schwarz, D. Growth, yield, and metabolic responses of temperature-stressed tomato to grafting onto rootstocks differing in cold tolerance. *J. Am. Soc. Hortic. Sci.* **2014**, *139*, 230–243. [CrossRef]
35. Tang, Y.; Wang, L.; Ma, C.; Liu, J.; Liu, B.; Li, H. The use of HPLC in determination of endogenous hormones in anthers of bitter melon. *J. Life Sci.* **2011**, *5*, 139–142.
36. Li, X.-J.; Yang, M.-F.; Chen, H.; Qu, L.-Q.; Chen, F.; Shen, S.-H. Abscisic acid pretreatment enhances salt tolerance of rice seedlings: Proteomic evidence. *Biochim. Biophys. Acta (BBA) Proteins Proteom.* **2010**, *1804*, 929–940. [CrossRef] [PubMed]
37. Campos, P.S.; nia Quartin, V.; chicho Ramalho, J.; Nunes, M.A. Electrolyte leakage and lipid degradation account for cold sensitivity in leaves of Coffea sp. plants. *J. Plant Physiol.* **2003**, *160*, 283–292. [CrossRef] [PubMed]

38. Zhang, Z.; Huang, R. Analysis of malondialdehyde, chlorophyll proline, soluble sugar, and glutathione content in Arabidopsis seedling. *Bio-Protocol* **2013**, *3*, e817. [CrossRef]
39. Mittova, V.; Volokita, M.; Guy, M.; Tal, M. Activities of SOD and the ascorbate-glutathione cycle enzymes in subcellular compartments in leaves and roots of the cultivated tomato and its wild salt-tolerant relative Lycopersicon pennellii. *Physiol. Plant* **2000**, *110*, 42–51. [CrossRef]
40. Aebi, H. [13] Catalase in vitro. In *Methods in Enzymology*; Elsevier: Amsterdam, The Netherlands, 1984; Volume105, pp. 121–126. ISBN 0076-6879.
41. Sharma, M.K.; Kumar, R.; Solanke, A.U.; Sharma, R.; Tyagi, A.K.; Sharma, A.K. Identification, phylogeny, and transcript profiling of ERF family genes during development and abiotic stress treatments in tomato. *Mol. Genet. Genom.* **2010**, *284*, 455–475. [CrossRef]
42. Chinnusamy, V.; Ohta, M.; Kanrar, S.; Lee, B.; Hong, X.; Agarwal, M.; Zhu, J.-K. ICE1: A regulator of cold-induced transcriptome and freezing tolerance in Arabidopsis. *Genes Dev.* **2003**, *17*, 1043–1054. [CrossRef] [PubMed]
43. Livak, K.J.; Schmittgen, T.D. Analysis of relative gene expression data using real-time quantitative PCR and the $2^{-\Delta\Delta CT}$ method. *methods* **2001**, *25*, 402–408. [CrossRef] [PubMed]
44. Peleg, Z.; Blumwald, E. Hormone balance and abiotic stress tolerance in crop plants. *Curr. Opin. Plant Biol.* **2011**, *14*, 290–295. [CrossRef]
45. Eremina, M.; Rozhon, W.; Poppenberger, B. Hormonal control of cold stress responses in plants. *Cell Mol. Life Sci.* **2016**, *73*, 797–810. [CrossRef] [PubMed]
46. Krishna, P. Brassinosteroid-mediated stress responses. *J. Plant Growth Regul.* **2003**, *22*, 289–297. [CrossRef] [PubMed]
47. Kuromori, T.; Seo, M.; Shinozaki, K. ABA transport and plant water stress responses. *Trends Plant Sci.* **2018**, *23*, 513–522. [CrossRef]
48. Xue-Xuan, X.; Hong-Bo, S.; Yuan-Yuan, M.; Gang, X.; Jun-Na, S.; Dong-Gang, G.; Cheng-Jiang, R. Biotechnological implications from abscisic acid (ABA) roles in cold stress and leaf senescence as an important signal for improving plant sustainable survival under abiotic-stressed conditions. *Crit. Rev. Biotechnol.* **2010**, *30*, 222–230. [CrossRef]
49. Ku, Y.-S.; Sintaha, M.; Cheung, M.-Y.; Lam, H.-M. Plant hormone signaling crosstalks between biotic and abiotic stress responses. *Int. J. Mol. Sci.* **2018**, *19*, 3206. [CrossRef]
50. Hu, Y.; Yu, D. BRASSINOSTEROID INSENSITIVE2 interacts with ABSCISIC ACID INSENSITIVE5 to mediate the antagonism of brassinosteroids to abscisic acid during seed germination in Arabidopsis. *Plant Cell* **2014**, *26*, 4394–4408. [CrossRef]
51. An, B.; Wang, Q.; Zhang, X.; Zhang, B.; Luo, H.; He, C. Comprehensive transcriptional and functional analyses of HbGASA genes reveal their roles in fungal pathogen resistance in Hevea brasiliensis. *Tree Genet. Genomes* **2018**, *14*, 41. [CrossRef]
52. Zhang, S.; Cai, Z.; Wang, X. The primary signaling outputs of brassinosteroids are regulated by abscisic acid signaling. *Proc. Natl. Acad. Sci. USA* **2009**, *106*, 4543–4548. [CrossRef]
53. Divi, U.K.; Rahman, T.; Krishna, P. Brassinosteroid-mediated stress tolerance in Arabidopsis shows interactions with abscisic acid, ethylene and salicylic acid pathways. *BMC Plant Biol.* **2010**, *10*, 151. [CrossRef]
54. Bajguz, A. Brassinosteroid enhanced the level of abscisic acid in Chlorella vulgaris subjected to short-term heat stress. *J. Plant Physiol.* **2009**, *166*, 882–886. [CrossRef] [PubMed]
55. Tanveer, M.; Shahzad, B.; Sharma, A.; Khan, E.A. 24-Epibrassinolide application in plants: An implication for improving drought stress tolerance in plants. *Plant Physiol. Biochem.* **2019**, *135*, 295–303. [CrossRef] [PubMed]
56. Anwar, A.; Liu, Y.; Dong, R.; Bai, L.; Yu, X.; Li, Y. The physiological and molecular mechanism of brassinosteroid in response to stress: A review. *Biol. Res.* **2018**, *51*, 46. [CrossRef] [PubMed]
57. Hardtke, C.S.; Dorcey, E.; Osmont, K.S.; Sibout, R. Phytohormone collaboration: Zooming in on auxin–brassinosteroid interactions. *Trends Cell Biol.* **2007**, *17*, 485–492. [CrossRef]
58. Tong, H.; Xiao, Y.; Liu, D.; Gao, S.; Liu, L.; Yin, Y.; Jin, Y.; Qian, Q.; Chu, C. Brassinosteroid regulates cell elongation by modulating gibberellin metabolism in rice. *Plant Cell* **2014**, *26*, 4376–4393. [CrossRef]
59. Kurepin, L.V.; Dahal, K.P.; Savitch, L.V.; Singh, J.; Bode, R.; Ivanov, A.G.; Hurry, V.; Huener, N. Role of CBFs as integrators of chloroplast redox, phytochrome and plant hormone signaling during cold acclimation. *Int. J. Mol. Sci.* **2013**, *14*, 12729–12763. [CrossRef]
60. Colebrook, E.H.; Thomas, S.G.; Phillips, A.L.; Hedden, P. The role of gibberellin signalling in plant responses to abiotic stress. *J. Exp. Biol.* **2014**, *217*, 67–75. [CrossRef]
61. Albacete, A.; Ghanem, M.E.; Martínez-Andújar, C.; Acosta, M.; Sánchez-Bravo, J.; Martínez, V.; Lutts, S.; Dodd, I.C.; Pérez-Alfocea, F. Hormonal changes in relation to biomass partitioning and shoot growth impairment in salinized tomato (Solanum lycopersicum L.) plants. *J. Exp. Bot.* **2008**, *59*, 4119–4131. [CrossRef]
62. Shibasaki, K.; Uemura, M.; Tsurumi, S.; Rahman, A. Auxin response in Arabidopsis under cold stress: Underlying molecular mechanisms. *Plant Cell* **2009**, *21*, 3823–3838. [CrossRef]
63. Sun, L.; Feraru, E.; Feraru, M.I.; Waidmann, S.; Wang, W.; Passaia, G.; Wang, Z.-Y.; Wabnik, K.; Kleine-Vehn, J. PIN-LIKES coordinate brassinosteroid signaling with nuclear auxin input in Arabidopsis thaliana. *Curr. Biol.* **2020**, *30*, 1579–1588. [CrossRef]
64. Goda, H.; Sawa, S.; Asami, T.; Fujioka, S.; Shimada, Y.; Yoshida, S. Comprehensive comparison of auxin-regulated and brassinosteroid-regulated genes in Arabidopsis. *Plant Physiol.* **2004**, *134*, 1555–1573. [CrossRef] [PubMed]
65. KIM, T.; Lee, S.M.; JOO, S.; Yun, H.S.; Lee, Y.E.W.; Kaufman, P.B.; Kirakosyan, A.R.A.; KIM, S.; Nam, K.H.; Lee, J.S. Elongation and gravitropic responses of Arabidopsis roots are regulated by brassinolide and IAA. *Plant Cell Environ.* **2007**, *30*, 679–689. [CrossRef] [PubMed]

66. Bai, M.-Y.; Shang, J.-X.; Oh, E.; Fan, M.; Bai, Y.; Zentella, R.; Sun, T.; Wang, Z.-Y. Brassinosteroid, gibberellin and phytochrome impinge on a common transcription module in Arabidopsis. *Nat. Cell Biol.* **2012**, *14*, 810–817. [CrossRef]

67. Gallego-Bartolomé, J.; Minguet, E.G.; Grau-Enguix, F.; Abbas, M.; Locascio, A.; Thomas, S.G.; Alabadí, D.; Blázquez, M.A. Molecular mechanism for the interaction between gibberellin and brassinosteroid signaling pathways in Arabidopsis. *Proc. Natl. Acad. Sci. USA* **2012**, *109*, 13446–13451. [CrossRef] [PubMed]

68. Li, Q.-F.; Wang, C.; Jiang, L.; Li, S.; Sun, S.S.M.; He, J.-X. An interaction between BZR1 and DELLAs mediates direct signaling crosstalk between brassinosteroids and gibberellins in Arabidopsis. *Sci. Signal.* **2012**, *5*, ra72. [CrossRef]

69. İşeri, Ö.D.; Körpe, D.A.; Sahin, F.I.; Haberal, M. Hydrogen peroxide pretreatment of roots enhanced oxidative stress response of tomato under cold stress. *Acta Physiol. Plant* **2013**, *35*, 1905–1913. [CrossRef]

70. Hayat, S.; Ali, B.; Hasan, S.A.; Ahmad, A. Brassinosteroid enhanced the level of antioxidants under cadmium stress in Brassica juncea. *Environ. Exp. Bot.* **2007**, *60*, 33–41. [CrossRef]

71. Sharma, I.; Ching, E.; Saini, S.; Bhardwaj, R.; Pati, P.K. Exogenous application of brassinosteroid offers tolerance to salinity by altering stress responses in rice variety Pusa Basmati-1. *Plant Physiol. Biochem.* **2013**, *69*, 17–26. [CrossRef] [PubMed]

72. Hayat, S.; Hayat, Q.; Alyemeni, M.N.; Wani, A.S.; Pichtel, J.; Ahmad, A. Role of proline under changing environments: A review. *Plant Signal. Behav.* **2012**, *7*, 1456–1466. [CrossRef] [PubMed]

73. Posmyk, M.M.; Janas, K.M. Effects of seed hydropriming in presence of exogenous proline on chilling injury limitation in Vigna radiata L. seedlings. *Acta Physiol. Plant* **2007**, *29*, 509–517. [CrossRef]

74. Çoban, Ö.; Baydar, N.G. Brassinosteroid effects on some physical and biochemical properties and secondary metabolite accumulation in peppermint (Mentha piperita L.) under salt stress. *Ind. Crops Prod.* **2016**, *86*, 251–258. [CrossRef]

75. Öktem, H.A.; Eyidoðan, F.; Demirba, D.; Bayraç, A.T.; Öz, M.T.; Özgür, E.; Selçuk, F.; Yücel, M. Antioxidant responses of lentil to cold and drought stress. *J. plant Biochem. Biotechnol.* **2008**, *17*, 15–21. [CrossRef]

76. Kamran, M.; Parveen, A.; Ahmar, S.; Malik, Z.; Hussain, S.; Chattha, M.S.; Saleem, M.H.; Adil, M.; Heidari, P.; Chen, J.-T. An Overview of Hazardous Impacts of Soil Salinity in Crops, Tolerance Mechanisms, and Amelioration through Selenium Supplementation. *Int. J. Mol. Sci.* **2020**, *21*, 148. [CrossRef]

77. Duan, M.; Feng, H.-L.; Wang, L.-Y.; Li, D.; Meng, Q.-W. Overexpression of thylakoidal ascorbate peroxidase shows enhanced resistance to chilling stress in tomato. *J. Plant Physiol.* **2012**, *169*, 867–877. [CrossRef] [PubMed]

78. Hu, Y.; Wu, Q.; Sprague, S.A.; Park, J.; Oh, M.; Rajashekar, C.B.; Koiwa, H.; Nakata, P.A.; Cheng, N.; Hirschi, K.D. Tomato expressing Arabidopsis glutaredoxin gene AtGRXS17 confers tolerance to chilling stress via modulating cold responsive components. *Hortic. Res.* **2015**, *2*, 1–11. [CrossRef]

79. Sharma, P.; Kumar, A.; Bhardwaj, R. Plant steroidal hormone epibrassinolide regulate–Heavy metal stress tolerance in Oryza sativa L. by modulating antioxidant defense expression. *Environ. Exp. Bot.* **2016**, *122*, 1–9. [CrossRef]

80. Wang, Q.; Ding, T.; Gao, L.; Pang, J.; Yang, N. Effect of brassinolide on chilling injury of green bell pepper in storage. *Sci. Hortic.* **2012**, *144*, 195–200. [CrossRef]

81. Hu, W.H.; Wu, Y.; Zeng, J.Z.; He, L.; Zeng, Q.M. Chill-induced inhibition of photosynthesis was alleviated by 24-epibrassinolide pretreatment in cucumber during chilling and subsequent recovery. *Photosynthetica* **2010**, *48*, 537–544. [CrossRef]

82. Wu, X.X.; He, J.; Zhu, Z.W.; Yang, S.J.; Zha, D.S. Protection of photosynthesis and antioxidative system by 24-epibrassinolide in Solanum melongena under cold stress. *Biol. Plant* **2014**, *58*, 185–188. [CrossRef]

83. Riechmann, J.L.; Heard, J.; Martin, G.; Reuber, L.; Jiang, C.-Z.; Keddie, J.; Adam, L.; Pineda, O.; Ratcliffe, O.J.; Samaha, R.R. Arabidopsis transcription factors: Genome-wide comparative analysis among eukaryotes. *Science* **2000**, *290*, 2105–2110. [CrossRef]

84. Faraji, S.; Filiz, E.; Kazemitabar, S.K.; Vannozzi, A.; Palumbo, F.; Barcaccia, G.; Heidari, P. The AP2/ERF Gene Family in Triticum durum: Genome-Wide Identification and Expression Analysis under Drought and Salinity Stresses. *Genes* **2020**, *11*, 1464. [CrossRef] [PubMed]

85. Nakano, T.; Suzuki, K.; Fujimura, T.; Shinshi, H. Genome-wide analysis of the ERF gene family in Arabidopsis and rice. *Plant Physiol.* **2006**, *140*, 411–432. [CrossRef]

86. Ahmadizadeh, M.; Heidari, P. Bioinformatics study of transcription factors involved in cold stress. *Biharean Biol.* **2014**, *8*, 83–86.

87. Kagale, S.; Divi, U.K.; Krochko, J.E.; Keller, W.A.; Krishna, P. Brassinosteroid confers tolerance in Arabidopsis thaliana and Brassica napus to a range of abiotic stresses. *Planta* **2007**, *225*, 353–364. [CrossRef] [PubMed]

88. Xie, Z.; Nolan, T.; Jiang, H.; Tang, B.; Zhang, M.; Li, Z.; Yin, Y. The AP2/ERF transcription factor TINY modulates brassinosteroid-regulated plant growth and drought responses in Arabidopsis. *Plant Cell* **2019**, *31*, 1788–1806. [CrossRef] [PubMed]

89. Dhaubhadel, S.; Chaudhary, S.; Dobinson, K.F.; Krishna, P. Treatment with 24-epibrassinolide, a brassinosteroid, increases the basic thermotolerance of Brassica napus and tomato seedlings. *Plant Mol. Biol.* **1999**, *40*, 333–342. [CrossRef]

90. Yin, W.; Dong, N.; Niu, M.; Zhang, X.; Li, L.; Liu, J.; Liu, B.; Tong, H. Brassinosteroid-regulated plant growth and development and gene expression in soybean. *Crop. J.* **2019**, *7*, 411–418. [CrossRef]

horticulturae

Article

Effect of 5-Aminolevulinic Acid (5-ALA) on Leaf Chlorophyll Fast Fluorescence Characteristics and Mineral Element Content of *Buxus megistophylla* Grown along Urban Roadsides

Hao Yang [1], Jianting Zhang [1], Haiwen Zhang [1], Yi Xu [2], Yuyan An [1] and Liangju Wang [1,*]

[1] College of Horticulture, Nanjing Agricultural University, Nanjing 210095, China; 2019804194@njau.edu.cn (H.Y.); 2019104027@njau.edu.cn (J.Z.); 2019104008@njau.edu.cn (H.Z.); anyuyan0447@njau.edu.cn (Y.A.)

[2] Changzhou Park Management Center, Changzhou 213001, China; hdtang@czie.edu.cn

* Correspondence: wlj@njau.edu.cn; Tel.: +86-25-84395265

Abstract: It is well known that trees grown on roadsides suffer from stressful environments, including poor soils, bad weather, and harmful gases from automobile exhaust. Improving the adaptability of roadside trees to adverse environments is important for urban management. An experiment was carried out with six-year-old *Buxus megistophylla* Levl. hedgerows, where 20 mg/L 5-aminolevulinic acids (5-ALA) solution was sprayed on the blade surface at the end of April. Three months later, plant morphology, chlorophyll fast fluorescence characteristics, antioxidant enzyme activities and the mineral element content were investigated. The results showed that leaf size and thickness were significantly greater with 5-ALA treatment, and the leaf color was also greener than those of the control. 5-ALA treatment significantly promoted the electron transfer activity of the PSII reaction center on the donor side, the reaction center itself and the receptor side. It reduced energy dissipation through the heat with increased photochemical quantum yields. The activities of antioxidant enzymes, including superoxide dismutase (SOD), catalase (CAT) and peroxidase (POD) in leaves and roots, were stimulated by 5-ALA treatment. The content of soluble sugars and free proline in leaves was significantly increased by 5-ALA treatment, as were the absorption and accumulation of several kinds of mineral nutrient elements, such as nitrogen, phosphate, calcium, magnesium, iron, copper and boron. Additionally, 5-ALA application significantly increased the content of cadmium, mercury, chromium and lead in the roots but decreased them in the leaves. This implies that 5-ALA may induce a mechanism in *B. megistophylla* in which toxic elements were intercepted in roots to avoid accumulation in leaves, which ensured healthy growth of the aboveground tissues. 5-ALA may regulate the absorption and utilization of mineral nutrient elements in soil with the interception of toxic heavy metal elements in roots, promote leaf photosynthetic performance, induce the accumulation of soluble sugars and free proline, and improve the antioxidant enzyme systems for plants to adapt to the stressful environment of urban roads. These results provide a basis for 5-ALA applications alongside city roads.

Keywords: 5-aminolevulinic acid; *Buxus megistophylla*; chlorophyll fast fluorescence characteristics; mineral nutrition; urban road greening

Citation: Yang, H.; Zhang, J.; Zhang, H.; Xu, Y.; An, Y.; Wang, L. Effect of 5-Aminolevulinic Acid (5-ALA) on Leaf Chlorophyll Fast Fluorescence Characteristics and Mineral Element Content of *Buxus megistophylla* Grown along Urban Roadsides. *Horticulturae* 2021, 7, 95. https://doi.org/10.3390/horticulturae7050095

Academic Editor: Douglas D. Archbold

Received: 8 March 2021
Accepted: 26 April 2021
Published: 2 May 2021

Publisher's Note: MDPI stays neutral with regard to jurisdictional claims in published maps and institutional affiliations.

1. Introduction

Road greening, or landscapes alongside roads, is an important part of urban garden management. The cultivation and maintenance of vibrant roadside green landscape belts have become an important feature of modern social civilization [1]. However, the soil used in road greening is often mixed with engineering construction waste, and even the foreign soil transported by large machinery is often deep subsoil, which lacks aggregate structure and has little fertility. East China is located in a subtropical monsoon climate, cold in winter and hot in summer, with four distinct seasons. The highest temperature in summer often

101

exceeds 40 °C, and the minimum temperature in winter can drop to −10 °C. In summer, it rains almost throughout the whole month of June and most of July, which floods the asphalt roads and the roadside trees for lengthy periods. In winter, it often snows, and a great deal of industrial salt (NaCl) is sprinkled on the road to prevent traffic accidents, which results in the salination of soils and underground water. Additionally, the plants are exposed to exhaust fumes from heavy traffic. Such severe environments subject the roadside plants to multiple stresses, including heat, cold, excess rain, stagnant water, drought, acid or saline–alkali soils, and air and/or soil pollution [2]. How to improve plants alongside heavily traffic roads to resist these multiple stresses and beautify modern cities has become an important problem to be solved in the process of developing a civilized society.

5-Aminolevulinic acid (5-ALA) is a δ-amino acid that does not participate in protein synthesis. It is the key biosynthetic precursor of all porphyrin compounds, such as chlorophyll, heme, cytochrome and so on [3,4]. At the same time, it is also effective at improving plant stress tolerance as a new natural, nontoxic, biodegradable and environmentally friendly plant growth regulator [5]. In agriculture, 5-ALA has been known to promote plant growth and improve the yield of many crops, such as rice, radish, barley, potato, garlic, broad bean and so on [6]. Under salt stress, 5-ALA promoted the growth of spinach seedlings [7]. Under low temperature and low light, 5-ALA improved the photosynthetic capacity of melon seedlings and promoted plant growth [8]. In addition, 5-ALA improved plant tolerance to biotic and abiotic stresses, such as high temperature [9], strong light [10], UV-B [11], drought [12], waterlogging [13], alkaline soil [14], heavy metal pollution [15], pesticides [16], herbicide damage [17], and fungal infection [18]. In garden plants, 5-ALA improved cold tolerance of camphor and rhododendron [19] and heat tolerance of *Ligustrum japonicum* and *Spiraea japonica* [20]. However, to date, there are no reports of 5-ALA applications to urban road greening tree or shrub species.

Buxus megistophylla is an evergreen shrub species [21], which is widely planted along roadsides in urban and rural public green spaces, roadside hedgerows, ecological ditches, and other related landscape constructions in many regions of China [22]. It has been proposed that the species is moderately sensitive to drought [23], chilling injury [24] and air pollution [25]. Yet, no attempts to improve its stress tolerance have been reported. In this study, hedgerows of *B. megistophylla* grown on the roadsides of the main transport highway were sprayed with exogenous 5-ALA solutions in the spring at the early stage of plant growth. The 5-ALA treatment significantly improved the growth of these hedgerow plants. Therefore, analyses were carried out to clarify the possible mechanisms contributing to the improved growth to provide a theoretical basis for applying this nonprotein amino acid to urban road greening plants.

2. Materials and Methods

2.1. Materials and Treatment

The materials selected in this experiment were *Buxus megistophylla* Levl. hedgerows, which were planted on both sides of Nenjiang Road, Changzhou City, Jiangsu Province in 2014. The planting density was 64 plantlet cuttings per m^2, and the hedgerows were pruned monthly to maintain a 70–80 cm height during the growing season. Our visual assessment was that the plants were growing poorly, exhibiting weak annual growth, which is why they were selected for this study.

5-Aminolevulinic acid (5-ALA) solution at a dosage of 20 mg/L was sprayed on the leaf surface of plants on 28 April 2020. Control plants were sprayed with the same volume of water. The width of the hedge was 1.8 m, and the length of one plot was about 100 m. There were 3 plots per treatment, arranged randomly. Ninety days after treatment, plant growth and morphology of the control and 5-ALA treated plants were investigated. Because the hedgerows were dense, it was difficult to dig a single plant with a complete root system. In addition, monthly pruning did not permit measurements of shoot growth. Therefore, leaf morphology and rapid chlorophyll fluorescence were measured in the roadside plants. Then, 5–8 plants with partial roots were randomly dug up from each plot. Mature leaf and fine root tissues were collected and prepared for physiological and biochemical analysis, rinsed in water, quickly frozen in liquid nitrogen, and then driven back to the lab and stored at $-80\,^{\circ}\text{C}$ until analysis.

2.2. Determination of Leaf Morphology and Relative Chlorophyll Content

The length, width and thickness of fresh plant leaves were measured with a ruler or vernier caliper. The relative chlorophyll content was determined with a SPAD-502 Plus produced by Konica Minolta. For these measurements, fifteen mature leaves of the control and the treated plants were randomly selected from different plants from each plot, and an average value was obtained from 45 observations.

2.3. Rapid Fluorescence Determination of Chlorophyll in Leaves and Analysis of Fluorescence Parameters

The fast fluorescence OKJIP curve and 820 nm reflection fluorescence absorption curve of the chlorophyll of leaves of *B. megistophylla* were measured in vivo with a Multifunctional plant efficiency analyzer (M-PEA, Hansatech Instrument Ltd., Norfolk, UK). Before determination, the leaves were dark-adapted for 30 min and then exposed to saturated pulsed light (3000 μmol m^{-2} s^{-1}) for 1 s. Fifteen leaves were measured repeatedly in the treatment and the control, respectively. The O-phase fluorescence (F_o) of the fast fluorescence curve was the initial fluorescence when the PSII reaction center was completely open. The fluorescence of the K phase (F_k), J phase (F_j), and I phase were measured at 300 μs, 2 ms, and 30 ms, respectively. F_m was the maximum fluorescence when the PSII reaction center was completely closed, and the time was about 300 ms. The absorption of modulated reflection fluorescence at 820 nm was expressed by its relative value, where the relative fluorescence $MR/MR_o = 1$ at 0.7 ms was set. The fluorescence parameters were calculated according to the method described previously [19,20]. The control and 5-ALA treatment measurements were both replicated 15 times from randomly selected plants, 5 per plot.

2.4. Determination of Soluble Sugars and Free Proline in Leaves

Frozen leaf samples were used for the measurements, where the content of soluble sugars was determined by anthrone colorimetry [26], while the free proline content was determined by the acid ninhydrin method [27]. Three biological repeats were conducted, and the average value was taken.

2.5. Determination of the Activities of Antioxidant Enzymes in Leaves and Roots

The frozen tissues were extracted according to the method described by Tan et al. [28]. The activities of superoxide dismutase (SOD), peroxidase (POD) and catalase (CAT) were determined in the crude extract according to the methods described by Zhang et al. [29] and Change and Maehly [30], respectively. All extractions were biologically repeated 3 times, and the average value was taken.

2.6. Determination of Mineral Nutrient Elements in Leaves and Roots

The mature leaves and fine roots of *B. megistophylla*, allowed to equilibrate at ambient temperature, were rinsed thoroughly with tap water and then deionized water, respectively, then dried and ground to a fine powder to pass a 100-mesh sieve. One hundred milligrams of the powdered samples were digested with concentrated nitric acid at 100 $^{\circ}$C for 90 min. After complete clarification, the volume of the solution was fixed to 20 mL. The content of mineral elements, including phosphorus, potassium, calcium, magnesium, iron, copper, zinc, manganese, sodium, aluminum, cadmium, mercury, chromium and lead, were determined using an ICP-OES2100 system. For the nitrogen content, 0.2 g of dry powder was combined with 3 g potassium sulfate, and 5 mL concentrated sulfuric acid. After the solution was clarified, 5 mL of 4% NaOH (w/v) and 30 mL 30% boric acid (w/v) were added and determined using a Kjeldahl nitrogen meter. There were 3 biologically replicates, 1 per plot, and the average value was taken.

2.7. Statistical Analysis

The data from control and treated plants were compared using the Student's *t*-test of SPSS 20.0 software. When $p \leq 0.05$, the effect was considered significant; when $p \leq 0.01$, the effect was considered extremely significant.

3. Results

3.1. Effect of 5-ALA Treatment on the Appearance of B. megistophylla on Both Sides of Urban Roads

We found that some leaves of the control plants appeared chlorotic, some older branches had dried up, and parts of shrubs were dead. However, there was no plant dead in the area sprayed with 5-ALA. Even if a few leaves of individual plants fell off, the lateral buds sprouted, and new leaves were developing. When the weak plants were dug up, there was almost no new root growth evident in control, but many new roots occurred in the treated plants. In addition, it can be seen from Table 1 that the leaf length and width of the treatment were 13.54% and 40.34% higher than those of the control, respectively, and the difference was extremely significant ($p \leq 0.01$). The leaf thickness was also 16.67% higher ($p \leq 0.05$), and the relative content of chlorophyll was 61.06% higher than that of the control ($p \leq 0.01$). These results showed that the 5-ALA treatment significantly promoted the growth of *B. megistophylla*, a popular species for urban greening, with deeper green leaves and a better greening effect.

Table 1. Effects of application with 5-ALA solution on leaf morphology of *B. megistophylla*.

Treatment	Leaf Length (cm)	Leaf Width (cm)	Leaf Thickness (cm)	SPAD
Control	3.25 ± 0.08 B [z]	2.38 ± 0.09 B	0.30 ± 0.002 b	41.70 ± 0.23 B
5-ALA	4.84 ± 0.03 A	3.34 ± 0.06 A	0.35 ± 0.002 a	67.16 ± 0.25 A

[z] The data in the table are means $\pm$ SE of 45 replicate measurements. The different capital or lowercase letters represent significant differences at $p \leq 0.01$ or 0.05, respectively, between the control and the treated plants by Student's *t*-test.

3.2. Effects of 5-ALA Treatment on Chlorophyll Fast Fluorescence and 820 nm Reflection Fluorescence Absorption Curve of B. megistophylla Leaves

Figure 1 shows the fast fluorescence OKJIP curve and 820 nm reflection fluorescence absorption curve of *B. megistophylla* leaf chlorophyll. It can be seen from Figure 1A that the fluorescence intensity of OKJIP curves in the 5-ALA treated leaves were significantly lower than those of the control in O ($t = 50$ μs), K ($t = 300$ μs) and J ($t = 3$ ms) phases, but significantly higher than the latter in I ($t = 30$ ms) and P ($t \approx 300$ ms) phases. These indicate that the heat consumption (O phase) of the photosystem II (PSII) reaction center of the control was higher than that of the 5-ALA treatment. Simultaneously, the activities of the donor side (K phase) and receptor side (J phase) were both inhibited in control. In contrast, after 5-ALA treatment, the electron transport activity of the donor side (oxygen evolved complex) and the receptor side increased, while the heat dissipation energy decreased significantly.

It can be seen from Figure 1B that the MR/MR_0 of the 820 nm reflection fluorescence absorption curve of *B. megistophylla* leaves decreased rapidly from 0.7 ms of JIP time, which decreased linearly until 3 ms, then slowly between 3 and 30 ms and down to the minimum. This dynamic reflects the loss of electrons in the photosynthetic system I (PSI) reaction center P_{700} to reduce the terminal electron receptor NADP and produce NADPH and ATP. The lower the minimum value of MR/MR_0, the stronger the ability of the PSI reaction center to reduce the terminal electron receptor. After this, the value of MR/MR_0 increased and reached the maximum at about 300 ms. This rise reflects the reducing activity of the PSI reaction center by electron transfer between PSII and PSI. Comparing the curve between the treatment and the control, the change in magnitude of the MR/MR_0 value of the control leaves was significantly smaller than that of the treatment. The minimum MR/MR_0 value of the treated leaves was significantly lower than that of the control at t = 30 ms, but the maximum MR/MR_0 value was significantly higher than that of the control at 300 ms. This indicates that the 5-ALA treatment improved the ability of the PSI reaction center to be oxidized by itself and then be reduced by the electrons transported from the PSII reaction center in the *B. megistophylla* leaves. Therefore, 5-ALA treatment had a significant promoting effect on the two photosynthetic systems in the leaves of *B. megistophylla*.

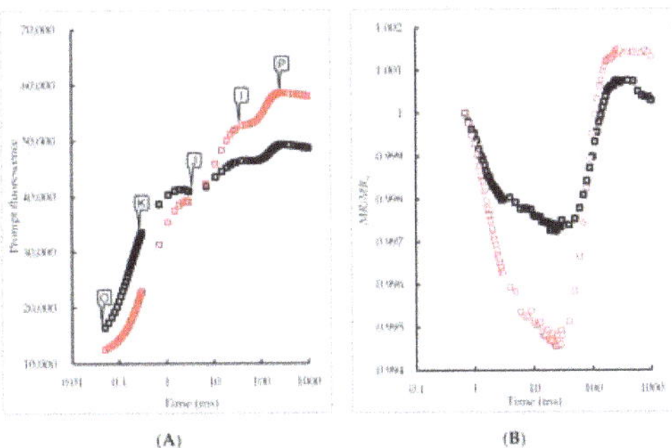

Figure 1. Effect of 5-ALA treatment on (**A**) leaf chlorophyll fast fluorescence and (**B**) 820 nm reflection absorption curve of *B. megistophylla*. Black lines indicate control, and the red lines indicate that 5-ALA treatment. O, K, J, I, and P represent different phases in the OKJIP curve.

3.3. Effect of 5-ALA on Chlorophyll Rapid Fluorescence Parameters in B. megistophylla Leaves

Table 2 shows that the initial fluorescence F_0 of leaves treated with 5-ALA was only 75.46% of that of the control, while the maximum fluorescence F_m and variable fluorescence F_v were 16.29% and 36.89% higher than that of the control, respectively. These differences reached an extremely significant level ($p \leq 0.01$).

Table 2. Effects of 5-ALA treatment on chlorophyll fluorescence parameters of *B. megistophylla*.

Fluorescence Parameters	Treatment		Fluorescence Parameters	Treatment	
	Control	5-ALA		Control	5-ALA
F_0	$(1.63 \pm 0.12) \times 10^4$ A [z]	$(1.23 \pm 0.03) \times 10^4$ B	ET_0/CS	$(2.79 \pm 0.19) \times 10^3$ B	$(4.51 \pm 0.19) \times 10^3$ A
F_m	$(4.91 \pm 0.22) \times 10^4$ B	$(5.71 \pm 0.17) \times 10^4$ A	DI_0/CS	$(5.83 \pm 0.83) \times 10^3$ A	$(2.65 \pm 0.01) \times 10^3$ B
F_v	$(3.28 \pm 0.22) \times 10^4$ B	$(4.49 \pm 0.15) \times 10^4$ A	ABS/RC	4.43 ± 0.40 A	2.18 ± 0.07 B
W_k	0.70 ± 0.04 A	0.43 ± 0.01 B	TR_0/RC	2.80 ± 0.14 A	1.70 ± 0.04 B
Ψ_0	0.28 ± 0.03 B	0.47 ± 0.02 A	ET_0/RC	0.73 ± 0.04 a	0.79 ± 0.02 a
M_0	2.06 ± 0.16 A	0.91 ± 0.05 B	DI_0/RC	1.64 ± 0.27 A	0.47 ± 0.02 B
φP_0	0.66 ± 0.03 B	0.78 ± 0.00 A	RC/CS	$(3.82 \pm 0.20) \times 10^3$ B	$(5.70 \pm 0.18) \times 10^3$ A
φE_0	0.19 ± 0.02 B	0.37 ± 0.01 A	V_{PSI}	$(8.72 \pm 3.22) \times 10^{-4}$ B	$(16.23 \pm 0.18) \times 10^{-4}$ A
φR_0	0.07 ± 0.01 B	0.11 ± 0.00 A	$V_{PSII\text{-}PSI}$	$(2.12 \pm 1.47) \times 10^{-5}$ B	$(6.72 \pm 1.38) \times 10^{-5}$ A
φD_0	0.34 ± 0.03 A	0.22 ± 0.00 B	PI_{ABS}	0.33 ± 0.09 B	1.62 ± 0.19 A
ABS/CS	$(1.63 \pm 0.12) \times 10^4$ A	$(1.23 \pm 0.03) \times 10^4$ B	PI_{total}	0.17 ± 0.04 B	0.66 ± 0.07 A
TR_0/CS	$(1.04 \pm 0.05) \times 10^4$ A	$(0.96 \pm 0.03) \times 10^4$ B			

[z] The data in the table are means $\pm$ SE of 15 replicate measurements. The different capital or lowercase letters represent significant differences at $p \leq 0.01$ or 0.05, respectively, between the control and the treated plants by Student's *t*-test.

W_k reflects the inhibition of the oxygen-evolving complex on the donor side of the PSII reaction center. The higher the W_k, the more serious the inhibition of the oxygen-evolving complex. The W_k of 5-ALA treated leaves was only 61.43% of that of the control (Table 2), indicating that 5-ALA treatment alleviated the inhibition of the oxygen-evolving complex on the donor side of the PSII reaction center. φ_0 represents the possibility of an exciton transferring electrons to the other electron receptors beyond Q_A^-. The φ_0 in the leaves treated with 5-ALA was 67.86% higher than that of the control (Table 2), indicating that 5-ALA improved the electron transport on the receptor side of the PSII reaction center. M_0 reflects the maximum closure rate when Q_A is completely reduced on the receptor side of the PSII reaction center. The higher M_o, the faster the PSII reaction center closes. As shown in Table 2, the M_o of leaves treated with 5-ALA was only 44.17% of that of the control, indicating that 5-ALA treatment slowed the closing rate of the PSII reaction center and facilitated the transfer of photosynthetic electrons to receptors farther away from Q_A^-. The differences in these parameters show that the 5-ALA treatment improved the photosynthetic electron transport activity on the donor side and receptor side of the PSII reaction center in the leaves of *B. megistophylla*.

Variables φP_o, φE_o, φR_o and φD_o represent the maximum photochemical efficiency of the PSII reaction center, the quantum yield of absorbed light energy for photosynthetic electron transfer, the maximum photochemical efficiency of the PSI reaction center and the quantum yield for heat dissipation, respectively. Among them, the higher the first three, the higher the photosynthetic energy utilization efficiency, and the lower the φD_o, the less the energy dissipation through heat. As shown in Table 2, φP_o, φE_o and φR_o of leaves treated with 5-ALA were 18.18%, 94.74% and 57.14% higher than those of the control, respectively, while φD_o was only 64.71% of that of the control, indicating that 5-ALA treatment significantly increased the maximum photochemical efficiency of PSII and PSI in the leaves of *B. megistophylla*, with increased photochemical quantum yield and reduced heat dissipation.

The calculated results of absorption (ABS/CS_o), trap (TR_o/CS), energy transfer (ET_o/CS) and heat dissipation (DI_o/CS) per excited cross-section in the treated leaves were 75.76%, 92.31%, 161.65% and 45.45% of the control (Table 2), respectively. This means that, except for ET_o/CS, which was significantly higher in the treated leaves than in the control, the other three parameters were lower in the treated leaves than those in control ($p \leq 0.01$). Furthermore, the calculated absorption (ABS/RC_o), trap (TR_o/RC), energy transfer (ET_o/RC) and heat dissipation per active reaction center (DI_o/RC) in the treated leaves were 49.13%, 60.94%, 107.54% and 28.95% of the control, respectively. Except for ET_o/RC, which was slightly higher in the treatment than the control, although without a significant difference, the other three parameters in the treated leaves were significantly lower than those of the control. These show that 5-ALA treatment alleviated photoinhibition by relaxing energy charge per cross-section or per reaction center. Furthermore, the density of active reaction centers per cross-section (RC/CS) of treated leaves was 49% higher than that of the control. This is an important reason for the improvement of the photosynthetic capacity of leaves treated with 5-ALA.

V_{PSI} and $V_{PSII\text{-}PSI}$ represent the oxidation rate of the PSI reaction center when electrons were lost and the reduction rate of the PSI reaction center when electrons were transferred between the two photosynthetic systems, respectively. V_{PSI} and $V_{PSII\text{-}PSI}$ of leaves treated with 5-ALA were 86.12% and 216.98% higher than those of the control, respectively (Table 2), indicating that 5-ALA treatment exhibited a good promoting effect on electron transfer in PSI reaction center and a much greater promoting effect on electron transfer between the two systems. Therefore, the photosynthetic performance index, PI_{abs}, based on absorption and PI_{total}, including two photosynthetic systems in leaves treated with 5-ALA, was 390.91% and 288.24% higher than those of the control, respectively (Table 2). These data show that the 5-ALA treatment greatly improved the photosynthetic performance of *B. megistophylla* leaves.

3.4. Effects of 5-ALA Treatment on the Content of Soluble Sugar and Free Proline in B. megistophylla Leaves

5-ALA treatment significantly increased the content of soluble sugars and free proline in the leaves of *B. megistophylla*, where the former increased by 122.41% and the latter by 150% (Table 3). These two substances can be used as osmotic solutes, and their increased content is beneficial for enhancing the adaptability of plants to stressful environments.

Table 3. Effects of 5-ALA treatment on leaf soluble sugar and free proline content of *B. megistophylla*.

Treatment	Soluble Sugar (mg g^{-1}FW)	Free Proline (mg g^{-1}FW)
Control	0.58 ± 0.07 b [z]	0.12 ± 0.05 b
5-ALA	1.29 ± 0.07 a	0.30 ± 0.07 a

[z] The data in the table are means ± SE of 3 replicate measurements. The different lowercase letters represent significant differences at $p \leq 0.05$ between the control and the treated plants by Student's *t*-test.

3.5. Effects of 5-ALA Treatment on the Activities of Antioxidant Enzymes in Leaves and Roots of B. megistophylla

The activities of superoxide dismutase (SOD), peroxidase (POD), and catalase (CAT) in leaves and roots of *B. megistophylla* sprayed with 5-ALA were all significantly higher than those of the control (Table 4). These indicate that the 5-ALA treatment improved the antioxidant capacity of leaves and roots of *B. megistophylla*.

Table 4. Effects of 5-ALA treatment on antioxidant enzyme activities of leaves and roots of *B. megistophylla*.

Treatments	SOD (U g^{-1}FW)		POD (U g^{-1}FW min^{-1})		CAT (U g^{-1}FW min^{-1})	
	Leaves	Roots	Leaves	Roots	Leaves	Roots
Control	36.05 ± 0.51 b [z]	34.35 ± 0.70 b	29.73 ± 5.45 B	11.09 ± 1.95 B	15.97 ± 0.72 B	39.85 ± 4.99 B
5-ALA	38.11 ± 0.07 a	36.95 ± 0.39 a	85.69 ± 2.74 A	37.44 ± 0.68 A	62.42 ± 1.58 A	63.76 ± 14.69 A

[z] The data in the table are means ± SE of 3 repeat measurements. The different capital or lowercase letters represent significant differences at $p \leq 0.01$ or 0.05, respectively, between the control and the treated plants by Student's *t*-test.

3.6. Effect of 5-ALA Treatment on the Mineral Element Content of B. megistophylla

The total nitrogen (N) content in roots of *B. megistophylla* was significantly higher than in leaves, and the N ratio between leaves and roots was 0.62 (Table 5), indicating that most of the N absorbed by plants remained in the roots and only a small part was transported to leaves. Compared with the control, the total N content of roots and leaves in the 5-ALA treatment increased significantly, of which leaves and roots increased by 95.58% and 26.04%, respectively. Based on the leaf and root averages, the total level of N increased by 52.68% following the 5-ALA treatment. Compared with the control, the root of N was 56.45% higher than that of the control ($p \leq 0.01$). The above results show that 5-ALA not only promoted the absorption of N but also promoted the transport and distribution of N to the leaves of *B. megistophylla*.

Table 5. Effect of 5-ALA treatment on the mineral element content in leaves and roots of *B. megistophylla*.

Element	Leaves		Roots		Ratio of Leaf to Root	
	Control	5-ALA	Control	5-ALA	Control	5-ALA
N (mg/g)	7.01 ± 0.06 B [z]	13.71 ± 0.82 A	11.29 ± 0.17 B	14.23 ± 0.21 A	0.62 ± 0.01 B	0.97 ± 0.07 A
P (mg/g)	3.20 ± 0.78 b	8.39 ± 1.44 a	0.32 ± 0.05 b	1.55 ± 0.21 a	11.50 ± 4.83 a	5.50 ± 0.82 a
K (mg/g)	10.45 ± 1.40 a	15.55 ± 3.23 a	4.67 ± 0.62 a	6.29 ± 0.62 a	2.26 ± 0.33 a	2.56 ± 0.67 a
Ca (mg/g)	27.40 ± 2.67 b	48.38 ± 5.18 a	17.73 ± 0.77 b	24.37 ± 0.81 a	1.55 ± 0.14 a	2.00 ± 0.29 a
Mg (mg/g)	0.76 ± 0.13 b	1.80 ± 0.25 a	2.40 ± 0.20 a	2.79 ± 0.34 a	0.31 ± 0.03 b	0.69 ± 0.18 a
Fe (mg/g)	0.11 ± 0.01 B	0.29 ± 0.01 A	4.39 ± 0.34 B	9.34 ± 0.47 A	0.03 ± 0.00 a	0.03 ± 0.00 a
Cu (mg/kg)	26.50 ± 0.50 b	32.30 ± 1.52 a	231.91 ± 71.74 b	463.19 ± 14.49 a	0.13 ± 0.03 a	0.07 ± 0.00 a
Zn (mg/kg)	28.21 ± 6.54 a	29.98 ± 6.08 a	66.78 ± 2.27 a	66.73 ± 1.22 a	0.43 ± 0.11 a	0.45 ± 0.09 a
Mn (mg/kg)	24.08 ± 5.23 a	23.52 ± 4.39 a	194.60 ± 37.62 a	90.67 ± 4.12 b	0.14 ± 0.01 a	0.26 ± 0.06 a
B (mg/kg)	28.07 ± 4.08 b	74.95 ± 10.61 a	121.24 ± 10.28 b	257.60 ± 18.56 a	0.23 ± 0.02 a	0.30 ± 0.07 a
Na (mg/g)	0.97 ± 0.00 a	1.04 ± 0.06 a	1.25 ± 0.04 b	1.45 ± 0.06 a	0.78 ± 0.03 a	0.72 ± 0.05 a
Al (mg/g)	0.05 ± 0.00 a	0.08 ± 0.01 a	5.11 ± 0.67 a	2.60 ± 0.27 b	0.01 ± 0.00 a	0.03 ± 0.00 a
Cd (mg/kg)	0.43 ± 0.04 a	0.28 ± 0.01 b	0.35 ± 0.10 b	0.76 ± 0.06 a	1.48 ± 0.51 a	0.37 ± 0.04 b
Hg (mg/kg)	69.45 ± 4.37 a	53.35 ± 1.46 b	213.78 ± 13.29 b	292.75 ± 20.48 a	0.32 ± 0.00 a	0.18 ± 0.01 b
Cr (mg/g)	0.36 ± 0.05 a	0.16 ± 0.01 b	1.29 ± 0.24 b	3.36 ± 0.39 a	0.30 ± 0.08 a	0.05 ± 0.01 b
Pb (mg/kg)	1.32 ± 0.10 a	0.90 ± 0.04 b	6.92 ± 1.15 b	19.11 ± 1.89 a	0.20 ± 0.04 a	0.05 ± 0.01 b

[z] The data in the table are means ± SE of 3 replicate measurements. The different capital or lowercase letters represent significant differences at $p \leq 0.01$ or 0.05, respectively, between the control and the treated plants by Student's *t*-test.

5-ALA treatment also significantly increased the content of phosphorus (P) in the leaves and roots of *B. megistophylla*. The content of P in the leaves of the control was 10 times higher than that in the roots (Table 5), indicating that most of the P absorbed by roots was transported to the shoot. After 5-ALA treatment, the P content in leaves increased by 162.19%, and the root content increased by 384.38%. After the average of the whole plant, the P level of the 5-ALA treatment was 182.39% higher than that of the control. 5-ALA treatment had no effect on the ratio of P in leaves to roots. These results showed that 5-ALA treatment promoted P uptake and accumulation but had no significant effect on P distribution.

The 5-ALA treatment had no significant effect on the level of potassium (K) in *B. megistophylla*. In addition, the K content in the leaves of *B. megistophylla* was 2.26 times higher than that in the roots, indicating that the K absorbed by the plants was mainly transported to the shoot. 5-ALA treatment had no significant effect on this allocation characteristic.

Calcium (Ca) and magnesium (Mg) are both essential medium elements for plant growth. In *B. megistophylla*, the leaf–root ratio of Ca in the control plant was 1.55, indicating that Ca mainly accumulated in leaves. 5-ALA treatment did not significantly affect this characteristic but increased the Ca content in leaves and roots by 76.57% and 37.45%, respectively. This indicates that 5-ALA treatment promoted the absorption and transport of Ca but did not affect the distribution in leaves and roots. Different from Ca, the leaf–root ratio of Mg in the control plant of *B. megistophylla* was 0.31, indicating that most of the magnesium was distributed in the root system. 5-ALA treatment significantly increased Mg content in leaves (by 136.84%) but had no effect on roots. Compared with the control, the leaf–root ratio of Mg in 5-ALA treatment increased by 122.58% ($p \leq 0.05$), indicating that 5-ALA promoted the absorption of Mg by the roots of *B. megistophylla* and the distribution of Mg to the leaves.

Iron (Fe), copper (Cu), zinc (Zn), manganese (Mn) and boron (B) are essential trace elements for plant growth. The content of Fe in roots of *B. megistophylla* was much higher than that in leaves, and the ratio of leaf to root was 0.03 (Table 5). With 5-ALA treatment, the Fe of leaves was 156.45% higher than that of the control, and that in the roots also increased by 113.10% ($p \leq 0.01$), but the leaf–root ratio of Fe did not change, indicating that 5-ALA significantly promoted the absorption and transport of Fe, but did not affect the distribution ratio. The leaf–root ratio of Cu in the control plant of *B. megistophylla* was 0.13, indicating that Cu was mainly distributed in the roots. 5-ALA treatment significantly promoted the content of Cu in leaves and doubled the content in roots but had no significant effect on the distribution ratio of Cu in leaves and roots. Zn of *B. megistophylla* was mainly distributed in the roots, and the leaf content was less than half of the roots (43–45%); 5-ALA had no significant effect on Cu content and distribution. Mn was mainly distributed in the roots of *B. megistophylla*, and the leaf–root ratio was only 0.14 in the control. 5-ALA treatment had no significant effect on leaf content but significantly decreased root Mn content (about half of the control). B was also mainly distributed in the roots of *B. megistophylla*. In the control and treatment, the leaf–root ratio of B was 0.23–0.30. 5-ALA treatment significantly increased B content in leaves and roots but had no significant effect on leaf root distribution (Table 5).

Sodium (Na) and aluminum (Al) are often regarded as harmful elements. If the Na content is too high, the soil may be salinized. If the Al content is too high, the soil tends to be acidified. The Na content in roots of *B. megistophylla* was higher than that in leaves, and the content in leaves and roots increased after 5-ALA treatment, but only the root increment reached a significant level. These indicate that 5-ALA promoted Na uptake but mainly retained in the roots, without a significant increase in the leaves. For Al, 5-ALA treatment reduced the Al content in roots of *B. megistophylla* by about half and had no effect on leaves. These suggest that 5-ALA treatment reduced the absorption of Al by *B. megistophylla* as a whole.

Cadmium (Cd), mercury (Hg), chromium (Cr) and lead (Pb) are considered to be toxic heavy metal elements. In the control plant, Cd was mainly distributed in the leaves, and the leaf–root ratio was 1.48, while the other three elements were mainly distributed in the roots, with the leaf–root ratios from 0.20 to 0.32. 5-ALA treatment reduced the leaf–root ratio of Cd to 0.37, which was only 25% of that of the control. For Hg, Cr and Pb, 5-ALA treatment significantly decreased the leaf content but increased the root content, so the leaf–root ratio decreased significantly. These results show that 5-ALA treatment can induce more toxic element storage in roots to avoid leaf accumulation. This effect is consistent with these four toxic heavy metal elements as a whole.

4. Discussion

Road greening is not only an important part of urban greening systems but also an important symbol of modern urban civilization. A good road greening system is conducive to driving safety, enjoyment of city landscapes, and environment protection [31]. However, while road traffic brings speed and convenience to people's lives, it also exerts new coercive factors on the growth of roadside plants. In addition to the conventional high or low temperatures, drought or waterlogging, salinization, and acid–alkali stress, automobile exhaust is a typical harmful factor on plant growth [32]. This is something rarely encountered by plants grown in a quiet garden. It has been proposed that the main characteristics of automobile exhaust were damage to cell membranes, decreased chlorophyll content and photosynthesis, leaf discoloration, leaf shedding and even death [33]. The plants of *B. megistophylla* used in this study were planted on both sides of the main road. However, automobile exhaust was not detected in the air. Traffic throughout the country was not particularly busy in the first half of 2020 due to the COVID-19 epidemic, so the exhaust levels were likely lower than usual. Even so, the growth of *B. megistophylla* on both sides of the road was restricted, and occasional death of shrubs occurred in the experimental site. In addition to the *B. megistophylla* reported in this paper, leaf yellowing, growth inhibition and plant death of another road greening species, *Gardenia jasminoides* Ellis, were also quite significant (data not listed). Therefore, improving the adaptability of plants used for road greening in such adverse environments, promoting plant growth, reducing mortality, and enhancing the greening effect has important practical significance.

5-Aminolevulinic acid (5-ALA) is a natural δ-amino acid commonly found in animals, plants, and microorganisms. It is not used to synthesize proteins but is a key biosynthetic precursor of porphyrin compounds, such as chlorophyll, heme, and cytochrome [3]. Therefore, it is closely related to plant photosynthesis and respiration [5]. A large number of studies have shown that 5-ALA can improve the resistance of plants to high temperature [9], low temperature [8], strong light [10], weak light [8], drought [12], waterlogging [13], salinization [7], heavy metal pollution [15], pesticide [16], herbicide toxicity [17], fungal diseases [18] and other biotic and abiotic stresses and has broad application potential in agricultural production [5]. However, whether 5-ALA can be used on urban road greening plants to improve environmental adaptability has never been reported. *Buxus megistophylla* is sensitive to automobile exhaust [34]. In urban roads with high concentrations of NO_2 and CO, the pH of leaf cell sap decreased, and the content of malondialdehyde (MDA), a product of membrane peroxidation, increased significantly. The results of the present work showed that 5-ALA promoted leaf growth of *B. megistophylla* (Table 1), promoted the occurrence of new roots, reduced the probability of plant death, and significantly improved urban road greening, indicating that it has an important application prospect in road greening.

5-ALA is the precursor of chlorophyll biosynthesis in plants. Exogenous application of 5-ALA increased the content of chlorophyll in leaves of *B. megistophylla* (Table 1). Chlorophyll a is the main pigment in the reaction center of the photosynthetic system (including PSI and PSII). It forms antenna pigment with chlorophyll b, carotenoids and lutein. A photosynthetic reaction center consists of about 300 chlorophyll molecules [35]. 5-ALA can increase the number of photosynthetic reaction centers by increasing the content of chlorophyll in the leaves of *B. megistophylla*. The active reaction center density (RC/CS) per cross-section increased significantly (Table 4).

The fast fluorescence OKJIP curve and 820 nm reflection fluorescence absorption curve of leaf chlorophyll were detected by M-PEA. The results showed that the O, K and J phase fluorescence of *B. megistophylla* in the leaves sprayed with 5-ALA was significantly lower than that of the control, while the I and P phase fluorescence was higher than that of the control (Figure 1A). These results showed that 5-ALA treatment increased the donor side activity, reaction center itself activity and receptor-side electron-transport activity of the PSII reaction center in leaves. Simultaneously, the minimum value of MR/MR_o decreased significantly at t = 30 ms, and the maximum value of MR/MR_o increased significantly at t $\approx$ 300 ms after 5-ALA treatment (Figure 1B). This means that the activity of the PSI reaction center and the ability to reduce terminal electron receptor NADP were promoted. Fluorescence parameter analysis (Table 2) showed that W_k and M_o decreased in the 5-ALA treated leaves, suggesting that 5-ALA alleviated the photoinhibition at the donor-side oxygen evolved complex and the receptor side, leading to a decrease in the maximum closure rate of the PSII reaction center. The increase of φ_o represents an increase in the probability of excitons transferring electrons to other electron receptors downstream of Q_A. 5-ALA treatment significantly promoted the activity of the PSII reaction center in *B. megistophylla* leaves (Table 2). V_{PSI} and $V_{PSII\text{-}PSI}$ represent the rate at which the PSI itself loses electrons when it is oxidized and the rate at which the PSI reaction center is reduced by electrons transferred by PSII [19,20]. After 5-ALA treatment, both V_{PSI} and $V_{PSII\text{-}PSI}$ increased significantly, indicating that 5-ALA significantly promoted electron transfer in PSI reaction centers and between PSII and PSI reaction centers. Furthermore, 5-ALA had a significant promoting effect on the absorption-based photosynthetic performance index PI_{abs} and the total photosynthetic performance index PI_{total}, including two photosynthetic systems (Table 2). These results are similar to the effects of 5-ALA treatment at low temperatures [19] or high temperatures [20]. In addition, from the point of view of photochemical energy conversion, the quantum yield (φE_o) and energy transfer per excited cross-section (ET_o/CS) of *B. megistophylla* leaves treated with 5-ALA were significantly increased. Although the promoting effect of energy transfer (ET_o/RC) based on unit reaction center did not reach a significant level, energy absorption (ABS/CS or ABS/RC), trap (TR_o/CS or TR_o/RC), and heat dissipation (DI_o/CS or DI_o/RC) decreased significantly. These indicate that 5-ALA treatment can reduce the energy charge pressure of the reaction center and reduce the possibility of photoinhibition. Recently, it was reported that 5-ALA treatment could upregulate the expression of core proteins D1 and D2 of the PSII reaction center in potato leaves [36]. This may be an important reason why the core proteins of the PSII reaction center maintain higher activity. Perhaps due to the improvement of leaf photosynthetic performance, the soluble sugars in the leaves of *B. megistophylla* treated with 5-ALA increased significantly (Table 3).

Soluble sugars can be used as osmotic solutes, which are related to plant stress resistance. However, more studies have shown that proline accumulation induced by 5-ALA can improve plant stress resistance [37]. In this experiment, it was observed that 5-ALA treatment increased the proline content in the leaves of *B. megistophylla* (Table 3). This may be one of the reasons for its resistance to environmental stress on urban roads. In addition, the increases in antioxidant enzyme activities induced by 5-ALA are also important reasons for the improvement of stress resistance. In this experiment, 5-ALA significantly increased the activities of superoxide dismutase (SOD), peroxidase (POD) and catalase (CAT) in leaves and roots of *B. megistophylla* (Table 4). These results are similar to previous reports [3,4], indicating that the increase of antioxidant enzyme activity is related to plant stress resistance.

There is currently no consensus on the effect of 5-ALA on the content of mineral elements in plants. Naeem et al. [38] suggested that 5-ALA alleviated the decrease of N, P, Ca, Mg, Zn and Fe content in rape (*Brassica napus* L.) leaves and roots under salt stress but had no effect on Mn and Cu contents. Zhang et al. [39] proposed that 5-ALA could increase Ca, Mg, Cu, Fe and Zn in apple (*Malus domestica Borkh.*) leaves but decrease the content of P, K and Na. Anwar et al. [40] observed that the content of N, P, K, Ca, Mg, Cu,

Fe, Mn and Zn in roots and leaves of cucumber (*Cucumis sativus* L.) seedlings increased significantly when 5-ALA alleviated the inhibition of low temperature on the growth of cucumber seedlings. It seems that the improvement of cold resistance of plants by 5-ALA is related to the maintenance of nutrient balance. However, Liu et al. [14] did not observe an increase of K, Fe and Mg content in leaves when they studied the improvement of alkaline tolerance of Swiss chard (*Beta vulgaris* L.) treated with 5-ALA. Therefore, the effect of 5-ALA on the content of mineral elements in plants may be related to species and types of stress. In the present work, it was observed that 5-ALA increased the level of N in leaves and roots of *B. megistophylla*. This effect was shown not only in the increase of N content in roots and leaves but also in the transport and distribution of N from roots to leaves (Table 5). It has been suggested that nitrate reductase (NR) activity and its coding gene expression was significantly upregulated in pakchoi (*Brassica campestris* ssp. *chinensis* var. *communis* Tsen et Lee) plants treated with 5-ALA, and the content of total N, amino acid and protein in leaves increased, while the content of nitrate nitrogen decreased [41]. The authors deduced that 5-ALA promoted N absorption and assimilation. Our result here is consistent with the previous report, indicating that 5-ALA can promote the absorption, transport, distribution and utilization of N in plants.

It was observed that 5-ALA increased P, Ca, Cu, Fe and B in roots and leaves of *B. megistophylla*. This is similar to the result of Naeem et al. [38]. However, 5-ALA treatment only increased leaf Mg content, had no significant effect on K and Zn content, and decreased root Mn content (Table 5). The reason for these differences is not clear. Yao et al. [42] reported that 5-ALA promoted the absorption and distribution of P in rice (*Oryza sativa* L.). We also observed that 5-ALA increased the P content of *B. megistophylla* plants (Table 5), indicating that the promoting effect of 5-ALA on P was universal. The transport of Ca by plants mainly depends on transpiration flow [43], while 5-ALA can promote stomatal opening [44] and increase transpiration rate. Therefore, it may also be common for 5-ALA treatment to promote the absorption and transport of Ca. For Mg, 5-ALA treatment led to an increase in leaf content (Table 5). This may be one reason for the increase of chlorophyll content in leaves (Table 1). For Cu, Naeem et al. [38] suggested that 5-ALA had no alleviating effect on the decrease of Cu content in rape plants under salt stress, but Anwar et al. [40] thought that 5-ALA could alleviate the decrease of Cu in cucumber induced by low temperature. Our results were similar to those of the latter. 5-ALA could significantly increase the Cu content in leaves and roots of *B. megistophylla* (Table 5). Lack of Fe will lead to leaf chlorosis. In this experiment, the leaves of the control plants were generally yellowing, while the leaves of the treated plants were green. This could not only be related to the higher N and Mg content mentioned above but to the 5-ALA treatment promoting an increase of Fe content in *B. megistophylla*. The content of Fe in both roots and leaves treated with 5-ALA was more than twice as high as that of the control. Such a huge increase is rarely reported. Similarly, 5-ALA induced a more than double increase in B content in leaves and roots of *B. megistophylla* (Table 5). These phenomena have rarely been reported before and are worthy of our attention and further study of the mechanism.

Na and Al are not essential elements for plant growth and development. Excessive accumulation will cause plant toxicity. The results showed that 5-ALA treatment increased the Na content of *B. megistophylla*, but mainly in the root system. This is similar to a previous report [45] that found 5-ALA upregulated expressions of genes, including *SOS1*, *NHX1* and *HKT1* in strawberry (*Fragaria* × *ananassa* Duch.) roots to intercept salt ions in root vacuoles and depress transport to the aboveground parts, avoiding leaf ion damage. The relationship between Al content and 5-ALA has not been reported before. In this experiment, the Al content in leaves was not affected by 5-ALA treatment, but it was decreased significantly in the roots, suggesting that 5-ALA treatment reduced the absorption of Al by plants. Due to the soil acidity of the experimental region, a decrease in Al content induced by 5-ALA may be related to the increase of acid tolerance of plants.

Cd, Hg, Cr and Pb are all toxic elements. Crops grown on both sides of roads are often poisoned by these heavy metal elements [46]. 5-ALA improved the tolerance of plants to

Pb [15] and Cr [47]. Recently, Xu et al. [48] proposed a mechanism of 5-ALA improving plant Cd tolerance, which is related to the upregulation of cellular antioxidant enzymes and metal carrier protein gene expression. In the present work, we observed that 5-ALA promoted the accumulation of heavy metals in the roots of *B. megistophylla* and decreased the content of leaves (Table 5). This mechanism is similar to that proposed by Wu et al. [45] that 5-ALA induces Na interception in strawberry roots under salt stress to reduce upward transport. Obviously, this mechanism is more ingenious, and it has yet to be proven. It has been reported that arbuscular mycorrhizal fungi (AMF) could improve plant tolerance against heavy metal pollution [49], where AMF can chelate part of metal elements in the rhizosphere soil [50]. Whether 5-ALA treatment improves the tolerance mediated by AMF has never been reported, but this deserves study in the future.

To sum up, 5-ALA promoted the growth of *B. megistophylla*, a road greening plant, improved plant survival, increased leaf color and enhanced the greening effect. The reason may be related to the increase of plant antioxidant enzyme activities, the improvement of plant mineral nutrition levels, and the inhibition of toxic and harmful heavy metal element accumulation in leaves. The toxic elements, such as Cd, Hg, Cr and Pb absorbed by the roots were trapped in the roots, the accumulation in leaves was reduced, the content of N, P, Ca, Mg, Cu, Fe and B in leaves were increased, and the nutritional status of plants was generally improved. The photosynthetic capacity of leaves also improved following 5-ALA treatment, the accumulation of soluble sugars and free proline was promoted, the activities of antioxidant enzymes were increased, and the stress resistance of plants was enhanced. Therefore, this nonprotein amino acid can use in urban landscapes to improve plant stress tolerance.

Author Contributions: Conceptualization, L.W. and Y.A.; methodology, H.Y. and J.Z.; validation, H.Y. and J.Z.; formal analysis, H.Y. and H.Z.; investigation, L.W., H.Y., J.Z. and Y.X.; data curation, L.W. and H.Y.; writing—original draft preparation, H.Y. and L.W.; writing—review and editing, L.W. and Y.A. All authors have read and agreed to the published version of the manuscript.

Funding: This research was supported by the National Natural Science Foundation of China (31772253), the Fundamental Research Funds for the Central Universities (KYYJ202004), a Project Funded by the Priority Academic Program Development of Jiangsu Higher Education Institutions (130-809005) and Jiangsu Agricultural Science and Technology Innovation Fund (CX (20)2023). The funders had no role in study design, data collection and analysis, decision to publish, or preparation of the manuscript.

Institutional Review Board Statement: Not applicable.

Informed Consent Statement: Not applicable.

Data Availability Statement: Data is contained within the article.

Acknowledgments: The authors thank YH Ma (Central Lab of College of Horticulture, Nanjing Agricultural University) for assistance in manipulating ICP-OES for mineral analysis.

Conflicts of Interest: The authors declare no conflict of interest.

References

1. Ma, J.; Luo, W. Importance and design method of urban road greening. *World Build. Mater.* **2013**, *34*, 50–52.
2. Gao, Y.D.; Liu, H.R. Physiological responses of road greening plants to traffic pollution. *Heilongjiang Agric. Sci.* **2016**, *3*, 93–97.
3. Wu, Y.; Liao, W.; Dawuda, M.M.; Hu, L.; Yu, J. 5-Aminolevulinic acid (ALA) biosynthetic and metabolic pathways and its role in higher plants: A review. *Plant Growth Regul.* **2019**, *87*, 357–374. [CrossRef]
4. Akram, N.A.; Ashraf, M. Regulation in plant stress tolerance by a potential plant growth regulator, 5-aminolevulinic acid. *J. Plant Growth Regul.* **2013**, *32*, 663–679. [CrossRef]
5. Wang, L.J.; Jiang, W.B.; Zhang, Z.; Yao, Q.H.; Matsui, H.; Ohara, H. Biosynthesis and physiological activity of 5-aminolevulinic acid and its potential application in agriculture. *Plant Physiol. Commun.* **2003**, *39*, 185–192.
6. Hotta, Y.; Tanaka, T.; Takaoka, H.; Takeuchi, Y.; Konnai, M. New physiological effects of 5-aminolevulinic acid in plants: The increase of photosynthesis, chlorophyll content, and plant growth. *Biosci. Biotechnol. Biochem.* **1997**, *61*, 2025–2028. [CrossRef] [PubMed]

7.   Nishihara, E.; Kondo, K.; Parvez, M.M.; Takahashi, K.; Watanabe, K.; Tanaka, K. Role of 5-aminolevulinic acid (ALA) on active oxygen-scavenging system in NaCl-treated spinach (*Spinacia oleracea*). *J. Plant Physiol.* **2003**, *160*, 1085–1091. [CrossRef]
8.   Wang, L.J.; Jiang, W.B.; Huang, B.J. Promotion of 5-aminolevulinic acid on photosynthesis of melon (*Cucumis melo*) seedlings under low light and chilling stress conditions. *Physiol. Plant* **2004**, *121*, 258–264. [CrossRef]
9.   Ma, N.; Qi, L.; Gao, J.; Chao, K.C.; Hu, Q.F.; Jiang, H.G.; Wang, L.J. Effects of 5-ALA on growth and chlorophyll fluorescence of *Ficus carica* cutting seedlings under high temperature. *J. Nanjing Agric. Univ.* **2015**, *38*, 546–553.
10. Liu, W.Q.; Kang, L.; Wang, L.J. Effect of ALA on photosynthesis of strawberry and its relationship with antioxidant enzymes. *Acta Bot. Boreal. Occident. Sin.* **2006**, *26*, 57–62.
11. Aksakal, O.; Algur, O.F.; Icoglu, A.F.; Aysin, F. Exogenous 5-aminolevulinic acid alleviates the detrimental effects of UV-B stress on lettuce (*Lactuca sativa* L.) seedlings. *Acta Physiol. Plant* **2017**, *39*, 55. [CrossRef]
12. Cai, C.Y.; He, S.S.; An, Y.Y.; Wang, L.J. Exogenous 5-aminolevulinic acid improves strawberry tolerance to osmotic stress and its possible mechanisms. *Physiol. Plant* **2020**, *168*, 948–962. [CrossRef]
13. An, Y.Y.; Qi, L.; Wang, L.J. ALA pretreatment improves waterlogging tolerance of fig plants. *PLoS ONE* **2016**, *11*, e0147202. [CrossRef]
14. Liu, L.Y.; El-Shemy, H.A.; Saneoka, H. Effects of 5-aminolevulinic acid on water uptake, ionic toxicity, and antioxidant capacity of Swiss chard (*Beta vulgaris* L.) under sodic-alkaline conditions. *J. Plant Nutr. Soil Sci.* **2017**, *180*, 535–543.
15. Singh, R.; Kesavan, A.K.; Landi, M.; Kaur, S.; Thakur, S.; Zheng, B.S.; Bhardwaj, R.; Sharma, A. 5-Aminolevulinic acid regulates Krebs cycle, antioxidative system and gene expression in *Brassica juncea* L. to confer tolerance against lead toxicity. *J. Biotechnol.* **2020**, *323*, 283–292. [CrossRef]
16. Taspinar, M.S.; Aydin, M.; Arslan, E.; Yaprak, M.; Agar, G. 5-Aminolevulinic acid improves DNA damage and DNA methylation changes in deltamethrin-exposed *Phaseolus vulgaris* seedlings. *Plant Physiol. Biochem.* **2017**, *118*, 267–273. [CrossRef]
17. Xu, L.; Islam, F.; Zhang, W.F.; Ghani, M.A.; Ali, B. 5-Aminolevulinic acid alleviates herbicide-induced physiological and ultrastructural changes in *Brassica napus*. *J. Integr. Agric.* **2018**, *17*, 579–592. [CrossRef]
18. Elansary, H.O.; El-Ansary, D.O.; Al-Mana, F.A. 5-Aminolevulinic acid and soil fertility enhance the resistance of rosemary to *Alternaria dauci* and *Rhizoctonia solani* and modulate plant biochemistry. *Plants* **2019**, *8*, 585. [CrossRef]
19. Chen, H.; Xu, L.; Li, X.; Wang, D.Y.; An, Y.Y.; Wang, L.J. Effect of 5-aminolevulinic acid on cold tolerance of *Rhododendron simsii* and *Cinnamomum camphora* leaves. *Plant Physiol. J.* **2017**, *53*, 2103–2113.
20. Wang, D.Y.; Li, X.; Xu, L.; An, Y.Y.; Wang, L.J. Effect of 5-aminolevulinic acid on leaf heat tolerance in *Ligustrum japonicum* and *Spiraea japonica*. *Bot. Res.* **2018**, *7*, 350–365.
21. Qiu, M.H.; Li, D.Z. Study on chemotaxonomy of Buxaceae. *Chin. J. Appl. Environ. Biol.* **2002**, *8*, 387–391.
22. Shen, Y.M.; Ma, G.S. Occurrence regularity and ecological control technology of diseases of hedgerow commonly used in greening in five cities of southern Jiangsu. *J. Landsc. Res.* **2019**, *11*, 141–144.
23. Tian, B. Effects of drought stress on protective enzymes and osmotic substances of *Buxus megistophylla* Levl. *Shandong For. Sci. Technol.* **2019**, *4*, 64–67.
24. Hou, L.Q. Effects of sustained low temperature stress on physiological characteristics of *Buxus megistophylla* Levl. *Shandong For. Sci. Technol.* **2019**, *4*, 68–71.
25. Li, J.X.; He, J.; Sun, Y.M.; Zhao, A.; Tain, Q. Physiological and ecological responses of ten garden plant functional traits to air pollution. *Ecol. Environ. Sci.* **2020**, *29*, 1205–1214.
26. Li, H.S. *Principles and Techniques of Plant Physiological and Biochemical Experiments*; Higher Education Press: Beijing, China, 2020.
27. Zhang, D.Z.; Wang, P.H.; Zhao, H.X. Determination of free proline content in wheat leaves. *Plant Physiol. Commun.* **1990**, *4*, 62–65.
28. Tan, W.; Liu, J.; Dai, T.; Jing, Q.; Cao, W.; Jiang, D. Alterations in photosynthesis and antioxidant enzyme activity in winter wheat subjected to post-anthesis water-logging. *Photosynthetica* **2008**, *46*, 21–27. [CrossRef]
29. Zhang, W.F.; Zhang, F.; Raziuddin, R.; Gong, H.J.; Yang, Z.M.; Lu, L.; Ye, Q.F.; Zhou, W.J. Effects of 5-aminolevulinic acid on oilseed rape seeding growth under herbicide toxicity stress. *J. Plant Growth Regul.* **2008**, *27*, 159–169. [CrossRef]
30. Change, B.; Maehly, A.C. Assay of catalases and peroxidase. *Methods Enzym.* **1995**, *2*, 764–775.
31. Wang, J.B. The role of urban road greening. *New Silk Road Horiz.* **2011**, *8*, 88–90.
32. Yang, X.X.; Feng, L.H.; Wei, P. Automobile exhaust pollution and its harm. *Front. Sci.* **2012**, *6*, 10–22.
33. Ling, Q.Y. Physiological response and resistance of landscape plants to automobile exhaust. *Green Technol.* **2015**, *6*, 206–207.
34. Chen, J.N.; Yuan, X.W.; Xiang, X.Z.; Yuan, X.H.; Li, H. Evaluation on the resistance of fourteen ground cover plants to automobile exhaust. *Resour. Environ. Arid Areas* **2018**, *32*, 176–181.
35. Li, L.B.; Kuang, T.Y. *An Overview of Primary Light Energy Conversion in Photosynthesis, Principle and Regulation of Primary Energy Transformation Process of Photosynthesis*; Kuang, T.Y., Ed.; Jiangsu Science and Technology Press: Nanjing, China, 2003; pp. 3–21.
36. Li, M.A.; Ma, L.; Hao, Q.; An, Y.Y.; Wang, L.J. Effect of 5-aminolevulinic acid on leaf photosynthetic characteristics, yield and quality of potato. *China Veg.* **2020**, *11*, 43–52.
37. Ou, C.; Yao, X.M.; Yao, X.J.; Ji, J.; Wang, W.G.; Guo, J. Effects of exogenous 5-aminolevulinic acid and PEG on photosynthetic and antioxidant characteristics of *Gardenia jasminoides* seedlings. *Agric. Res. Arid Areas* **2016**, *34*, 235–242.
38. Naeem, M.S.; Jin, Z.L.; Wan, G.L.; Liu, D.; Liu, H.B.; Yoneyama, K.; Zhou, W.J. 5-Aminolevulinic acid improves photosynthetic gas exchange capacity and ion uptake under salinity stress in oilseed rape (*Brassica napus* L.). *Plant Soil* **2010**, *332*, 405–415. [CrossRef]

39. Zhang, L.Y.; Feng, X.X.; Gao, J.J.; An, Y.Y.; Tian, F.; Li, J.; Zhang, Z.P.; Wang, L.J. Effects of ALA solution on leaf physiological characteristics and fruit quality of apple under rhizosphere irrigation. *Jiangsu J. Agric. Sci.* **2015**, *1*, 158–165.
40. Anwar, A.; Yan, Y.; Liu, Y.; Li, Y.; Yu, X. 5-Aminolevulinic acid improves nutrient uptake and endogenous hormone accumulation, enhancing low-temperature stress tolerance in cucumbers. *Int. J. Mol. Sci.* **2018**, *19*, 3379. [CrossRef]
41. Wei, Z.Y.; Zhang, Z.P.; Lee, M.R.; Sun, Y.P.; Wang, L.J. Effect of 5-aminolevulinic acid on leaf senescence and nitrogen metabolism of pakchoi under different nitrates. *J. Plant Nutr.* **2012**, *35*, 49–63. [CrossRef]
42. Yao, S.M.; Wang, W.J.; Chen, G.X. Effects of 5-aminolevulinic acid on the uptake and distribution of ^{32}P in rice plants. *Acta Phytonutri Fertil* **2006**, *12*, 70–75.
43. Montanaro, G.; Dichio, B.; Lang, A.; Mininni, A.; Xiloyannis, C. Fruit calcium accumulation coupled and uncoupled from its transpiration in kiwifruit. *J. Plant Physiol.* **2015**, *181*, 67–74. [CrossRef]
44. An, Y.; Xiong, L.J.; Hu, S.; Wang, L.J. PP2A and microtubules function in 5-aminolevulinic acid-mediated H_2O_2 signaling in Arabidopsis guard cells. *Physiol. Plant* **2020**, *168*, 709–724. [CrossRef] [PubMed]
45. Wu, W.W.; He, S.S.; An, Y.Y.; Cao, R.X.; Sun, Y.P.; Tang, Q.; Wang, L.J. Hydrogen peroxide as a mediator of 5-aminolevulinic acid-induced Na^+ retention in roots for improving salt tolerance of strawberries. *Physiol. Plant* **2019**, *167*, 5–20. [CrossRef] [PubMed]
46. Chen, B. Study on Heavy Metal Distribution and Pollution Characteristics of Crops in Highway Area—Taking Xisan and Xiyan Expressways as Examples. Master's Thesis, Chang'an University, Xi'an, China, 2012.
47. Ali, B.; Wang, B.; Ali, S.; Ghani, M.A.; Hayat, M.T.; Yang, C.; Xu, L.; Zhou, W.J. 5-Aminolevulinic acid ameliorates the growth, photosynthetic gas exchange capacity, and ultrastructural changes under cadmium stress in *Brassica napus* L. *J. Plant Growth Regul.* **2013**, *32*, 604–614. [CrossRef]
48. Xu, L.; Li, J.J.; Najee, U.; Li, X.; Pan, J.M.; Huang, Q.; Zhou, W.J.; Liang, Z.S. Synergistic effects of EDDS and ALA on phytoextraction of cadmium as revealed by biochemical and ultrastructural changes in sunflower (*Helianthus annuus* L.) tissues. *J. Hazard. Mater.* **2020**, *407*, 124764. [CrossRef]
49. Zhang, X.H.; Guo, Y.L.; Lin, A.J.; Huang, Y.Z. Effects of arbuscular mycorrhizal fungi colonization on toxicity of soil contaminated by heavy metals to *Vicia faba*. *Chin. J. Environ. Eng.* **2008**, *2*, 274–278.
50. Wang, H.X.; Han, L.L.; Li, Y.; Zhang, L.L.; Li, X.; Yuan, Z.L. Study on mitigation of cadmium stress in wheat seedlings by inoculation with arbuscular mycorrhizal fungi. *J. Henan Agric. Univ.* **2021**, *55*, 29–34.

 *horticulturae*

Article

Exogenous Application of Chitosan Alleviate Salinity Stress in Lettuce (*Lactuca sativa* L.)

Geng Zhang [1,2,†], **Yuanhua Wang** [1,2,†], **Kai Wu** [3], **Qing Zhang** [1,2], **Yingna Feng** [1,2], **Yu Miao** [1] and **Zhiming Yan** [1,2,*]

[1] Department of Agronomy and Horticulture, Jiangsu Vocational College of Agriculture and Forestry, Jurong 212400, China; gengzhang@jsafc.edu.cn (G.Z.); wangyuanhua@jsafc.edu.cn (Y.W.); zhangqing@jsafc.edu.cn (Q.Z.); fyn@jsafc.edu.cn (Y.F.); miaoyu0721@163.com (Y.M.)

[2] Engineering and Technical Center for Modern Horticulture, Jiangsu Vocational College of Agriculture and Forestry, Jurong 212400, China

[3] China Agriculture Press, Beijing 100125, China; kkw0214@163.com

* Correspondence: yanzhim@jsafc.edu.cn

† These authors contributed equally to this work.

Abstract: Soil salinity is one of the major factors that affect plant growth and decrease agricultural productivity worldwide. Chitosan (CTS) has been shown to promote plant growth and increase the abiotic stress tolerance of plants. However, it still remains unknown whether the application of exogenous CTS can mitigate the deleterious effects of salt stress on lettuce plants. Therefore, the current study investigated the effect of foliar application of exogenous CTS to lettuce plants grown under 100 mM NaCl saline conditions. The results showed that exogenous CTS increased the lettuce total leaf area, shoot fresh weight, and shoot and root dry weight, increased leaf chlorophyll a, proline, and soluble sugar contents, enhanced peroxidase and catalase activities, and alleviated membrane lipid peroxidation, in comparison with untreated plants, in response to salt stress. Furthermore, the application of exogenous CTS increased the accumulation of K^+ in lettuce but showed no significant effect on the K^+/Na^+ ratio, as compared with that of plants treated with NaCl alone. These results suggested that exogenous CTS might mitigate the adverse effects of salt stress on plant growth and biomass by modulating the intracellular ion concentration, controlling osmotic adjustment, and increasing antioxidant enzymatic activity in lettuce leaves.

Keywords: antioxidant enzymes; chitosan (CTS); lettuce; proline; salinity; soluble sugars

Citation: Zhang, G.; Wang, Y.; Wu, K.; Zhang, Q.; Feng, Y.; Miao, Y.; Yan, Z. Exogenous Application of Chitosan Alleviate Salinity Stress in Lettuce (*Lactuca sativa* L.). *Horticulturae* **2021**, *7*, 342. https://doi.org/10.3390/horticulturae7100342

Academic Editors: Agnieszka Hanaka, Jolanta Jaroszuk-Ściseł and Małgorzata Majewska

Received: 24 August 2021
Accepted: 22 September 2021
Published: 24 September 2021

Publisher's Note: MDPI stays neutral with regard to jurisdictional claims in published maps and institutional affiliations.

1. Introduction

Saline stress is a harmful form of abiotic stress that restricts the growth and function of plants and thus can cause a 10%–25% decrease in the yield of many agricultural crops [1]. More than 20% of global farmland is affected by various degrees of salinity, and the farmland area (approximately 20,000 km^2 per year) affected by salinity is increasing each year, which severely limits agricultural productivity [2,3]. Salinity in soils can occur naturally or as a result of human activities. Weathering of rock minerals and flooding by seawater causes inherent soil salinity. Irrigation water with a high salt concentration, excessive chemical fertilization, and poor soil management are the main reasons for an increase in the area of saline–alkali land [4,5]. In some semi-arid and arid areas (e.g., Sahara in North Africa, Saudi Arabia, large parts of Iran and Iraq, parts of Asia, California in the USA, South Africa, and most of Australia), high temperatures and uneven distribution of rainfall result in higher evapotranspiration rates than the size of the leaching fraction, which causes an accumulation of soluble salts in the plough layer [6]. To increase the output of salinized agricultural land, the salt tolerance of plants must be increased and the conditions of saline–alkali land must be improved.

Salt stress negatively influences several processes in plant growth and production by causing ion toxicity, hyperosmotic stress, nutritional imbalance, oxidative damage,

115

metabolic disorders, and photoinhibition [4,7]. Because of their sessile nature, plants must evolve several mechanisms to adapt to high-salinity environments. The various physiological and biochemical processes that plants use to adapt to salt stress can be grouped into three categories: Osmotic stress, ionic stress, and detoxification responses [8]. Plants' primary response to salt stress is osmotic stress. Plants adjust their osmotic balance by accumulating organic and inorganic osmolytes, such as proline, glycine betaine, soluble sugar, soluble protein, and sodium (Na^+) and potassium (K^+) ions, to maintain cell turgor [2,9,10]. Na^+ and chloride (Cl^-) enter the root system through nonselective cation channels, K^+ transporters, and Cl^- transporters [11]. The excessive accumulation of Na^+ and Cl^- in plant cells and tissues can adversely affect the growth and development of plants by disturbing the water structure, inhibiting enzymes, and creating nutritional imbalance [12,13]. Plants usually maintain a balanced cytosolic Na^+/K^+ ratio through Na^+ and K^+ transporters and channels [12]. Salt stress also causes the accumulation of reactive oxygen species (ROS), which results in oxidative-stress-induced toxic effects in plants. ROS sources, such as superoxide radical ($O_2{}^-$), hydrogen peroxide (H_2O_2), and hydroxyl radical ($\cdot OH$), are generated by plants' photosynthetic and respiratory electron transport chains, xanthine oxidase, and nicotinamide adenine dinucleotide phosphate oxidase [14]. Normally, cellular ROS levels are regulated by enzymatic (e.g., ROS scavenging enzymes) and nonenzymatic scavengers (e.g., ascorbic acid [AsA], glutathione, and carotenoids) to mitigate the ROS-induced damage caused by salt stress [15,16]. However, although plants adopt these strategies to reduce the harmful effect of salt stress, survival in a salty environment is difficult, let alone producing a good yield.

Chitosan (poly[1,4]-2-amino-2-deoxy-D-glucose; CTS) is a biopolymer obtained through the deacetylation of nontoxic and biofunctional chitin from the exoskeleton of crustaceans [17]. Chitosan has three types of functional groups on its backbone: The amino/acetamido group, and primary and secondary hydroxyl groups, which enhance its affinity for ions and various pollutants [18]. CTS is a natural, low-toxicity, biodegradable, environmentally friendly, renewable, and inexpensive resource and has many applications in the agriculture sector [19]. Since the discovery of CTS by Rouget in 1859 [20], several studies have proven its role in enhancing plant growth and increasing plants' abiotic stress tolerance [21], including rice [19], maize [22], safflower and sunflower [23], and creeping bentgrass [24]. The beneficial role of CTS in stress mitigation is broadly attributable to the alleviation of oxidative stress [25] and the increase in water use efficiency [26], mineral nutrient uptake [27], chlorophyll (Chl) content, and photosynthesis [28] caused by CTS. The application of exogenous CTS increases plants' tolerance to several forms of stress, such as drought, salt, osmotic, and low-temperature stress [19,22,23,29]. Certain concentrations of exogenous CTS have been used to increase plants' resistance to several biotic and abiotic stresses by increasing water use efficiency, enhancing antioxidant activity, and regulating the content of osmotic regulation substances and defense gene expression [17,26,30–32].

Lettuce (*Lactuca sativa* L.) is a leafy vegetable mainly consumed raw and in salad mixes [33]. The production and cultivation area of lettuce has increased because of its marketability, sensory characteristics, and health-promoting properties [5,34]. According to the last available FAO data (http://www.fao.org/faostat/en/#data/QC, accessed on 19 September 2021), the global cultivation area and yield of lettuce were 243.97 thousand hectares and 16.31 million tonnes in 2019, an increase of 0.54% and 2.3% over 2018, respectively. The majority of lettuce comes from China, the United States, and India—the world's top three lettuce producers. Lettuce is a moderately to highly salt-sensitive vegetable [35]. Salinity reduces the seed germination rate, leaf number, photosynthesis, and cell growth and increases the accumulation of ROS, which negatively affects lettuce growth and yield [36,37]. Although the negative effects of salinity on lettuce have been studied [36,38], information regarding the effects of CTS on lettuce growth and production under saline conditions is lacking. Therefore, the present study evaluated the effectiveness of exogenous CTS in mitigating the adverse effects of salinity on the growth and physiological attributes of lettuce plants. In addition, this study identified the effects of exogenous CTS on the accumulation of osmolytes, the biosynthesis of antioxidants, and the activity of antioxidant enzymes in lettuce under saline conditions.

2. Materials and Methods

2.1. Plant Materials and Treatments

Romaine lettuce (*Lactuca sativa* var. *longifolia* L. cv. Romana, Takii seed, Japan) was used as the test material, and an experiment was conducted from November 2020 to January 2021. Lettuce seeds were sown in urethane cubes ($2.3 \times 2.3 \times 2.7$ cm^3), and the seedlings were cultivated in a 22 °C growth chamber at 200 ± 10 μmol·m^{-2}·s^{-1} photosynthetic photon flux for 12 h by using cool white fluorescent lamps (Figure 1A). At 21 days after sowing (DAS), uniform seedlings were selected and transplanted into a cultivation room of the Engineering and Technical Center for Modern Horticulture in Jurong. The plants were grown in a deep-flow hydroponic system in an Enshi formula nutrient solution (electrical conductivity [EC]: 1.5 ± 0.2, pH: 6.9 ± 0.2) [39]. Air pumps were used to oxygenate the nutrient solution and supply a constant stream of oxygen. The lighting for plant cultivation was provided by light-emitting diode lights (Figure 1B). The photosynthetic photon flux density of the light was 200 ± 10 μmol·m^{-2}·s^{-1}, and its photoperiod was 16 h. The air temperature was maintained as 25 °C during the day and 20 °C at night, with the relative humidity being maintained as $65\% \pm 5\%$.

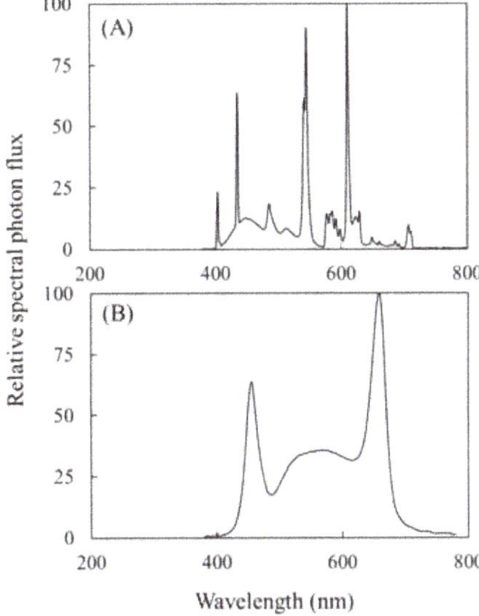

Figure 1. The relative spectral photon flux of (**A**) cool white fluorescent lamps and (**B**) LED lights. The wavelengths of light sources were recorded at 380–800 nm with a spectrometer.

All treatments were performed at 7 days after transplanting (28 DAS) into the cultivation room, so that the lettuce seedlings were well adapted to the environment of the cultivation room. The treatments were divided into four groups: (1) The control group, in which the plants were grown in a nutrient solution with water sprayed on the leaf surface; (2) the CTS group, in which the plants were grown under the same conditions as those of the control group and 100 mg/L of CTS (instead of water) was sprayed on the leaf surface; (3) the NaCl group, in which the plants were grown in a nutrient solution containing 100 mM NaCl and water was sprayed on the leaf surface; and (4) the NaCl + CTS group, in which the plants were grown under the same conditions as those of the NaCl group and 100 mg/L of CTS (instead of water) was sprayed on the leaf surface. CTS or water was sprayed on the leaves of the hydroponically grown lettuce continuously for 5 days from

28 DAS. Approximately 30 mL of CTS or water solution was sprayed on both adaxial and abaxial surfaces of leaves for each lettuce plant. For groups (3) and (4), NaCl was dissolved into the nutrient solution after the 5-day CTS induction treatment. Three replications were performed for each treatment, and each replication comprised six plants. Plants of each treatment were sampled at 28 and 49 DAS for further examinations.

2.2. Plant Growth Analysis

At 28 and 49 DAS, the plants were sampled to determine total leaf area, shoot and root fresh weight (FW), and shoot and root dry weight (DW). Total leaf area was determined using a Li-3100 leaf area meter (Li-Cor, Lincoln, NE, United States). Shoot and root DW was obtained after the plant tissues were dried at 80 °C to a constant weight. Chl was extracted in N,N-dimethylformamide from fresh lettuce leaves, and the Chl content was determined spectrophotometrically according to the method of Porra et al. [40].

Growth analysis parameters, namely the relative growth rate (RGR), the net assimilation rate (NAR), and the leaf area ratio (LAR), were estimated using the method of Ohtake et al. [41] by using the following equations:

$$RGR = (1/W)(\Delta W/\Delta t) = [\ln(W_2) - \ln(W_1)]/(t_2 - t_1), \tag{1}$$

where W_1 and W_2 are the total DWs of a plant at times t_1 (28 DAS) and t_2 (49 DAS).

$$NAR = (1/L)(\Delta W/\Delta t) = [(W_2 - W_1)/(t_2 - t_1)] \times [\ln(L_2) - \ln(L_1)]/(L_2 - L_1), \tag{2}$$

where L_1 and L_2 are the total leaf areas of a plant at times t_1 and t_2.

$$LAR = L/W = (L_1/W_1 + L_2/W_2)/2 \tag{3}$$

2.3. Estimation of Leaf Relative Water Content and Electrolyte Leakage

Relative water content (RWC) was measured using the method of Yamasaki and Dillenburg [42] by adopting the following equation:

$$RWC\ (\%) = (FW - DW)/(\text{turgid weight} - DW) \times 100 \tag{4}$$

Electrolyte leakage (EL) was measured using the method described by Ahmad et al. [43]. Leaf disks with a diameter of 13 mm were produced from the leaves in each treatment group and submerged in deionized water to measure EC_a. Then, tubes containing the leaf disks were incubated in a water bath at 50–60 °C for 25 min to determine the EC_b value for each treatment. Finally, the tubes with the leaf disks were boiled at 100 °C for 10 min to measure EC_c. The EL was calculated using the following equation:

$$EL\ (\%) = (EC_b - EC_a)/EC_c \times 100 \tag{5}$$

2.4. Determination of the Potassium and Sodium Contents in Lettuce Leaves

The potassium and sodium contents in lettuce leaves were determined through inductively coupled plasma optical emission spectrometry (Thermo Fisher Scientific, Cambridge, United Kingdom) by using the method of Zhang et al. [44].

2.5. Estimation of the Proline, Soluble Sugar, and Ascorbic Acid Contents

The proline content of the leaf samples was determined according to the method described by Bates et al. [45]. The soluble sugar content of the leaves was measured using the anthrone–sulfuric acid method [46]. The ascorbic acid (AsA) content of the leaves was determined according to the method of Kampfenkel et al. [47].

2.6. Examination of H_2O_2 Content, O_2^- Generation, and Malondialdehyde Content

The H_2O_2 concentration of the leaf samples was determined using the method of Patterson et al. [48]. The O_2^- generation was assayed spectrophotometrically by measuring the reduction of nitroblue tetrazolium by using the method of Averina et al. [49]. The malondialdehyde (MDA) concentration of the leaf samples was estimated using the method of Heath et al. [50]. The absorbance of the leaf samples was measured at 450, 532, and 600 nm by using a spectrophotometer.

2.7. Enzyme Assays

The fresh leaf samples were homogenized in phosphate buffer saline (50 mM, pH 7.8), and the homogenate was centrifuged at $10,000 \times g$ and 4 °C for 15 min. The supernatants were collected to determine the superoxide dismutase (SOD; EC: 1.15.1.1), peroxidase (POD; EC: 1.11.1.7), and catalase (CAT; EC: 1.11.1.6) activity [51]. Protein content was determined using the method of Bradford [52]. The activity of the enzymes was expressed in units per milligram of protein.

2.8. Statistical Analysis

The data are presented as the means $\pm$ standard errors (SEs) of the three replications for each treatment. Statistical analysis was performed using one-way analysis of variance with Tukey's HSD test (SPSS v. 18.0, IBM Inc., Chicago, IL, USA). p values of ≤ 0.05 were considered significant.

3. Results

3.1. Exogenous CTS Improved the Growth and Biomass of Lettuce under NaCl Stress

Compared with the control plants, the lettuce plants exposed to NaCl stress exhibited considerably inhibited plant growth in terms of a lower total leaf area, shoot FW, and shoot DW (Table 1). The total leaf area, shoot FW, root FW, shoot DW, and root DW of the NaCl group were 67.3%, 60.3%, 73.8%, 66.5%, and 51.6% lower, respectively, than those of the control group (Table 1). The application of 100 mg/L of exogenous CTS mitigated the lettuce growth inhibition caused by the salinity stress (Figure 2A). The total leaf area, shoot FW, root FW, shoot DW, and root DW of the NaCl + CTS group were 141.2%, 127.3%, 72.3%, 95.0%, and 60.0% higher, respectively, than those of the NaCl group (Table 1). The total leaf area, shoot FW, root FW, shoot DW, and root DW of the NaCl + CTS group were 21.2%, 40.4%, 31.5%, 34.7%, and 22.6% lower, respectively, than those of the control group (Table 1). However, no significant change in the aforementioned growth parameters was observed between the CTS and control groups (Table 1). The Chl a, Chl b, and total Chl contents of the NaCl group were 14.4%, 20.6%, and 16.1% lower, respectively, than those of the control group (Table 1). The Chl a and total Chl contents of the NaCl + CTS group were 10.1% and 8.6% higher, respectively, than those of the NaCl group (Table 1). The Chl b and total Chl contents of the NaCl + CTS group were 17.3% and 8.9% lower, respectively, than those of the control group (Table 1). In addition, the Chl a content of the plants subjected to exogenous CTS treatment alone were significantly higher than that of the control group (Table 1).

Plant biomass is strongly and positively correlated to RGR, and plant growth analysis decomposes RGR into NAR and LAR. In order to determine how physiological and morphological traits contribute to the plant biomass of each group, the growth analysis parameters of each group were estimated using total plant DW and total leaf area as described in the aforementioned text. The RGR of the NaCl group was significantly lower than that of the control group and corresponded to the lowest DW (Figure 2B, Table 1). The decrease in the RGR of the NaCl group was mitigated by the application of 100 mg/L of exogenous CTS (Figure 2B). A similar trend was observed in NAR, and the lowest NAR was observed for the NaCl group. Moreover, an insignificant difference in NAR was observed between the NaCl + CTS and NaCl groups (Figure 2C). No significant difference in LAR was observed among all the groups (Figure 2D). Similar to the growth parameters, no significant changes in the aforementioned growth analysis parameters were observed between the CTS and control groups.

Table 1. Effects of chitosan (CTS) on total leaf area, shoot and root fresh weight, shoot and root dry weight, and leaf chlorophyll content of lettuce plants under salt stress.

Treatments	Total Leaf Area (m^2)	Shoot FW $(g \cdot plant^{-1})$	Shoot DW $(g \cdot plant^{-1})$	Root FW $(g \cdot plant^{-1})$	Root DW $(g \cdot plant^{-1})$	Chl a $(\mu g \cdot ml^{-1})$	Chl b $(\mu g \cdot ml^{-1})$	Total Chl $(\mu g \cdot ml^{-1})$
CK	0.052 ± 0.004 a	48.8 ± 0.9 a	2.39 ± 0.21 a	8.73 ± 0.63 a	0.31 ± 0.02 a	8.68 ± 0.15 b	3.30 ± 0.23 a	11.98 ± 0.13 a
CTS	0.054 ± 0.010 a	45.2 ± 1.9 a	2.11 ± 0.19 ab	9.80 ± 1.23 a	0.28 ± 0.02 ab	9.24 ± 0.14 a	3.29 ± 0.11 a	12.53 ± 0.24 a
NaCl	0.017 ± 0.003 b	12.8 ± 1.7 c	0.80 ± 0.05 c	3.47 ± 0.51 b	0.15 ± 0.01 c	7.43 ± 0.14 c	2.62 ± 0.04 b	10.05 ± 0.16 c
NaCl + CTS	0.041 ± 0.007 a	29.1 ± 4.5 b	1.56 ± 0.19 b	5.98 ± 0.47 b	0.24 ± 0.03 b	8.18 ± 0.20 b	2.73 ± 0.15 b	10.91 ± 0.35 b

Data presented are the means ± SEs ($n = 3$). Different letters in each column indicate significant differences ($p < 0.05$). CK (control) = 0 mM NaCl + 0 mg/L CTS; CTS = 0 mM NaCl + 100 mg/L CTS; NaCl = 100 mM NaCl + 0 mg/L CTS; NaCl + CTS = 100 mM NaCl +100 mg/L CTS. FW, fresh weight; DW, dry weight; Chl, chlorophyll.

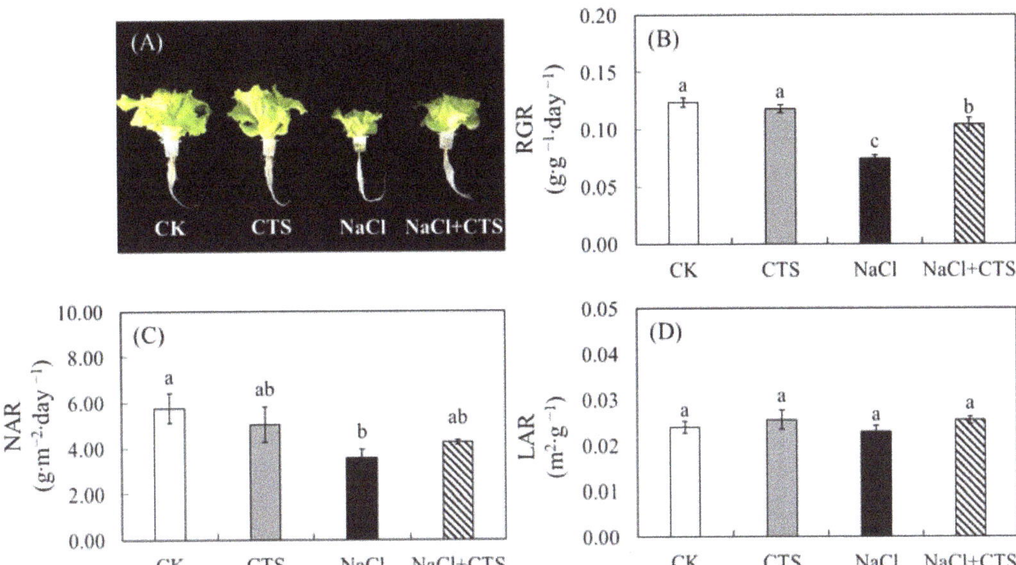

Figure 2. Effects of exogenous chitosan (CTS) on (**A**) plant morphology and (**B–D**) plant growth analysis parameters of lettuce plants under salt stress. Data presented are the means ± SEs (*n* = 3). Different letters on top of bars indicate a significant difference (*p* < 0.05) according to Tukey's HSD test. CK (control) = 0 mM NaCl + 0 mg/L CTS; CTS = 0 mM NaCl + 100 mg/L CTS; NaCl = 100 mM NaCl + 0 mg/L CTS; NaCl + CTS = 100 mM NaCl +100 mg/L. RGR, relative growth rate; NAR, net assimilation rate; LAR, leaf area ratio.

3.2. Effects of NaCl and CTS on the RWC, EL, and the Contents of Potassium and Sodium of the Lettuce Leaves

The leaf RWC of the NaCl group was 15.9% lower than that of the control group (Table 2). This decrease in leaf RWC due to salinity was mitigated by the application of exogenous CTS. The leaf RWC of the NaCl + CTS group was 15.7% higher than that of the NaCl group (Table 2). The leaf EL of the NaCl group was 160.9% higher than that of the control group (Table 2). Moreover, the leaf EL of the NaCl + CTS group was 21.2% lower than that of the NaCl group (Table 2). The leaf RWC and EL of the lettuce plants subjected to the exogenous CTS treatment alone were not significantly different from those of the control group (Table 2).

Table 2. Effects of chitosan (CTS) on leaf RWC, EL, and the contents of potassium and sodium in leaves of lettuce plants under salt stress.

Treatments	RWC (%)	EL (%)	Potassium (mg·g^{-1} DW)	Sodium (mg·g^{-1} DW)	K$^+$/Na$^+$ Ratio
CK	74.8 ± 1.3 ab	16.1 ± 3.0 c	73.96 ± 1.30 a	1.09 ± 0.04 c	68.23 ± 3.73 a
CTS	79.0 ± 2.9 a	15.5 ± 2.5 c	72.02 ± 2.25 a	1.05 ± 0.07 c	69.38 ± 6.97 a
NaCl	62.9 ± 1.6 c	42.0 ± 3.1 a	52.92 ± 1.55 c	22.61 ± 1.31 a	2.35 ± 0.07 b
NaCl + CTS	72.8 ± 0.4 b	33.1 ± 0.5 b	61.27 ± 1.23 b	12.21 ± 1.36 b	5.11 ± 0.42 b

Data presented are the means ± SEs (*n* = 3). Different letters in each column indicate significant differences (*p* < 0.05). CK (control) = 0 mM NaCl + 0 mg/L CTS; CTS = 0 mM NaCl + 100 mg/L CTS; NaCl = 100 mM NaCl + 0 mg/L CTS; NaCl + CTS = 100 mM NaCl +100 mg/L CTS. RWC, relative water content; EL, electrolyte leakage; FW, fresh weight; K$^+$, potassium; Na$^+$, sodium.

The potassium content of the NaCl group was 28.4% lower than that of the control group (Table 2). The addition of exogenous CTS to the plants treated with NaCl significantly mitigated the inhibition of potassium accumulation in the lettuce leaves (Table 2). The sodium contents of the NaCl and NaCl + CTS groups were 19.7 and 11.1 times higher, respectively, than that of the control group, whereas the accumulation of sodium was 46.0% lower in the NaCl + CTS group than in the NaCl group (Table 2). The K^+/Na^+ ratio of the NaCl group was significantly lower than that of the control group, and no significant change in the K^+/Na^+ ratio was observed between the NaCl and NaCl + CTS groups. Non-significant differences were noted in the potassium sodium contents and the K^+/Na^+ ratio between the CTS and control groups (Table 2).

3.3. Effects of NaCl and CTS on the Proline Content, MDA Content, O_2^- Generation, H_2O_2 Content, Soluble Sugar Content, and AsA Content of the Lettuce Leaves

Proline biosynthesis was triggered by salinity stress, which resulted in the NaCl group having a proline content 2.1 times higher than the control group (Figure 3A). Moreover, the proline content of the NaCl + CTS group was 66.5% higher than that of the NaCl group (Figure 3A). MDA, the product of membrane lipid peroxidation caused by ROS, can be used to evaluate the degree of membrane injury under stress [44]. The MDA content of the NaCl group was 127.6% higher than that of the control group (Figure 3B). Moreover, the MDA content of the NaCl + CTS group was 14.3% lower than that of the NaCl group (Figure 3B). The exposure of the lettuce plants to salinity stress induced the production of ROS in cells. The lettuce plants treated with NaCl generated ROS such as H_2O_2 and O_2^-. The H_2O_2 content and O_2^- of the NaCl group were 3 and 1.8 times higher than those of the control group, respectively (Figure 3C,D). The application of exogenous CTS to the plants treated with NaCl slowed the generation of H_2O_2 and O_2^- (Figure 3C,D). The soluble sugar content of the NaCl group was considerably higher than that of the control group (Figure 3E). The soluble sugar content of the NaCl + CTS group was 40.8% higher than that of the NaCl group (Figure 3E). No significant difference in AsA content was observed among the groups (Figure 3F). No significant change in the proline content, MDA content, H_2O_2 content, superoxide radical production rate, soluble sugar content, and AsA content was observed between the CTS and control groups (Figure 3).

3.4. Effects of NaCl and CTS on the Antioxidant Enzyme Activity in the Lettuce Leaves

Figure 4 displays the results regarding the effects of NaCl and CTS on the antioxidant enzyme activity in the lettuce leaves. SOD, POD, and CAT exhibited various responses to the different treatments. SOD activity was consistent among all the groups (Figure 4A). However, the POD and CAT activities of the NaCl group were 43.3% lower and 181.9% higher than those of the control group, respectively (Figure 4B,C). The CTS + NaCl group had considerably higher POD and CAT activities than the NaCl group (Figure 4B,C). No significant change in the SOD, POD, and CAT activities was observed between the CTS and control groups (Figure 4).

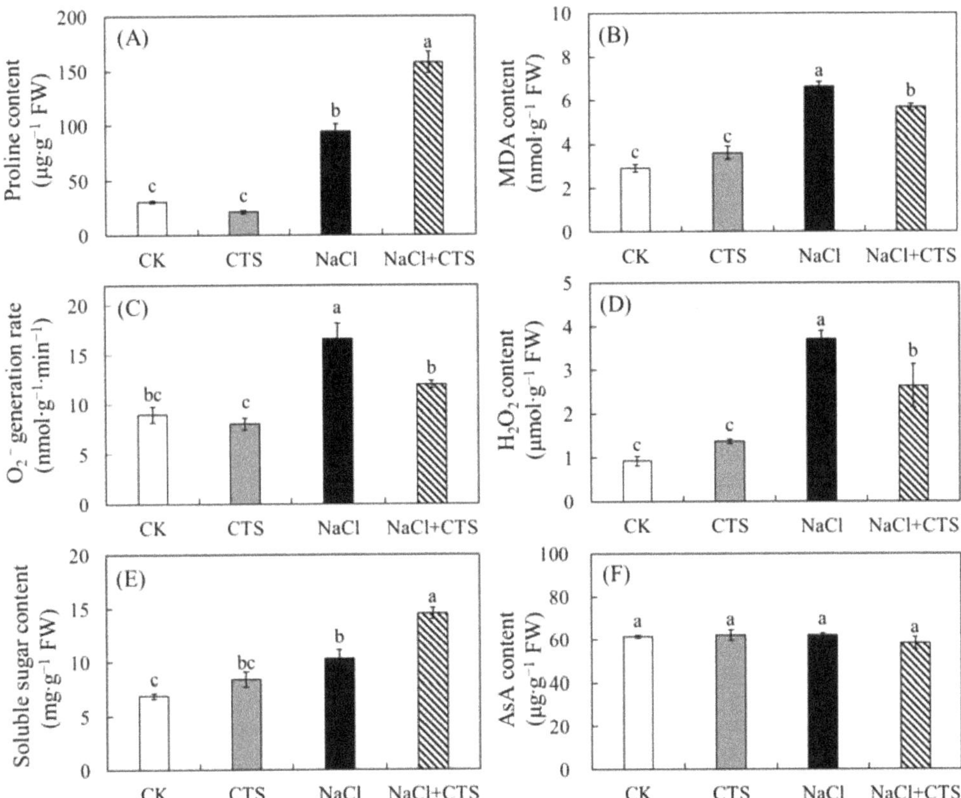

Figure 3. Effects of exogenous chitosan (CTS) on (**A**) proline content, (**B**) MDA content, (**C**) O_2^- generation rate, (**D**) H_2O_2 content, (**E**) soluble sugar content, and (**F**) AsA content in leaves of lettuce plants under salt stress. Data presented are the means ± SEs (*n* = 3). Different letters on top of bars indicate a significant difference (*p* < 0.05) according to Tukey's HSD test. CK (control) = 0 mM NaCl + 0 mg/L CTS; CTS = 0 mM NaCl + 100 mg/L CTS; NaCl = 100 mM NaCl + 0 mg/L CTS; NaCl + CTS = 100 mM NaCl +100 mg/L. MDA, malondialdehyde; O_2^-, superoxide radical; H_2O_2, hydrogen peroxide; AsA, ascorbic acid.

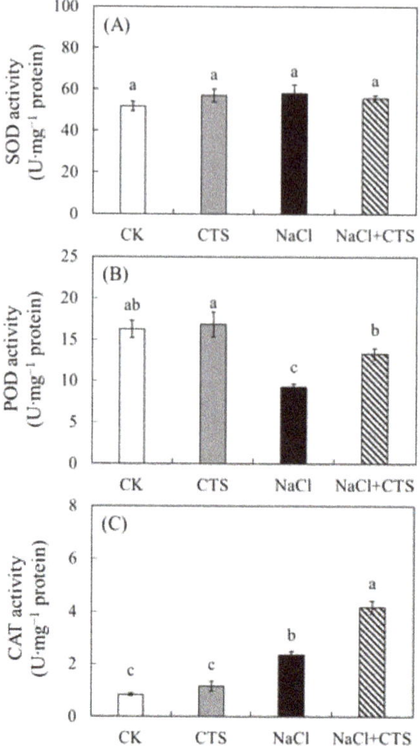

Figure 4. Effects of exogenous chitosan (CTS) on activities of (**A**) SOD, (**B**) POD, and (**C**) CAT in leaves of lettuce plants under salt stress. Data presented are the means ± SEs (n = 3). Different letters on top of bars indicate a significant difference ($p < 0.05$) according to Tukey's HSD test. CK (control) = 0 mM NaCl + 0 mg/L CTS; CTS = 0 mM NaCl + 100 mg/L CTS; NaCl = 100 mM NaCl + 0 mg/L CTS; NaCl + CTS = 100 mM NaCl +100 mg/L. SOD, superoxide dismutase; POD, peroxidase; CAT, catalase.

4. Discussion

Limited plant growth and productivity are common responses to salinity stress [5]. The data obtained in this study indicate that salinity adversely affected the growth and biomass of lettuce plants and resulted in a significant decrease in their total leaf area, FW, and DW (Table 1). These results are consistent with those of studies on other crops, including tomatoes [53], peppers [54], and chickpeas [43]. CTS, which is a derivative of chitin, has many applications in the agricultural sector because it regulates plant growth and development and increases plants' resistance to a wide range of abiotic and biotic stresses [19,21,55]. The application of exogenous CTS can mitigate the effects of salt stress on plant growth in many crops, such as ajowan [56], maize [57], and wheat [58]. A similar finding was obtained in this study, which indicates that the application of exogenous CTS alleviated the inhibition of the growth of lettuce caused by saline conditions (Table 1, Figure 2A). Growth analysis is a widely used analytical method for characterizing plant growth [59]. Plant biomass is strongly and positively correlated with RGR. We observed that the RGR of the NaCl group was significantly lower than that of the control group (Figure 2B), which indicates that the lettuce exposed to saline conditions accumulated less biomass than the lettuce grown under normal conditions during the growth stage. The NAR (average growth per unit leaf) of the NaCl group was significantly lower than that of the control group (Figure 2C). However, the LAR of the lettuce was relatively consistent between the saline and normal conditions (Figure 2D). These results indicate that the decrease in the RGR of the lettuce under salt stress was mainly associated with NAR and less associated with LAR. The exogenous application of CTS markedly increased the RGR of the lettuce under salt stress (Figure 2B), which proves that exogenous CTS can mitigate the inhibition of plant growth caused by salinity. However, we observed only a minor variation in NAR between the NaCl + CTS and NaCl groups (Figure 2C). Thus, the increase in the RGR of the NaCl + CTS group may have been related to both NAR and LAR. The Chl content of leaves is commonly considered a reliable predictor of the health and photosynthesis capacity of plants during growth [60,61]. Chlorophyll degradation under salt stress is usually related to the accumulation of ROS, which causes lipid peroxidation of chloroplast membranes [62]. In this study, the Chl content of the NaCl group was considerably lower than that of the control group (Table 1), which is in agreement with previous findings [63]. This result might be ascribable to the impaired biosynthesis or accelerated degradation of Chl pigment under saline conditions [43,64]. The application of exogenous CTS to the lettuce under saline conditions increased the Chl a and total Chl contents (Table 1). Similarly, Zou et al. [62] found that exogenous polysaccharides increased the Chl a content in wheat seedling leaves under salt stress. Chl a is responsible for the absorption of light and the initiation of photosynthesis. Excessive amounts of salt accumulated in chloroplasts can exert a direct toxic effect on photosynthesis through the destruction of pigment–protein complexes [65]. Thus, the increases in Chl a content in NaCl + CTS group are possibly because exogenous CTS protected Chl a from degradation in salt-stressed lettuce leaves, leading to high efficiency in photosynthesis. Moreover, the NaCl + CTS group did not exhibit a significantly higher Chl b content than the NaCl group possibly because Chl b tends to transform into Chl a when plants are subjected to saline conditions [66].

Leaf RWC is an accurate measure of the status of water in plants and indicates the water content of a leaf relative to the maximum amount of water that the leaf can contain under full turgidity [67]. Our study indicates that salinity decreased leaf RWC substantially; however, the exogenous application of CTS positively affected the leaf RWC of the lettuce exposed to saline conditions (Table 2). Geng et al. [24] also found that exogenous CTS application significantly increased the leaf RWC and water use efficiency in creeping bentgrass under salt stress, which contributes to maintaining a better water status in plants exposed to salinity stress. Thus, the results obtained from the current study were possibly due to the regulation of the balance between the water supply and leaf transpiration in the lettuce of the NaCl + CTS group. Under normal conditions, the Na^+ content was very low in lettuce leaves, and no significant change was observed between the CK and CTS groups (Table 2). Salt stress significantly increased Na^+ content but decreased the K^+ content in lettuce leaves (Table 2). The application of exogenous CTS increased the K^+ content and decreased the Na^+ content in the leaves of lettuce exposed to saline conditions, but the K^+ content in leaves was still lower than that of the control group (Table 2). A similar trend was also found in salt-stressed wheat seedlings applied with polysaccharides [62]. The accumulation of high levels of Na^+ in plant tissues subjected to saline conditions has a devastating effect on the metabolism of cytoplasm and organelles. As cytosolic Na^+ is noxious to cells, so too is chloroplastic Na^+ accumulation [12]. Excess Na^+ will cause an imbalance in cellular Na^+ and K^+ homeostasis, which often leads to a low K^+/Na^+ ratio [68]. Hereafter, plants will suffer from K^+ deficiencies stemming from the competitive inhibition of its uptake by Na^+ in plants exposed to salt stress. Potassium is an essential nutrient for plant growth and production, it is involved in the balance of osmotic pressure and the regulation of stomatal closure, and it affects photosynthesis and enzymatic activity [44]. Maintaining a high shoot K^+/Na^+ ratio is an important trait of plant salt tolerance [69]. In the current study, the shoot K^+/Na^+ ratio decreased dramatically under saline conditions due to excessive Na^+ accumulation in leaves. However, K^+ accumulation in NaCl + CTS-group plants was accompanied by a higher K^+/Na^+ ratio but not a significant difference compared to that of the NaCl group.

These results, combined with the mitigation of growth inhibition, suggest that exogenous CTS might facilitate plant growth by regulating the nutritional balance and reducing ion toxicity. A previous study has shown that polysaccharides enhanced salt tolerance in wheat by maintaining a high K^+/Na^+ ratio through regulation of several Na^+/K^+ transporter genes, coordinating the efflux and compartmentation of Na^+ [58]. The study of Geng et al. [24] found that CTS enhanced salt overly sensitive pathways and upregulated the expression of *AsHKT1* and genes encoding Na^+/H^+ exchangers under saline conditions, thus inhibiting Na^+ transport to the photosynthetic tissues. In the present study, although the lower Na^+ content in leaves was observed in the NaCl + CTS group at harvest time, the Na^+ levels in shoot and root were not evaluated in a time-response manner, and the gene expressions related to Na^+ transport still have to be analyzed. To further understand the CTS-induced salt tolerance in lettuce, future studies need to focus on examining the accumulation pattern of Na^+ in shoot and root and investigating the gene expressions involved in Na^+ transport.

Osmotic regulation is a major adaptation mechanism for plants to resist salt stress. High levels of osmolytes, such as proline, soluble sugars, and soluble proteins, are accumulated in the cytosol and other organelles to adjust osmotic pressure [17,44]. The high accumulation of proline in the NaCl group indicated the crucial nature of this osmolyte in the osmotic adjustment under saline conditions (Figure 3A). Under salt stress, proline can regulate osmotic potential, stabilize the cellular structure, reduce damage to the photosynthetic apparatus, and induce the expression of salt stress-responsive genes, consequently enhancing the adaption of the plant to saline conditions [58]. The application of exogenous CTS increased the proline level and Chl a content of the lettuce leaves subjected to saline conditions (Figure 3A), which suggests that exogenous CTS not only balances osmosis in cells but also protects their photosynthetic machinery. It had been proved that proline accumulation might be a result of a salt-induced increase in N metabolism [70]. A previous study found in wheat that exogenous CTS could effectively enhance N metabolism [71]. Thus, it is interesting to further study whether CTS increases proline levels by modulating N metabolism in lettuce plants during salt stress. The quantity of another vital osmolyte, namely soluble sugar, increased in the lettuce under saline conditions regardless of whether the leaves were treated with exogenous CTS; however, the soluble sugar levels of lettuce leaves treated with exogenous CTS were considerably higher than those of lettuce leaves treated with NaCl alone (Figure 3E). Because soluble sugar plays a key role in many physiological and biochemical processes, including photosynthesis, ROS scavenging, and the induction of adaptive pathway destructive conditions, a substantial increase in the total soluble sugar content may effectively protect lettuce plants exposed to NaCl stress [72,73].

Salt stress triggers the generation of a large number of ROS, such as O_2^-, H_2O_2, and $\cdot OH$, which poses challenges to plant cells [74,75]. The excessive generation of ROS causes the oxidation of lipids and proteins and the breakage of nucleic acids and limits the effectiveness of enzymes, which results in abnormalities at the cellular level and thus the inhibition of plant growth [43,76]. MDA, which is the final product of membrane lipid peroxidation caused by ROS, is generally an indicator of the degree of cell membrane damage in plants subjected to stress [77,78]. The results of our experiment reveal that the O_2^-, H_2O_2, MDA contents, and EL of the NaCl group were considerably higher than those of the control group, which indicates that the integrity and stability of the cell membrane decreased due ROS-induced oxidative damage (Figure 3B–D; Table 2). The O_2^-, H_2O_2, and MDA contents and EL of the CTS + NaCl group were lower than those of the NaCl group (Figure 3B–D; Table 2). These results suggest that CTS can mitigate oxidative damage and regulate the stability of the cell membrane system under saline conditions. Moreover, the observed significant increases in Chl a content and reductions in the MDA level in lettuce leaves of NaCl + CTS group also proved that exogenous CTS reduced lipid peroxidation and mitigated the salt-induced reduction in chlorophyll content (Figure 3B; Table 1). Similarly, Turk [57] reported that the application of exogenous CTS decreases ROS levels and lipid peroxidation in peppers under salt stress.

Plants possess a wide range of radical scavenging systems to manage oxidative damage, including antioxidative enzymes, such as SOD, POD, and CAT, and nonenzymatic compounds, such as proline and AsA [44,79–81]. SOD is a major O_2^- scavenger that catalyzes O_2^- to H_2O_2 and O_2 [82]. Thereafter, the toxic H_2O_2 can be removed by POD, CAT, or ascorbate peroxidase [83,84]. In the current study, the O_2^-, H_2O_2, and MDA contents of the NaCl group were considerably higher than those of the control group (Figure 3B–D). Moreover, the CAT and POD activities of the NaCl group were lower and higher, respectively, than those of the control group (Figure 4). The exogenous application of CTS significantly decreased the O_2^-, H_2O_2, and MDA levels and considerably increased the POD and CAT activities (Figure 3B–D and Figure 4). These results are consistent with those of studies on maize [57] and suggest that the exogenous application of CTS can mitigate the damage to the cell membrane system by salt stress by increasing the POD and CAT activities in lettuce leaves. The enzymatic activities of POD and CAT may have played a more crucial role than that of SOD in scavenging the overproduction of ROS in the plants treated with NaCl alone because no significant difference was observed in SOD activity among all the groups (Figure 4). However, the current study did not investigate the expression patterns of *SOD*, *CAT*, and *POD* genes, and additional studies should be conducted on this topic. In addition, studies have indicated that proline and AsA are potent antioxidants that can scavenge various types of ROS and shield the cell from oxidative damage [85,86]. In the current study, the application of exogenous CTS significantly increased the proline content of the lettuce leaves (Figure 3A), indicating that lettuce may accumulate high levels of proline to scavenge ROS, decrease oxidative damage, and safeguard cell membranes from the adverse effects of salt stress. No significant difference was observed in the AsA contents of the CTS + NaCl and NaCl groups (Figure 3F), which suggests that AsA may not play an important role in maintaining the strong antioxidant capacity of lettuce under saline conditions. A current study also found that exogenous oligo-alginate in NaCl-treated plants did not change the AsA content [87].

5. Conclusions

CTS, a natural polysaccharide, has many applications in the agriculture sector as an exogenous additive substance, being both safe and cheap. In the present study, the effects of exogenous CTS on lettuce plants under salt stress were investigated. The results showed that exogenous CTS could improve plant growth and biomass under salt stress. Exogenous CTS application increased proline and soluble sugar accumulations and enhanced peroxidase and catalase activities, thereby reducing oxidative damage to leaves. The CTS also curbed the accumulation of sodium but enhanced the accumulation of potassium in the leaves of NaCl-treated plants. These outcomes may help optimize the production technology of lettuce under saline conditions. However, the mechanism of CTS on alleviating salinity damage is still not fully understood. Future studies should focus on analyzing Na^+/K^+ transporter gene expressions and possible signal transduction pathways involved in CTS-regulated increased tolerance of lettuce plants to salt stress. As a biopolymer, the presence of amine and hydroxyl groups in CTS may also prevent Na^+ from reaching the photosynthetic tissue by chelating part of it at the root/lower tissue level, which needs further exploration depending on the CTS application method.

Author Contributions: Conceptualization, G.Z. and Y.W.; methodology, G.Z., Z.Y. and Y.W.; validation, Q.Z. and Y.F.; data analysis, G.Z., Y.W. and Q.Z.; writing—original draft preparation, G.Z. and Y.M.; writing—review and editing, K.W. All authors have read and agreed to the published version of the manuscript.

Funding: This research was funded by the Natural Science Foundation of the Jiangsu Higher Education Institutions of China, 19KJB210011; Innovation Training Program for College Students of Jiangsu Vocational College of Agriculture and Forestry, 202113103005y.

Institutional Review Board Statement: Not applicable.

Informed Consent Statement: Not applicable.

Data Availability Statement: The data presented in this study are available on request from the corresponding author.

Acknowledgments: The authors would like to thank Cullen Pitney and Randy Johnson for many helpful comments and suggestions.

Conflicts of Interest: The authors declare no conflict of interest.

References

1. Shahid, S.A.; Zaman, M.; Heng, L. Introduction to Soil Salinity, Sodicity and Diagnostics Techniques. In *Guideline for Salinity Assessment, Mitigation and Adaptation Using Nuclear and Related Techniques*; Springer Nature: Vienna, Switzerland, 2018.
2. Deinlein, U.; Stephan, A.B.; Horie, T.; Luo, W.; Schroeder, J.I. Plant salt-tolerance mechanisms. *Trends Plant Sci.* **2014**, *19*, 371–379. [CrossRef]
3. Ke, Q.; Wang, Z.; Ji, C.Y.; Jeong, J.C.; Lee, H.S.; Li, H.; Xu, B.; Deng, X.; Kwak, S.S. Transgenic poplar expressing codA exhibits enhanced growth and abiotic stress tolerance. *Plant Physiol. Biochem.* **2016**, *100*, 75–84. [CrossRef]
4. Chen, M.; Yang, Z.; Liu, J.; Zhu, T.; Wei, X.; Fan, H.; Wang, B. Adaptation mechanism of salt excluders under saline conditions and its applications. *Int. J. Mol. Sci.* **2018**, *19*, 3668. [CrossRef]
5. Azarmi-Atajan, F.; Sayyari-Zohan, M.H. Alleviation of salt stress in lettuce (*Lactuca sativa* L.) by plant growth-promoting rhizobacteria. *J. Hortic. Postharvest Res.* **2020**, *3*, 67–78.
6. Sohaid, M.; Zahir, Z.; Khan, M.; Ans, M.; Asghar, H.; Yasin, S.; Al-Barakah, F. Comparative evaluation of different carrier-based multi-strain bacterial formulations to mitigate the salt stress in wheat. *Saudi J. Biol. Sci.* **2020**, *27*, 777–787.
7. Zhu, J.K. Regulation of ion homeostasis under salt stress. *Curr. Opin. Plant Biol.* **2003**, *6*, 441–445. [CrossRef]
8. Yang, Y.; Guo, Y. Unraveling salt stress signaling in plants. *J. Integr. Plant Biol.* **2018**, *60*, 58–66. [CrossRef] [PubMed]
9. Hasegawa, P.M. Plant cellular and molecular responses to high salinity. *Annu. Rev. Plant Physiol. Plant Mol. Biol.* **2000**, *51*, 463–499. [CrossRef]
10. Wu, G.Q.; Shui, Q.Z.; Wang, C.M.; Zhang, J.L.; Yuan, H.J.; Li, S.J.; Liu, Z.J. Characteristics of Na^+ uptake in sugar beet (*Beta vulgaris* L.) seedlings under mild salt conditions. *Acta Physiol. Plant.* **2015**, *37*, 70. [CrossRef]
11. Zhang, J.L.; Wetson, A.M.; Wang, S.M.; Gurmani, A.R.; Bao, A.K.; Wang, C.M. Factors associated with determination of root $^{22}Na^+$ influx in the salt accumulation halophyte *Suaeda maritima*. *Biol. Trace Elem. Res.* **2011**, *139*, 108–117. [CrossRef] [PubMed]
12. Assaha, D.V.M.; Ueda, A.; Saneoka, H.; Al-Yahyai, R.; Yaish, M.W. The role of Na^+ and K^+ transporters in salt stress adaptation in glycophytes. *Front. Physiol.* **2017**, *8*, 509. [CrossRef] [PubMed]
13. Isayenkov, S.V.; Maathuis, F.J.M. Plant salinity stress: Many unanswered questions remain. *Front. Plant Sci.* **2019**, *10*, 80. [CrossRef]
14. Lv, X.; Chen, S.; Wang, Y. Advances in understanding the physiological and molecular responses of sugar beet to salt stress. *Front. Plant Sci.* **2019**, *10*, 1431. [CrossRef] [PubMed]
15. Li, J.; Liu, J.; Wang, G.; Cha, J.Y.; Li, G.; Chen, S.; Li, Z.; Guo, J.; Zhang, C.; Yang, Y.; et al. A chaperone function of NO CATALASE ACTIVITY1 is required to maintain catalase activity and for multiple stress responses in Arabidopsis. *Plant Cell* **2015**, *27*, 908–925. [CrossRef]
16. Del, L.A.D.; López-Huertas, E. ROS generation in peroxisomes and its role in cell signaling. *Plant Cell Physiol.* **2016**, *57*, 1364–1376. [CrossRef]
17. Zhao, J.; Pan, L.; Zhou, M.; Yang, Z.; Meng, Y.; Zhang, X. Comparative physiological and transcriptomic analyses reveal mechanisms of improved osmotic stress tolerance in annual ryegrass by exogenous chitosan. *Genes* **2019**, *10*, 853. [CrossRef] [PubMed]
18. Betchem, G.; Johnson, N.; Wang, Y. The application of chitosan in the control of post-harvest diseases: A review. *J. Plant Dis. Protect.* **2019**, *126*, 495–507. [CrossRef]
19. Pongprayoon, W.; Roytrakul, S.; Pichayangkura, R.; Chadchawan, S. The role of hydrogen peroxide in chitosan-induced resistance to osmotic stress in rice (*Oryza sativa* L.). *Plant Growth Regul.* **2013**, *70*, 159–173. [CrossRef]
20. Morin-Crini, N.; Lichtfouse, E.; Torri, G.; Crini, G. Applications of chitosan in food, pharmaceuticals, medicine, cosmetics, agriculture, textiles, pulp and paper, biotechnology, and environmental chemistry. *Environ. Chem. Lett.* **2019**, *17*, 1667–1692. [CrossRef]
21. Malerba, M.; Cerana, R. Chitosan effects on plant systems. *Int. J. Mol. Sci.* **2016**, *17*, 996. [CrossRef]
22. Guan, Y.J.; Hu, J.; Wang, X.J.; Shao, C.X. Seed priming with chitosan improves maize germination and seedling growth in relation to physiological changes under low temperature stress. *J. Zhejiang Univ. Sci. B* **2009**, *10*, 427–433. [CrossRef]
23. Jabeen, N.; Ahmad, R. The activity of antioxidant enzymes in response to salt stress in safflower (*Carthamus tinctorius* L.) and sunflower (*Helianthus annuus* L.) seedlings raised from seed treated with chitosan. *J. Sci. Food Agric.* **2012**, *93*, 1699–1705. [CrossRef]
24. Geng, W.; Li, Z.; Hassan, M.; Peng, Y. Chitosan regulates metabolic balance, polyamine accumulation, and Na^+ transport contributing to salt tolerance in creeping bentgrass. *BMC Plant Biol.* **2020**, *20*, 506. [CrossRef] [PubMed]
25. Cho, M.H.; No, H.K.; Prinyawiwatkul, W. Chitosan treatments affect growth and selected quality of sunflower sprouts. *J. Food Sci.* **2010**, *73*, S70–S77. [CrossRef] [PubMed]

26. Bittelli, M.; Flury, M.; Campbell, G.S.; Nichols, E.J. Reduction of transpiration through foliar application of chitosan. *Agric. For. Meteorol.* **2001**, *107*, 167–175. [CrossRef]

27. Dzung, N.A. Chitosan and their derivatives as prospective biosubstances for developing sustainable eco-agriculture. *Adv. Chitin Sci.* **2007**, *10*, 453–459.

28. Dzung, N.A.; Thang, N.T. Effect of oligoglucosamine on the growth and development of peanut (Arachis hypogea L.). In Proceedings of the 6th Asia-Pacific on Chitin, Chitosan Symposium, Singapore, 23–26 May 2004; Khor, E., Hutmacher, D., Yong, L.L., Eds.; ISBN 981-05r-r0904-9.

29. Li, Z.; Zhang, Y.; Zhang, X.; Merewitz, E.; Peng, Y.; Ma, X.; Huang, L.; Yan, Y. Metabolic pathways regulated by chitosan contributing to drought resistance in white clover. *J. Proteome Res.* **2017**, *16*, 3039–3052. [CrossRef]

30. Yang, F.; Hu, J.; Li, J.; Wu, X.; Qian, Y. Chitosan enhances leaf membrane stability and antioxidant enzyme activities in apple seedlings under drought stress. *Plant Growth Regul.* **2009**, *58*, 131–136. [CrossRef]

31. Hadwiger, L.A. Anatomy of a nonhost disease resistance response of pea to *Fusarium solani*: PR gene elicitation via DNase, chitosan and chromatin alterations. *Front. Plant Sci.* **2015**, *6*, 373. [CrossRef]

32. Aleksandrowicz-Trzcińska, M.; Bogusiewicz, A.; Szkop, M.; Drozdowski, S. Effect of chitosan on disease control and growth of scots pine (*Pinus sylvestris* L.) in a forest nursery. *Forests* **2015**, *6*, 3165–3176. [CrossRef]

33. Viacava, G.E.; Gonzalez-Aguilar, G.G.; Roura, S.I. Determination of phytochemicals and antioxidant activity in butterhead lettuce related to leaf age and position. *J. Food Biochem.* **2014**, *38*, 352–362. [CrossRef]

34. Aksakal, O.; Tabay, D.; Esringu, A.; Icoglu, A.F.; Esim, N. Effect of proline on biochemical and molecular mechanisms in lettuce (*Lactuca sativa* L.) exposed to UV-B radiation. *Photochem. Photobiol. Sci.* **2017**, *16*, 246–254. [CrossRef] [PubMed]

35. Yildirim, E.; Ekinci, M.; Turan, M.; Dursun, A.; Kul, R.; Parlakova, F. Roles of glycine betaine in mitigating deleterious effect of salt stress on lettuce (*Lactuca sativa* L.). *Arch. Agron. Soil Sci.* **2015**, *61*, 1673–1689. [CrossRef]

36. Fernández, J.A.; Niirola, D.; Ochoa, J.; Orsini, F.; Pennisi, G.; Gianquinto, G.; Egea-Gilabert, C. Root adaptation and ion selectivity affects the nutritional value of salt-stressed hydroponically grown baby-leaf *Nasturtium officinale* and *Lactuca sativa*. *Agr. Food Sci.* **2016**, *25*, 230–239. [CrossRef]

37. Ahmed, S.; Ahmed, S.; Roy, S.K.; Woo, S.H.; Sonawane, K.D.; Shohael, A.M. Effect of salinity on the morphological, physiological and biochemical properties of lettuce (*Lactuca sativa* L.) in Bangladesh. *Open Agric.* **2019**, *4*, 361–373. [CrossRef]

38. Garrido, Y.; Tudela, J.A.; Marín, A.; Mestre, T.; Martínez, V.; Gil, M.I. Physiological, phytochemical and structural changes of multi-leaf lettuce caused by salt stress. *J. Sci. Food Agric.* **2014**, *94*, 1592–1599. [CrossRef] [PubMed]

39. Joshi, J.; Zhang, G.; Shen, S.; Supaibulwatana, K.; Watanabe, C.K.A.; Yamori, W. A combination of downward lighting and supplemental upward lighting improves plant growth in a closed plant factory with artificial lighting. *HortScience* **2017**, *52*, 831–835. [CrossRef]

40. Porra, R.J.; Thompson, W.A.; Kriedemann, P.E. Determination of accurate extinction coefficients and simultaneous equations for assaying chlorophylls a and b extracted with four different solvents: Verification of the concentration of chlorophyll standards by atomic absorption spectroscopy. *BBA Bioenerg.* **1989**, *975*, 384–394. [CrossRef]

41. Ohtake, N.; Ishikura, M.; Suzuki, H.; Yamori, W.; Goto, E. Continuous irradiation with alternating red and blue light enhances plant growth while keeping nutritional quality in lettuce. *HortScience* **2018**, *53*, 1804–1809. [CrossRef]

42. Yamasaki, S.; Dillenburg, L.R. Measurements of leaf relative water content in *Araucaria angustifolia*. *Rev. Bras. Fisiol. Veg.* **1999**, *11*, 69–75.

43. Ahmad, P.; Latef, A.A.A.; Hashem, A.; Abd_Allah, E.F.; Gucel, S.; Tran, L.P. Nitric oxide mitigates salt stress by regulating levels of osmolytes and antioxidant enzymes in chickpea. *Front. Plant Sci.* **2016**, *7*, 347. [CrossRef]

44. Zhang, G.; Yan, Z.; Wang, Y.; Feng, Y.; Yuan, Q. Exogenous proline improve the growth and yield of lettuce with low potassium content. *Sci. Hortic.* **2020**, *271*, 109469. [CrossRef]

45. Bates, L.S.; Waldren, R.P.; Teare, I.D. Rapid determination of free proline for water-stress studies. *Plant Soil* **1973**, *39*, 205–207. [CrossRef]

46. Yemm, E.W.; Willis, A.J. The estimation of carbohydrates in plant extracts by anthrone. *Biochem. J.* **1954**, *57*, 508–514. [CrossRef] [PubMed]

47. Kampfenkel, K.; Van Montagu, M.; Inzé, D. Extraction and determination of ascorbate and dehydroascorbate from plant tissue. *Anal. Biochem.* **1995**, *225*, 165–167. [CrossRef]

48. Patterson, B.D.; MacRae, E.A.; Ferguson, I.B. Estimation of hydrogen peroxide in plant extracts using titanium (IV). *Anal. Biochem.* **1984**, *139*, 487–492. [CrossRef]

49. Averina, N.G.; Nedved, E.L.; Shcherbakov, R.A.; Vershilovskaya, I.V.; Yaronskaya, E.B. Role of 5-aminolevulinic acid in the formation of winter rape resistance to sulfonylurea herbicides. *Russ. J. Plant Physiol.* **2014**, *61*, 679–687. [CrossRef]

50. Heath, R.L.; Packer, L. Photoperoxidation in isolated chloroplasts: I. Kinetics and stoichiometry of fatty acid peroxidation. *Arch. Biophys. Biophys.* **1968**, *125*, 189–198. [CrossRef]

51. Jin, P.; Wang, H.; Zhang, Y.; Huang, Y.; Wang, L.; Zheng, Y. UV-C enhances resistance against gray mold decay caused by *Botrytis cinerea* in strawberry fruit. *Sci. Hortic.* **2017**, *225*, 106–111. [CrossRef]

52. Bradford, M.M. A rapid and sensitive method for the quantitation of microgram quantities of protein utilizing the principaldye binding. *Anal. Biochem.* **1976**, *72*, 248–254. [CrossRef]

53. Latef, A.A.H.A.; Chaoxing, H. Effect of arbuscular mycorrhizal fungi on growth, mineral nutrition, antioxidant enzymes activity and fruit yield of tomato grown under salinity stress. *Sci. Hortic.* **2011**, *127*, 228–233. [CrossRef]

54. Latef, A.A.H.A.; Chaoxing, H. Does the inoculation with Glomus mosseae improve salt tolerance in pepper plants? *J. Plant Growth Regul.* **2014**, *33*, 644–653. [CrossRef]

55. Hemantaranjan, A.; Deepmala, K.; Bharti, S.; Nishant, B.A. A future perspective in crop protection: Chitosan and its oligosaccharides. *Adv. Plants Agric. Res.* **2014**, *1*, 00006. [CrossRef]

56. Mahdavi, B.; Rahimi, A. Seed priming with chitosan improves the germination and growth performance of ajowan (*Carum copticum*) under salt stress. *Eurasia. J. Biosci.* **2013**, *7*, 69–76. [CrossRef]

57. Turk, H. Chitosan-induced enhanced expression and activation of alternative oxidase confer tolerance to salt stress in maize seedlings. *Plant Physiol. Bioch.* **2019**, *141*, 415–422. [CrossRef]

58. Zou, P.; Li, K.C.; Liu, S.; Xing, R.G.; Qin, Y.K.; Yu, H.H.; Zhou, M.M.; Li, P.C. Effect of chitooligosaccharides with different degrees of acetylation on wheat seedlings under salt stress. *Carbohydr. Polym.* **2015**, *126*, 62–69. [CrossRef]

59. Hoffmann, W.A.; Poorter, H. Avoiding bias in calculations of relative growth rate. *Ann. Bot.* **2002**, *80*, 37–42. [CrossRef]

60. Rady, M.M.; El-Shewy, A.A.; El-Yazal, M.A.; Abdelaal, K.E.S. Response of salt-stressed common bean plant performances to foliar application of phosphorus (MAP). *Int. Lett. Nat. Sci.* **2018**, *72*, 7–20. [CrossRef]

61. Qi, H.; Wu, Z.; Zhang, L.; Li, J.; Zhou, J.; Jun, Z.; Zhu, B. Monitoring of peanut leaves chlorophyll content based on drone-based multispectral image feature extraction. *Comput. Electron. Agr.* **2021**, *187*, 106292. [CrossRef]

62. Zou, P.; Lu, X.; Jing, C.; Yuan, Y.; Lu, Y.; Zhang, C.; Meng, L.; Zhao, H.; Li, Y. Low-molecular-weightt polysaccharides from *Pyropia yezoensis* enhance tolerance of wheat seedlings (*Triticum aestivum* L.) to Salt Stress. *Front. Plant Sci.* **2018**, *9*, 427. [CrossRef]

63. Sen, S.; Chouhan, D.; Das, D.; Ghosh, R.; Mandal, P. Improvisation of salinity stress response in mung bean through solid matrix priming with normal and nano-sized chitosan. *Int. J. Biol. Macromol.* **2020**, *145*, 108–123. [CrossRef] [PubMed]

64. Rasool, S.; Ahmad, A.; Siddiqi, T.O.; Ahmad, P. Changes in growth, lipid peroxidation and some key antioxidant enzymes in chickpea genotypes under salt stress. *Acta Physiol. Plant.* **2013**, *35*, 1039–1050. [CrossRef]

65. Negrão, S.; Schmöckel, S.M.; Tester, M. Evaluating physiological responses of plants to salinity stress. *Ann. Bot.* **2017**, *119*, 1–11. [CrossRef] [PubMed]

66. Eckardt, N.A. A new chlorophyll degradation pathway. *Plant Cell* **2009**, *21*, 700. [CrossRef] [PubMed]

67. Suriya-arunroj, D.; Supapoj, N.; Toojinda, T.; Vanavichit, A. Relative leaf water content as an efficient method for evaluating rice cultivars for tolerance to salt stress. *ScienceAsia* **2004**, *30*, 411–415. [CrossRef]

68. Mekawy, A.M.; Assaha, D.V.; Yahagi, H.; Tada, Y.; Ueda, A.; Saneoka, H. Growth, physiological adaptation, and gene expression analysis of two Egyptian rice cultivars under salt stress. *Plant Physiol. Biochem.* **2015**, *87*, 17–25. [CrossRef]

69. Cuin, T.A.; Betts, S.A.; Chalmandrier, R.; Shabala, S. A root's ability to retain K^+ correlates with salt tolerance in wheat. *J. Exp. Bot.* **2008**, *59*, 2697–2706. [CrossRef]

70. Rocha, I.M.A.; Vitorello, V.A.; Silva, J.S.; Ferreira-Silva, S.L.; Viegas, R.A.; Silva, E.N.; Silveira, J.A.G. Exogenous ornithine is an effective precursor and the delta-ornithine amino transferase pathway contributes to proline accumulation under high N recycling in salt-stressed cashew leaves. *J. Plant Physiol.* **2012**, *169*, 41–49. [CrossRef]

71. Zhang, X.Q.; Li, K.C.; Xing, R.G.; Liu, S.; Li, P.C. Metabolite profiling of wheat seedlings induced by chitosan: Revelation of the enhanced carbon and nitrogen metabolism. *Front. Plant Sci.* **2017**, *8*, 13. [CrossRef]

72. Gupta, A.K.; Kaur, N. Sugar signalling and gene expression in relation to carbohydrate metabolism under abiotic stresses in plants. *J. Biosci.* **2005**, *30*, 761–776. [CrossRef]

73. Bolouri-Moghaddam, M.R.; Le, R.K.; Xiang, L.; Rolland, F.; Van den Ende, W. Sugar signalling and antioxidant network connections in plant cells. *FEBS J.* **2010**, *277*, 2022–2037. [CrossRef]

74. Hayat, S.; Yadav, S.; Wani, A.S.; Irfan, M.; Alyemini, M.N.; Ahmad, A. Impact of sodium nitroprusside on nitrate reductase, proline and antioxidant systemin *Solanum lycopersicum* under salinity stress. *Hort. Environ. Biotechnol.* **2012**, *53*, 362–367. [CrossRef]

75. Butt, M.; Ayyub, C.M.; Amjad, M.; Ah-mad, R. Proline application enhances growth of chilli by improving physiological and biochemical attributes under salt stress. *Pak. J. Agr. Sci.* **2016**, *53*, 43–49.

76. Miller, G.; Suzuki, N.; Ciftci-Yilmaz, S.; Mittler, R. Reactive oxygen species homeostasis and signalling during drought and salinity stresses. *Plant Cell Environ.* **2010**, *33*, 453–467. [CrossRef] [PubMed]

77. Li, J.T.; Qiu, Z.B.; Zhang, X.W.; Wang, L.S. Exogenous hydrogen peroxide can enhance tolerance of wheat seedlings to salt stress. *Acta Physiol. Plant.* **2011**, *33*, 835–842. [CrossRef]

78. Wu, G.Q. Exogenous application of proline alleviates salt-induced toxicity in sainfoin seedlings. *J. Anim. Plant Sci.* **2017**, *27*, 246–251.

79. Asada, K. Ascorbate peroxidase—A hydrogen peroxide-scavenging enzyme in plants. *Physiol. Plant.* **1992**, *85*, 235–241. [CrossRef]

80. Alscher, R.G.; Donahue, J.L.; Cramer, C.L. Reactive oxygen species and antioxidants: Relationships in green cells. *Physiol. Plant.* **1997**, *100*, 224–233. [CrossRef]

81. Orsini, F.; Pennisi, G.; Mancarella, S.; Nayef, M.A.; Sanoubar, R.; Nicola, S.; Gianquinto, G. Hydroponic lettuce yields are improved under salt stress by utilizing white plastic film and exogenous applications of proline. *Sci. Hortic.* **2018**, *233*, 283–293. [CrossRef]

82. Apel, K.; Hirt, H. Reactive oxygen species: Metabolism, oxidative stress, and signal transduction. *Annu. Rev. Plant Biol.* **2004**, *55*, 373–399. [CrossRef]

83. Foyer, C.H.; Noctor, G. Oxidant and antioxidant signalling in plants: A re-evaluation of the concept of oxidative stress in a physiological context. *Plant Cell Environ.* **2005**, *28*, 1056–1071. [CrossRef]

84. Kusvuran, S.; Bayat, R.; Ustun, A.S.; Ellialtioglu, S.S. Exogenous proline improves osmoregulation, physiological and biochemical responses of eggplant under salt stress. *Fresenius Environ. Bull.* **2020**, *29*, 152–161.

85. Thomas, C.E.; Mclean, L.R.; Parkar, R.A.; Ohleweiler, D.F. Ascorbate and phenolic antioxidant interactions in prevention of liposomal oxidation. *Lipids* **1992**, *27*, 543–550. [CrossRef]

86. Iqbal, N.; Umar, S.; Khan, N.A.; Khan, M.I.R. A new perspective of phytohormones in salinity tolerance: Regulation of proline metabolism. *Environ. Exp. Bot.* **2014**, *100*, 34–42. [CrossRef]

87. Piotr, S.; Monika, G.; Edward, M.; Marcin, S. Oligo-alginate with low molecular mass improves growth and physiological activity of eucomis autumnalis under salinity stress. *Molecules* **2018**, *23*, 812.

Article

Adaptive Morphophysiological Features of *Neottia ovata* (Orchidaceae) Contributing to Its Natural Colonization on Fly Ash Deposits

Maria Maleva [1,*,†], Galina Borisova [1,†], Nadezhda Chukina [1], Olga Sinenko [1], Elena Filimonova [2], Natalia Lukina [2] and Margarita Glazyrina [2]

[1] Department of Experimental Biology and Biotechnology, Ural Federal University, 620002 Ekaterinburg, Russia; G.G.Borisova@urfu.ru (G.B.); nadezhda.chukina@urfu.ru (N.C.); Olga.Sinenko@urfu.ru (O.S.)
[2] Laboratory of Anthropogenic Dynamics of Ecosystems, Ural Federal University, 620002 Ekaterinburg, Russia; Elena.Filimonova@urfu.ru (E.F.); natalia.lukina@urfu.ru (N.L.); Margarita.Glazyrina@urfu.ru (M.G.)
* Correspondence: maria.maleva@urfu.ru
† Equal contribution by both the authors.

Citation: Maleva, M.; Borisova, G.; Chukina, N.; Sinenko, O.; Filimonova, E.; Lukina, N.; Glazyrina, M. Adaptive Morphophysiological Features of *Neottia ovata* (Orchidaceae) Contributing to Its Natural Colonization on Fly Ash Deposits. *Horticulturae* **2021**, *7*, 109. https://doi.org/10.3390/ horticulturae7050109

Academic Editors: Jolanta Jaroszuk-Ściseł, Małgorzata Majewska and Agnieszka Hanaka

Received: 7 April 2021
Accepted: 4 May 2021
Published: 11 May 2021

Publisher's Note: MDPI stays neutral with regard to jurisdictional claims in published maps and institutional affiliations.

Abstract: In previous decades, some species of the Orchidaceae family have been found growing in man-made habitats. *Neottia ovata* is one of the most widespread orchids in Europe, however it is quite rare in Russia and is included in several regional Red Data Books. The purpose of this study was to compare the chemical composition and morphophysiological parameters of *N. ovata* from two forest communities of the Middle Urals, Russia: natural and transformed (fly ash dump of Verkhnetagil'skaya Thermal Power Station) for determining orchid adaptive features. The content of most of the studied metals in the underground parts (rhizome + roots) of *N. ovata* was considerably higher than in the leaves, which diminished the harmful effect of toxic metals on the aboveground organs. The adaptive changes in the leaf mesostructure of *N. ovata* such as an increase in epidermis thickness, the number of chloroplasts in the cell, and the internal assimilating surface were found for the first time. The orchids from the fly ash deposits were characterized by a higher content of chlorophyll *b* and carotenoids than plants from the natural forest community that evidenced the compensatory response on the decrease in chlorophyll *a*. The ability of *N. ovata* from the transformed habitat to maintain a relatively favorable water balance and stable assimilation indexes further contribute to its high viability. The study of orchid adaptive responses to unfavorable factors is necessary for their successful naturalization and introduction into a new environment.

Keywords: orchid; transformed ecosystems; fly ash; metals; adaptive responses; water exchange; leaf mesostructure; photosynthetic pigments; photosynthesis; plant introduction

Highlights

Neottia ovata successfully colonize the fly ash dump (FAD) due to less phytocoenotic stress.
N. ovata plants from transformed habitat demonstrate high viability.
Sequestration of metals mainly in underground organs reduced harmful effect on orchid plants.
Natural orchid colonization of FAD was facilitated by adaptive structural and functional changes.
The FAD plants were characterized by the higher chlorophyll *b* and carotenoids content.
N. ovata from FAD maintained a relatively favorable water balance and stable assimilation indexes.

1. Introduction

The Orchidaceae family has a broad variety of more than 28,000 species distributed in about 763 genera and widespread from the Arctic tundra to tropical Brazilian rainforests [1,2]. It includes species with complex adaptations to pollination by specific insect

Horticulturae **2021**, *7*, 109. https://doi.org/10.3390/horticulturae7050109   https://www.mdpi.com/journal/horticulturae

species and very different life strategies: from epiphytic to terrestrial, from evergreen to completely chlorophyll-free [3,4]. Studies have been conducted on orchids' taxonomy, morphology, ecology, breeding [5–7], pollination [8–10], genetics [11,12], mycorrhizal association [13–15], etc. At the same time, the physiological parameters of Orchidaceae species are still less studied and need much more attention [16].

Changing natural habitats have caused the extinction of many orchid species [17]. However, some orchids, especially in temperate regions of Europe and North America, have been found in anthropogenically disturbed territories, such as industrial dumps formed after the excavation and extraction of coal, iron, and some trace elements, and the fly ash dumps of thermal power plants [13,14,18–21].

The monitoring of vegetation restoration on disturbed lands in the Middle Urals, Russia, has shown that dumps from mining and processing industries are often colonized by some rare orchid species at the initial stages of the forest phytocoenoses formation [22–24]. Low competition in man-made habitats contributes to the conservation of the gene pool of Orchidaceae species. The local populations of *Neottia ovata* (L.) Bluff and Fingerh. (syn. *Listera ovata* (L.) R. Br. or Common twayblade) are of particular interest, as they have been found in recent years in disturbed territories of the Middle Urals, including fly ash dumps [22].

Common twayblade is one of the widespread orchids in Europe, especially in the British Isles [25]. However, this species is quite seldom encountered in Russia and has the status of a "rare species" in many regional Red Data Books, including the Red Book of Sverdlovsk Region [26]. This is a short-rhizome herbaceous perennial, mesophyte, European–West Asian, boreal-immoral species [3,25,27]. Like other orchids, *N. ovata* is characterized by low competitiveness. *N. ovata* grow on both fertile and poor soils. Sometimes it is found in disturbed habitats, along roadsides and railways, and in abandoned limestone quarries [4]. The *N. ovata* is a typical entomophile, the spectrum of pollinators is very wide [9]. This species reproduces well both by seeds and vegetatively [28]. For the germination of seeds in the first years of life, the presence of mycorrhiza is necessary [4].

Fly ash is considered a problematic form of solid waste throughout the world [29,30]. It is well known that fly ash substrates are characterized by low microbiological activity, an insufficient supply of nutrients, especially nitrogen, and adverse physicochemical properties [31]. Moreover, fly ash may also contain toxic concentrations of As, Cd, Cr, Pb, Co, Cu, etc. [29–31].

Different plant species growing in stressful environments show great variation in their tolerance mechanisms [29]. Unfavorable abiotic factors affect photosynthesis, respiration, water regime, and mineral nutrition, leading to impaired growth and development [21–24]. Photosynthesis is the main fundamental process that determines the productivity of plants [32,33]. The leaves are the primary photosynthetic organs, serving as key sites where the absorption of light and CO_2 assimilation take place. The internal organization of the leaf is influenced by environmental factors such as the chemical properties of soil substrates, light, temperature, and water availability. Leaf structure is known to be highly plastic in response to growing conditions, varying greatly in morphology, anatomy, and physiology [32–34]. The investigation of plant leaf traits and its responses to environmental change has increasingly gained more attention in recent decades [34–37]. Maintaining the functional activity of the photosynthetic apparatus in stressful conditions is one of the essential prerequisites that allow plants to colonize transformed territories.

The purpose of this study was a comparative analysis of the chemical composition and the structural and functional parameters of *N. ovata* from natural and transformed (fly ash dump) habitats. This will allow us to identify the adaptive morphophysiological characteristics of this species that contribute to its colonization on infertile technogenic substrates.

2. Materials and Methods

2.1. Study Area

The research was conducted in the Middle Urals, Russia (subzone of the southern taiga). The southern taiga subzone is characterized as moderately cold in terms of heat supply and over-humidified in terms of moisture availability. The average annual temperature is 1.9 °C, the annual precipitation is almost 570 mm, and the hydrothermal coefficient is about 1.5 [32]. The fly ash dump formed by brown coal ash is located on an area of 1.25 km². The fly ash deposits formed after mining (1968–1970) were left for colonization by natural forest [22,31,38]. The investigation was carried out in the vicinity of Verkhniy Tagil town (Sverdlovsk region) during the period of orchid blooming (mid-July 2018–2019). All samples were collected during the same period (from 15 to 18 July) under similar weather conditions (temperature during the daytime was 23 ± 3 °C and the relative humidity was about 60%). Two naturally growing orchid populations were studied: P-1 (57°20′13″ N 60°01′43″ E) from the natural forest community (NFC) near Belorechka village and P-2 (57°20′45″ N 59°56′46″ E) from the fly ash dump (FAD) near Verkhnetagil'skaya Thermal Power Station, VTTPS (Figures 1 and 2). The studied area of each site was about 400 m².

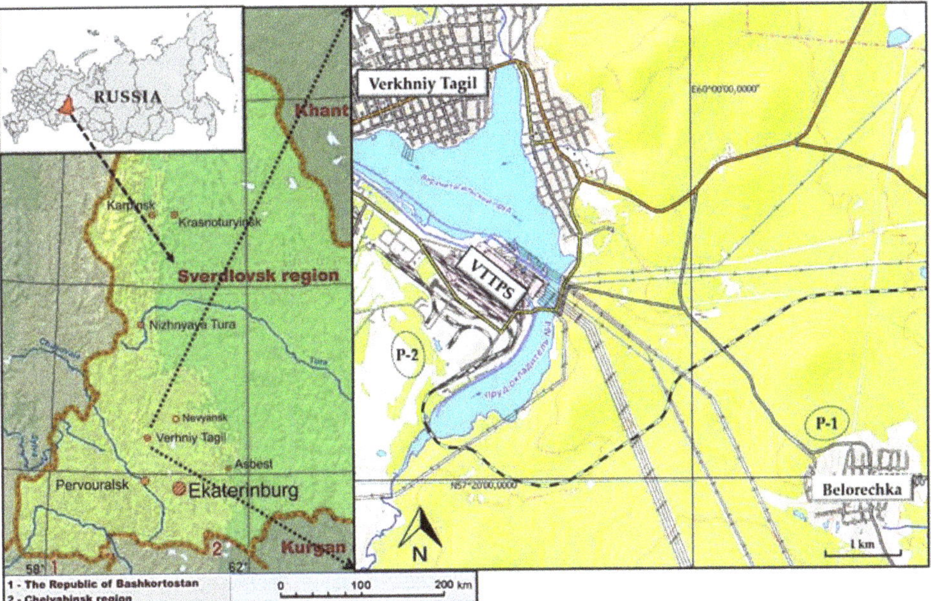

Figure 1. Map of Sverdlovsk region, Russia. Detailed presentation of the locations of studied *N. ovata* populations: P-1—from natural forest community near Belorechka village; P-2—from the fly ash dump near Verkhnetagil'skaya Thermal Power Station (VTTPS).

The natural forest community was represented by a mixed forest. The soil of this site was is sod-podzolic. The age of trees was between 80 and 100 years and the tree crown density was between 0.5 and 0.6. The height of the first layer of trees was 10–24 m and the second layer was 6–12 m. Coniferous species (*Picea obovata* Ledeb., *Larix sibirica* Ledeb., *Pinus sylvestris* L.) predominated in this forest. *N. ovata* individuals in P-1 grew mainly under the deciduous species *Betula pendula* Roth and *B. pubescens* Ehrh., as well as in small glades and meadows. The total projective shrub cover was 30–50% and contained *Tilia cordata* Mill., *Sorbus aucuparia* L., *Padus avium* Mill., *Rosa majalis* Herm., *R. acicularis* Lindl., and *Rubus idaeus* L. The total projective herbaceous cover was 70–80%, which reached up to 100% in the glades. Dominant among the herbaceous species were

Calamagrostis arundinacea (L.) Roth, *Aegopodium podagrária* L., *Círsium heterophyllum* (L.) Hill., *Brachypodium pinnatum* (L.) Beauv., *Geranium sylvaticum* L., *Lathyrus vernus* (L.) Bernh., *Anthoxanthum odoratum* L., *Alchemilla vulgaris* L., *Poa pratensis* L., *Melica nutans* L., *Vicia sepium* L., and *Ranunculus acris* L. The moss–lichen layer was weak. In total, more than 90 species grew on the investigated site and the species richness was 31 species per 100 m^2.

Figure 2. Flowering *N. ovata* plants from: (**a**) natural forest community (P-1) and (**b**) fly ash dump (P-2); (**c**) naturally colonized fly ash dump of VTTPS; *N. ovata* (**d**) inflorescence and (**e**) flower.

The orchid population P-2 was found in the young forest community formed during the natural revegetation of the fly ash deposits. Soil formation was proceeding according to zonal type in the fly ash substrate under the forest communities [38]. The tree crown density was 0.4, which reached up to 0.6 in some places. The 35–40-year-old forest community was dominated by *B. pendula*, *Populus tremula* L., *P. sylvestris*, *P. obovata*, and, less often, *B. pubescens*. In the undergrowth, there were singular instances of *P. obovata*, *L. sibirica*, and *Abies sibirica* Ledeb. The total projective shrub cover was 10–30% and contained *Salix myrsinifolia* Sm., *Sorbus aucuparia* L., *Padus avium* Mill., *Viburnum opulus* L., and *Chamaecytisus ruthenicus* (Fisch. ex Wołoszcz.) Klásková. The total projective herbaceous cover was 20–25%, in some places reaching 70%. The dominant species were *Platanthera bifolia* (L.) Rich., *Calamagrostis epigejos* (L.) Roth, *Amoria repens* (L.) C. Presl, *Pyrola rotundifolia* L., *Orthilia secunda* (L.) House, *Poa pratensis* L., *Festuca rubra* L., and *Equisetum arvense* L. The moss–lichen layer was weak. In total, more than 60 species grew on this site and the species richness was 19 species per 100 m^2.

2.2. Plant and Soil Substrate Sample Collection, Preparation and Analysis

From each site, no more than 20% of the total number of orchids in the studied populations were randomly selected to minimize the damage to these populations. Since the populations differed in terms of the number of individuals (39 and 194 in P-1 and P-2, respectively), four flowering orchid plants from natural forest population (P-1), and eight from the fly ash deposits (P-2), were carefully excavated with part of the soil substrate to preserve the underground organs (rhizome and roots). The studied plants from both the sites were in the range of 50–60 cm in length. The samples were placed into separate sterile 10 L bags to minimize dehydration and transferred to the laboratory for further analysis.

The soil substrate samples (about 2 kg) were taken from each orchid root zone at a depth of 0–15 cm, and a composite sample was formed for analysis. Subsequently, the soil samples were air dried, homogenized, passed through a sieve (<2 mm), and preserved for physicochemical analysis (pH, electrical conductivity, total dissolved solids, total and available metal concentrations).

Before chemical analysis, the plants were carefully washed by ultrasonication (UM-4, Unitra Unima, Olsztyn, Poland) and finally with deionized water (Milli-Q system, Millipore SAS, Molsheim, France). The leaves, rhizome, and roots from each individual plant were separated and then dried for 24 h at 75 °C along with soil samples. Afterwards, the dried samples were weighed and digested with concentrated HNO_3 (analytical grade) using a MARS 5 Digestion Microwave System (CEM, Matthews, NC, USA). The available form of metals was analyzed by extracting the soil sample (5 g) with 10 mL of 0.5 M nitric acid solution as described earlier [24]. All the samples were prepared using deionized Millipore water. The concentrations of K, Ca, Mg, Fe, Zn, Mn, Pb, Cu, Ni, Cr, and Co in all the samples were determined using an atomic absorption spectrometer AA240FS (Varian Australia Pty Ltd., Mulgrave, Victoria, Australia) [24]. Standard reference materials (JSC Ural Chemical Reagents Plant, Verkhnyaya Pishma, Russia) were used for the preparation and calibration of each analytical batch. The calibration coefficients were maintained at a high level of no less than 0.99.

The bioconcentration factor (BCF) was calculated as the ratio of the metal concentration in the underground/aboveground organs to the available concentration in the soil. The translocation factor (TF) was calculated as the ratio of metal concentration in the leaves to the concentration in the rhizome + roots.

The pH, electrical conductivity (EC), and total dissolved solids (TDS) of the soil–water suspensions (1:2.5; *w/v*) were measured using a portable multivariable analyzer HI98129 Combo (Hanna Instruments GmbH, Graz, Austria). The total nitrogen and phosphorus content in the *N. ovata* leaves and rhizome + roots were measured spectrophotometrically at 440 and 640 nm, respectively, after wet digestion with an acid mixture of $HClO_4$ and H_2SO_4 (1:10; *v/v*). The total nitrogen was measured after the reaction with Nessler's reagent [39], whereas the total phosphorus was determined by standard method using ammonium molybdate in the acid medium [40].

2.3. Morphological, Anatomical Parameters and Mycorrhiza Assay

Twenty flowering plants from each population were used to study shoot and inflorescence length, the number of flowers, and the total leaf area under in situ conditions. To calculate the leaf area, each leaf was photographed on graph paper and digital image analysis was performed using special MesoPlant software (OOO SIAMS, Ekaterinburg, Russia). From the lower leaf of 10 plants about 30 leaf discs (0.7 cm diameter) were fixed in 3.5% glutaraldehyde solution in a phosphate buffer (pH 7.0, *v/v*) in order to study the mesostructural parameters: leaf mesophyll and epidermis thickness (μm), number of cells per unit leaf area (thousand cm^{-2}), chloroplasts per mesophyll cell (pieces), and cell and chloroplast volumes ($μm^3$). The transverse sections of the leaf discs were obtained using a freezing microtome MZ-2 (JSC Kharkov plant "Tochmedpribor", Kharkov, Ukraine). All measurements were carried out in 30 replicates using a Meiji MT 4300 L light microscope (Meiji Techno, San Jose, CA, USA). The quantitative parameters of the mesophyll were determined with a computer-assisted protocol based on MesoPlant software (OOO SIAMS, Ekaterinburg, Russia). The number of cells per unit of leaf surface area was counted in a Goryaev cytometer after tissue maceration in a 20% KOH solution (*v/v*) with heating at 80–90 °C. All other measurements were carried out on leaf discs preliminarily macerated with 5% chromic acid dissolved in 1 N HCl (*v/v*) [34].

The quantitative indices of the leaf mesophyll were determined according to Mokronosov [31], modified by Ivanova and P'yankov [34]. The cell volume per chloroplast (CVC,

μm^3) was calculated as the ratio of cell volume to the number of chloroplasts per cell. The chloroplast membrane index (CMI, cm^2 cm^{-2}) was calculated as the ratio of the total surface area of the outer membranes of chloroplasts to the unit of leaf surface area [35].

Fresh roots of *N. ovata* from both the studied sites were used for investigating mycorrhizal association. Root tips up to 1.5 cm were cut into 20 μm cross sections with a freezing microtome and analyzed under a Meiji MT 4300 L light microscope (Meiji Techno, San Jose, CA, USA) [24].

2.4. Physiological and Biochemical Parameters Assay

To measure photosynthesis and transpiration, freshly dug-up plants with bulk soil (as described earlier in Section 2.2) were transported to the laboratory and studied no later than 3 h after collection to minimize the dehydration.

The gas exchange (μM CO_2 m^{-2} s^{-1}) and transpiration rate (mM H_2O m^{-2} s^{-1}) were measured with the lower leaf of four plants using a LI-6400XT portable infrared gas analyzer (LI-COR, Lincoln, NE, USA) with a LED Light Source chamber (3×4 cm). The following parameters were set: operating at an ambient concentration of CO_2 and humidity, the temperature was +23 °C and the saturating light intensity of 1600 μM m^{-2} s^{-1}. This value of light intensity was experimentally established earlier by constructing average light curves. The CO_2 uptake was recalculated to mg of CO_2 per unit leaf area (dm^2), per mg of chlorophyll (Chl $a + b$), and per chloroplast (10^8) per hour; the transpiration rate was recalculated to g of H_2O per unit leaf area (dm^2) per hour. Subsequently, fresh leaf cuttings (0.7 cm diameter) from these leaves were used to measure the water exchange and photosynthetic pigment content.

The relative water content (RWC, %) and water saturation deficit (WSD, %) of the plant tissue were measured by floating disc method and calculated according to Hellmuth [41]. The fresh leaf cutting was immediately weighed to obtain FW and then saturated by submerging the sample in distilled water for 3 h. Afterwards, the surface water was blotted carefully, the discs were weighed to obtain the saturated weight, and later dried 24 h at 75 °C to determine the dry weight. The fresh weight (FW) to dry weight (DW) ratios were used for further calculations. Simultaneously, the part of the leaf discs was immediately used for photosynthetic pigment determination. Three discs from each plant (about 40–50 mg of FW) were homogenized in 2 mL of a cold 80% acetone solution (v/v) with addition of a small amount of $CaCO_3$ to protect the pigments from oxidation, and centrifuged at $8000 \times g$ for 10 min. The homogenate was decanted, acetone solution was added to the precipitate and stirred again; this procedure was repeated threefold until the precipitate was completely discolored. The content of chlorophyll a (Chl a), chlorophyll b (Chl b), and carotenoids (CAR) was determined spectrophotometrically ("APEL" PD-303 UV) at wavelengths of 470, 647, and 663 nm, respectively, and calculated according to Lichtenthaler [42], and expressed as mg g^{-1} DW. The physiological and biochemical parameters were determined in four biological and three analytical replicates.

2.5. Statistical Analysis

The values are presented as mean values of 5 replicates for the physicochemical analysis of the soil and the elemental analysis of plant samples, 30 replicates for structural characteristics, and 12 replicates for physiological and biochemical plant parameters with standard error (SE). After checking the normality by Shapiro–Wilk test and the homogeneity of variance by Levene's test, the differences between the studied orchid populations were determined with the nonparametric Mann–Whitney U-test, $p < 0.05$. The relationship between different parameters was determined by Spearman's rank correlation coefficient. Asterisks in the tables and figures indicate significant differences between the studied populations.

3. Results

3.1. Brief Description of Studied Populations

In NFC, *N. ovata* plants were found both as single individuals and in groups of up to three individuals. The total number of *N. ovata* plants in P-1 was 39, with a density of 0.10 individuals per 1 m^2; the age spectrum was dominated by flowering plants (71%).

The distribution of *N. ovata* plants in P-2 was uneven. The orchid population from FAD was young and vegetatively oriented, with pregenerative individuals predominating (78%). The total number of *N. ovata* plants in P-2 was 194 individuals and the density was 0.49 individuals per 1 m^2.

Orchid mycorrhiza, represented by pelotons, were found in the root cells of *N. ovata* from both studied populations. The intensity of mycorrhizal association was high and ranged from 96 to 98%.

3.2. Soil and Plant Composition

3.2.1. Physicochemical Characteristics and Metal Content in Soil Substrates

The pH of the soil substrates varied between acidic and slightly acidic for both sites. At the same time, the pH of the fly ash substrate was slightly higher than that of the natural forest soil (Table 1). The EC and TDS values in the soil substrate from NFC were higher (1.7-fold on average) than in FAD.

Table 1. The pH (water solution), electrical conductivity and total dissolved solids in soil substrates (0–15 cm depth) from the natural forest community (NFC) and the fly ash dump (FAD).

Site	pH (H$_2$O)	Electrical Conductivity (EC), μS cm^{-1}	Total Dissolved Solids (TDS), mg L^{-1}
NFC	5.80 ± 0.04 [1] (5.62–6.34) [2]	162.8 ± 11.3 (108.0–280.0)	89.9 ± 6.1 (51.0–145.0)
FAD	6.19 ± 0.03 * (5.79–6.32)	96.6 ± 7.2 * (46.0–146.0)	48.0 ± 3.7 * (23.0–73.0)

[1] Data is presented as mean ± SE (n = 5); [2] In the brackets are the minimum and maximum values. Asterisks (*) indicate significant differences between the studied habitats according to Mann–Whitney U-test ($p < 0.05$).

The total metal contents in the NFC soil were found in the following order: Fe > Ca > Mg > Mn > K > Zn > Pb > Cu > Cr > Ni > Co (Table 2). A similar trend was noted in the metal content distribution in the FAD substrate, with the exception of Mn and K that switched places. The largest difference between the two sites was found for Mn; its total and available content in the NFC soil exceeded its concentration in the FAD substrate by 3.3 times on average. The total contents of Ca, Mg, Zn, and Cu were also higher in the soil of the natural habitat than in the disturbed one, but the differences between the sites were less noticeable (1.3–2.0 times). Whereas, for K, Pb, and Cr, their content (both total and available) was higher in the FAD substrate (on average by 1.4 times). At the same time, there were no significant differences between the total Fe, Ni, and Co contents in both sites. The available concentration of most of the studied metals (namely, Ca, Mg, Fe, Zn, Mn, Cu, and Ni) in the NFC soil was higher compared to the FAD substrate (on average 2.0 times, Table 2). As for Co, there were no significant differences in the content between the sites studied.

Table 2. Total and available metal content in the soil substrates (0–15 cm depth) from the natural forest community (NFC) and the fly ash dump (FAD).

Metal	Total Content, mg kg^{-1} DW		Available Content, mg kg^{-1} DW	
	NFC	FAD	NFC	FAD
K	948.8 ± 87.1 [1] (735.8–1162.1) [2]	1256.2 ± 75.8 * (1080.8–1450.6)	211.6 ± 5.6 (198.0–225.4)	310.9 ± 3.8 * (290.6–326.3)
Ca	16,779.6 ± 856.1 (14,690.2–18,876.0)	8369.1 ± 437.2 * (7286.3–10676.0)	6644.8 ± 534.3 (5395.9–7823.9)	4439.9 ± 480.9 * (3098.7–6351.4)
Mg	5321.0 ± 329.9 (4515.1–6127.2)	2708.0 ± 602.6 * (999.8–4368.0)	1264.1 ± 47.2 (1149.0–1379.2)	792.4 ± 139.6 * (411.7–1243.8)
Fe	36,273.4 ± 1472.1 (32,670.0–39877.8)	33,087.7 ± 3008.1 (24,256.6–41777.9)	4810.1 ± 214.8 (4288.5–5332.0)	2592.9 ± 358.0 * (1601.5–3705.0)
Zn	269.9 ± 25.4 (208.8–332.6)	192.1 ± 10.2 * (159.0–224.1)	158.4 ± 14.4 (123.3–193.3)	92.7 ± 4.7 * (77.8–111.6)
Mn	2043.4 ± 153.4 (1694.0–2393.2)	572.7 ± 114.6 * (270.8–957.0)	1388.2 ± 123.8 (1086.4–1690.2)	439.9 ± 93.0 * (166.6–712.5)
Pb	152.6 ± 7.8 (133.8–171.3)	180.0 ± 5.1 * (158.8–196.7)	69.8 ± 7.2 (58.2–81.3)	86.2 ± 3.0 * (74.3–98.2)
Cu	110.3 ± 5.4 (98.1–122.4)	69.2 ± 8.4 * (42.1–98.2)	83.1 ± 5.4 (70.2–96.5)	39.5 ± 3.7 * (26.8–51.9)
Ni	26.2 ± 4.1 (16.8–35.5)	22.9 ± 4.9 (9.1–37.5)	8.5 ± 0.6 (6.9–10.0)	4.5 ± 0.6 * (2.8–7.8)
Cr	41.7 ± 1.9 (37.6–45.9)	63.3 ± 3.1 * (49.5–72.8)	4.8 ± 0.4 (4.1–5.6)	8.1 ± 0.5 * (5.9–10.1)
Co	5.4 ± 0.5 (3.9–7.2)	7.5 ± 0.9 (4.3–11.5)	3.4 ± 0.4 (2.3–4.3)	4.5 ± 0.4 (3.1–6.1)

[1] Data is presented as mean ± SE (n = 5); [2] In the brackets are the minimum and maximum values. Asterisks (*) indicate significant differences between studied habitats according to Mann–Whitney U-test ($p < 0.05$).

3.2.2. Macronutrient and Metal Content in *N. ovata*

The leaves of *N. ovata* from P-2 contained a smaller amount of total nitrogen, while the content of total phosphorus was higher than in the P-1 plants (by 1.2 times, Table 3).

Table 3. Macronutrient content in the aboveground and underground organs of *N. ovata* from the natural forest community (P-1) and the fly ash dump (P-2).

Macronutrient	Leaves, mg g^{-1} DW		Rhizome + Roots, mg g^{-1} DW	
	P-1	P-2	P-1	P-2
N	42.84 ± 1.30 [1]	23.08 ± 2.32 *	29.05 ± 3.92	28.15 ± 3.13
P	4.48 ± 0.04	5.48 ± 0.15 *	3.96 ± 0.03	2.92 ± 0.25 *
K	33.00 ± 0.61	46.97 ± 0.73 *	9.13 ± 0.36	12.70 ± 1.37 *
Ca	19.17 ± 0.28	18.13 ± 0.98	16.41 ± 0.62	15.57 ± 0.85
Mg	2.22 ± 0.08	1.54 ± 0.08 *	1.58 ± 0.09	1.32 ± 0.05

[1] Data is presented as mean ± SE (n = 5). Asterisks (*) indicate significant differences between the studied populations according to Mann–Whitney U-test ($p < 0.05$).

As for content of these nutrients in the *N. ovata* rhizome + roots, there were no significant differences in the nitrogen content between the studied orchid populations, although lower phosphorus content was noted in the plants growing on the fly ash substrate.

As expected, among the studied metals the *N. ovata* plants accumulated K, Ca, and Mg in the greatest amounts (Table 3). The potassium content in the both the aboveground and underground organs of *N. ovata* was 1.4 times higher than in the plants from the FAD while the studied populations did not significantly differ in terms of Ca and Mg content.

As for the other metals, they accumulated to the greatest extent in the *N. ovata* underground organs (Figure 3). The content of toxic elements such as Pb and Cr was higher in the rhizome + roots of the plants colonizing FAD (by 43 and 26%, respectively). In contrast, the content of essential Cu and Fe was lower (1.3 and 5.8 times, respectively). A similar tendency was noted for Mn. The differences between the studied populations in terms of metal content in their leaves were less noticeable, with the exceptions of Pb and Fe (Figure 3).

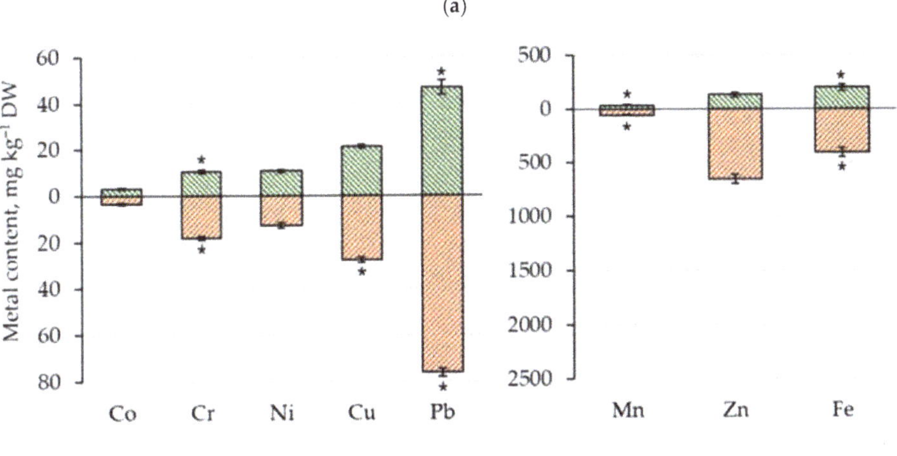

Figure 3. Heavy metal content in the aboveground and underground organs of *N. ovata* from: (**a**) the natural forest community (P-1) and (**b**) the fly ash dump (P-2). Data is presented as mean ± SE ($n = 5$). Asterisks (*) indicate significant differences between the studied populations according to Mann–Whitney U-test ($p < 0.05$).

The concentration of some trace metals (Zn, Cu, Cr and Co) in the underground organs correlated with their total content in the soil (on average $r = 0.64$; Supplementary Table S1). Moreover, for copper and chromium a significant correlation was also noted with regard to the available concentration in the soil (on average $r = 0.66$, Supplementary Table S2).

The BCF for macronutrients was found in the following order at both sites: K > Ca > Mg (Table 4). Potassium was released: its average BCF in the rhizome + roots of the orchid was 42, while in the leaves was 153. The plants from FAD had increased BCF values for zinc, which were several times higher than 1 and significantly higher than in NFC. The BCF for Ni and Cr were also greater than 1, while for other studied metals were ≤1.

Table 4. Bioconcentration factors (BCF) and translocation factor (TF) from underground to aboveground organs of *N. ovata* from natural forest community (P-1) and fly ash dump (P-2).

Metal	BCF(Aboveground)		BCF(Underground)		TF(Aboveground/Underground)	
	P-1	**P-2**	**P-1**	**P-2**	**P-1**	**P-2**
K	155.94	151.09	43.13	40.86	3.62	3.70
Ca	2.89	4.08	2.47	3.51	1.17	1.16
Mg	1.75	1.95	1.25	1.67	1.41	1.17
Fe	0.05	0.08	0.49	0.16	0.11	0.48
Zn	0.85	1.39	4.50	7.11	0.19	0.20
Mn	0.04	0.08	0.15	0.13	0.31	0.62
Pb	0.49	0.54	0.71	0.82	0.70	0.66
Cu	0.23	0.54	0.44	0.70	0.53	0.77
Ni	1.17	2.40	1.62	2.87	0.72	0.84
Cr	1.34	1.30	2.99	2.24	0.45	0.58
Co	0.84	0.82	1.04	0.94	0.81	0.88

3.3. Morphological and Anatomical Characteristics of N. ovata

The orchid plants growing on the fly ash substrate (P-2) had a lower shoot and inflorescence length, and number of flowers (by 30, 27, and 20%, respectively) compared to P-1, but at the same time they had a 1.4-fold larger leaf area (Table 5).

Table 5. Morphological characteristics of the flowering individuals of the *N. ovata* populations from the natural forest community (P-1) and the fly ash dump (P-2).

Parameters	Populations	
	P-1	**P-2**
Shoot length, cm	60.2 ± 7.2 [1] (46.0–69.5) [2]	45.0 ± 2.1 * (23.0–69.0)
Inflorescence length, cm	18.8 ± 4.9 (13.0–28.5)	14.8 ± 1.1 (7.2–30.0)
Number of flowers, pcs.	29.0 ± 8.3 (17.0–45.0)	23.7 ± 2.1 (7.0–46.0)
Upper leaf area, cm^2	38.9 ± 8.7 (12.9–60.6)	48.1 ± 5.8 (22.8–87.5)
Lower leaf area, cm^2	33.7 ± 11.9 (15.1–63.6)	53.6 ± 7.2 (24.1–101.4)

[1] Data is presented as mean ± SE ($n = 20$); [2] In the brackets are the minimum and maximum values. Asterisks (*) indicate significant differences between the studied populations according to Mann–Whitney U-test ($p < 0.05$).

The *N. ovata* leaves have a homogeneous mesophyll structure. The study showed that the orchids colonizing the fly ash substrate were distinguished by a thicker epidermis (by 14%) and lower mesophyll thickness (by 6%), compared to individuals from the natural habitat (Figure 4a,b).

There were no significant differences between the studied populations in terms of the number of mesophyll cells (Figure 4c), and their surface area and volume (Supplementary Table S5), whereas an increased number of chloroplasts both per cell (by 18%) and per unit cell area (by 13%), was noted for P-2 plants, compared to their P-1 counterparts (Figure 4d; Supplementary Table S5). The cell volume per chloroplast in the leaves of *N. ovata* from FAD was lower (Figure 4e) than in plants from NFC, while the reverse trend was observed for the chloroplast membrane index (Figure 4f).

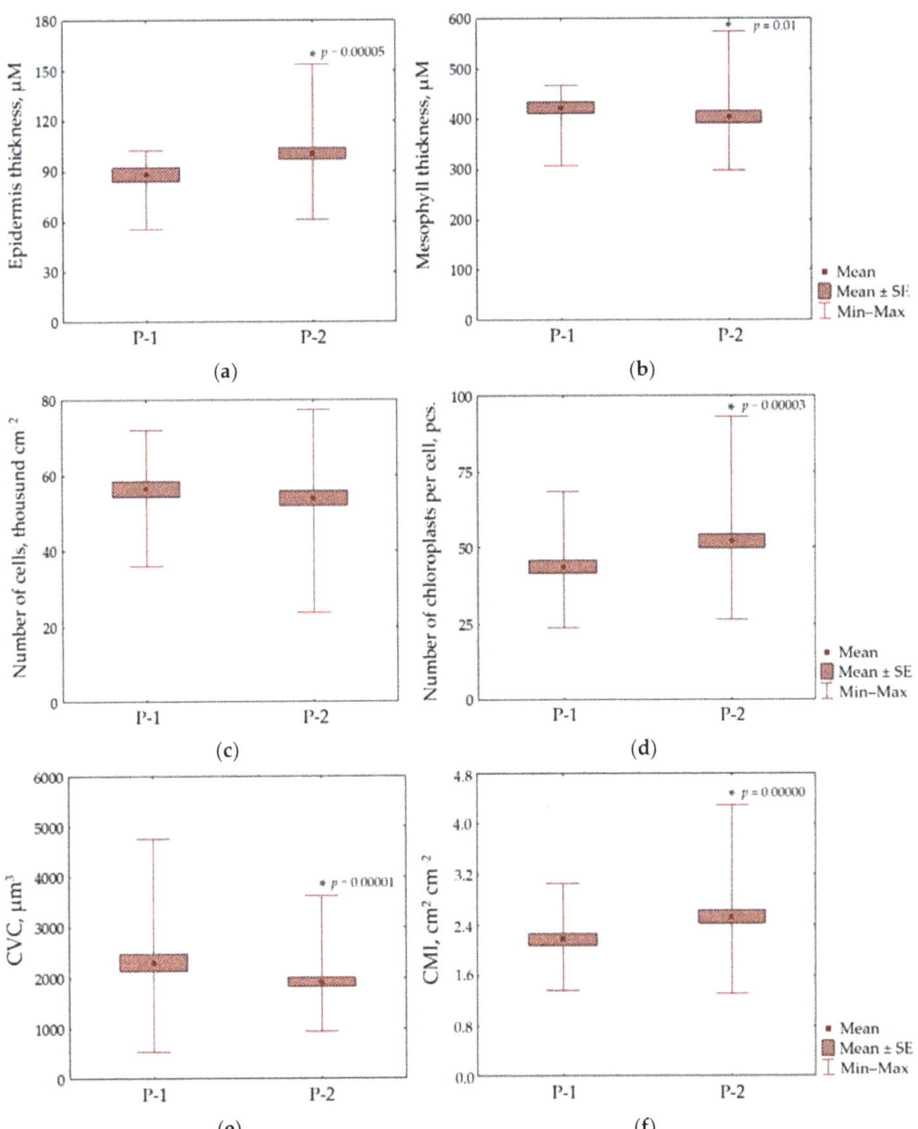

Figure 4. The mesostructure parameters of the lower leaf of *N. ovata* from the natural forest community (P-1) and the fly ash dump (P-2): (**a**) epidermis thickness; (**b**) mesophyll thickness; (**c**) number of cells; (**d**) number of chloroplasts per cell; (**e**) cell volume per chloroplast (CVC); (**f**) chloroplast membrane index (CMI). The small solid square indicates mean values ($n = 30$); boxes present mean $\pm$ SE; the whiskers are the minimum and maximum values. Asterisks (*) indicate significant differences between the studied populations according to Mann–Whitney U-test ($p < 0.05$).

3.4. Physiological and Biochemical Parameters of N. ovata

As shown in Figure 5a, the leaves of *N. ovata* from the natural habitat contained 2.5 times higher Chl *a* than Chl *b*. The Chl *a* content in the P-2 plants was 1.6 times lower than in the P-1 plants. In contrast, the Chl *b* and carotenoid content in the orchids on the fly ash substrates was higher (14% and 33%, respectively). A 1.9-fold decrease in Chl *a/b* and

Chl $(a + b)$/CAR ratios was observed in plants from FAD while the (CAR+ Chl b)/Chl a ratio increased almost 2 times compared to plants from the NPC site.

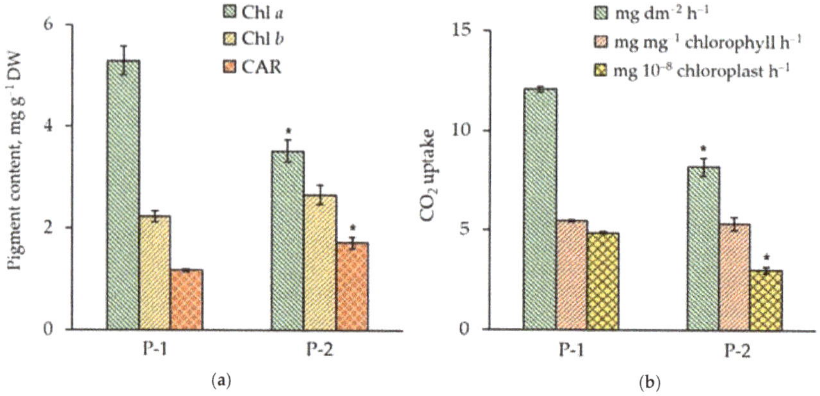

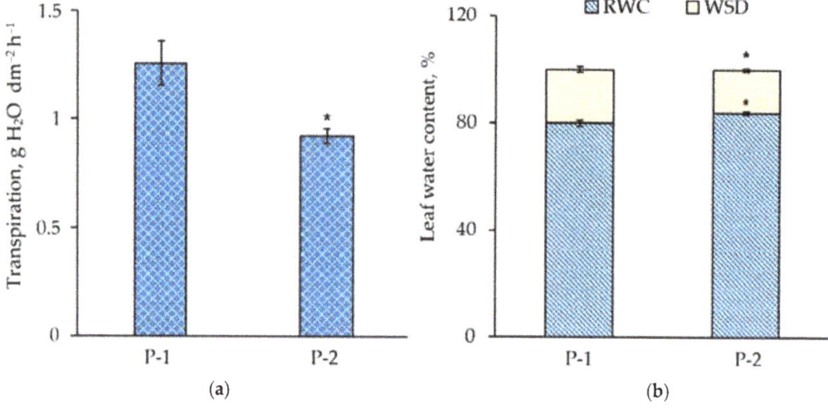

Figure 5. The photosynthetic parameters of the leaves of *N. ovata* from the natural forest community (P-1) and the fly ash dump (P-2): (**a**) photosynthetic pigment content; (**b**) intensity of CO_2 assimilation. Data is presented as mean $\pm$ SE ($n = 12$). Asterisks (*) indicate significant differences between the studied populations according to Mann–Whitney U-test ($p < 0.05$).

In addition, a high positive correlation between Chl a and available content of Cu, Ni, Fe, Mn, Mg, and Ca in soil (on average, $r = 0.86$, Supplementary Table S3) and negative correlation between Chl a and total Pb and Cr content in the leaves (on average, $r = -0.86$; Supplementary Table S4) were found for *N. ovata* plants.

The data on CO_2 assimilation intensity (Figure 5b) showed that its uptake per unit area and per chloroplast per hour in P-2 plants was 1.7 times lower than in P-1 plants. At the same time, there were no significant differences between the populations in terms of the CO_2 uptake per mg of chlorophyll per hour (Figure 5b).

The *N. ovata* plants on the fly ash substrate had a lower intensity of transpiration compared to the plants from the natural habitat (by 1.4 times, Figure 6a). The relative water content and water saturation deficit indexes entered the range of values of most plants and did not differ much between sites (Figure 6b). However, P-1 plants experienced a greater lack of moisture than P-2.

Figure 6. The water exchange parameters in the leaves of *N. ovata* from the natural forest community (P-1) and the fly ash dump (P-2): (**a**) transpiration intensity; (**b**) relative water content (RWC) and water saturation deficit (WSD). Data is presented as mean $\pm$ SE ($n = 12$). Asterisks (*) indicate significant differences between the studied populations according to Mann–Whitney U-test ($p < 0.05$).

4. Discussion

The present study aims to identify the adaptive responses of the rare orchid *N. ovata* that contribute to natural colonization under the adverse conditions of a fly ash deposits. No similar investigations have been carried out in the Middle Urals. A comparative analysis of the structural and functional characteristics of *N. ovata* plants in disturbed (fly ash dump of VTTPS) and natural forest ecosystems are vital for achieving this goal.

It is well known that fly ash substrates are characterized by unfavorable physico-chemical properties, which depend on the type and origin of the coal, the conditions of combustion, the type of emission control devices, and the storage and handling methods [29–31,43]. The pH values of fly ash can vary from 4.5 to 12, depending on the coal type [29]. As a rule, fly ash formed during the combustion of brown coal is alkaline [29]. The lowered pH value of the FAD substrate is obviously explained by the fact that soil formation in this area had proceeded according to the zonal type (under the conditions of a flushing water regime) [38]. According to Gajic' et al. [29], unweathered fly ash had high values of electrical conductivity (150–352 μS cm^{-1}), which indicate a large amount of soluble salts, while EC values usually decrease (101–217 μS cm^{-1}) in weathered fly ash. In general, plants growing during the weathering of fly ash improve the physicochemical properties of the fly ash substrate. It was found that EC of the FAD substrate was reliably lower compared to the NFC substrate. This is associated with significantly lower available concentrations of most of the studied metals in the FAD substrate, which is also confirmed by the low TDS values.

As noted, toxic concentrations of Cr, Pb, Cd, As and other metal(loid)s are often a limiting factor that reduce the rate of natural colonization of fly ash dumps [30]. The studied metal content in the soil of both sites did not exceed the maximum permissible concentrations [44]. All the studied metals, with the exception of macronutrients (Ca, Mg, and K), accumulated to the greatest degree in the roots of *N. ovata*. The increased level of trace metal accumulation in *N. ovata* roots indicates the functioning of barrier mechanisms and contributes to the implementation of an ontogenetic program [34]. Nevertheless, increased Co, Cr, Ni, and Pb content in the leaves was noted (above the normal level), which corresponds to an excessive or toxic concentration [44].

The plants growing on FAD showed the greatest difference in terms of potassium in both the aboveground and underground organs. This is due to the higher concentration of potassium in the fly ash substrate. Potassium plays a vital role in such important processes as photosynthesis, growth, assimilate transport, water exchange, etc. [45]. Thus, the high ability of *N. ovata* to absorb potassium and translocate it to leaves is one of its adaptive responses.

Fly ash substrates are known to be very poor in nitrogen content [29,30,38]. Nevertheless, its content in the orchid plants from FAD was within normal limits (at the level of average values) [45]. These results confirm the existing view that orchids can effectively assimilate nitrogen even from soils poor in this element [46]. In contrast, the total phosphorus content was higher in the leaves of *N. ovata* from FAD compared with those from NFC. Since phosphorus is involved in many biochemical, energy, and physiological processes [45], its accumulation in leaves can obviously be regarded as an adaptive response.

The *N. ovata* individuals from FAD were characterized by lower values of shoot height, inflorescence length, and the number of generative organs. These were compensated by an increase in the area of the assimilative organs.

Most orchids, especially species with thin leaves, assimilate carbon dioxide through a C_3-pathway [16]. The intensity of photosynthesis depends on the activity of photosynthetic enzymes and pigment concentration [32]. On the other hand, this is largely related to the leaf blade's anatomical and morphological characteristics, which determine the optical properties and diffusion rate of CO_2 to the carboxylation centers [33]. Under the stress factors, changes in the mesostructure of the photosynthetic apparatus can take place as an adaptive reaction [34,35].

The studied *N. ovata* populations revealed a lack of significant differences in leaf blade thickness. At the same time, the properties of the substrate affected the thickness of the mesophyll and epidermis: a thinner mesophyll and a thicker epidermis were characteristic of plants from FAD. This is a protective response which is probably associated with increased atmospheric dust at dump sites [29].

An increased number of chloroplasts per cell in the *N. ovata* growing in a transformed habitat can be regarded as a compensatory adaptive reaction to the lack of chlorophylls, the content of which was significantly lower in plants from the fly ash deposits. The cell volume per chloroplast is a parameter that indicates the size of the cell volume, which is provided by metabolites as well as energy substrates due to the activity of one chloroplast [34,35]. A significant decrease in this indicator in *N. ovata* from FAD, compared with NFC, is explained by an increase in the number of chloroplasts in its mesophyll cells, while the cell volume remained practically unchanged.

The chloroplast membrane index is an integral indicator of the photosynthetic apparatus [34]. Obviously, the more significant CMI observed for P-2 plants is explained by the significant increase in the number of plastids in mesophyll cells. Thus, the greater development of the leaf's internal assimilation surface in orchid plants growing on the fly ash substrate is apparently associated with lower values of tree crown density and the total projective cover of the grass-shrub layer and as a consequence, greater lighting.

It is well reported that the photosynthetic apparatus of plants ensures their vital activity in various environmental conditions [47]. The pigment complex of plants is highly sensitive to the effects of adverse factors and capable of adaptive changes. Therefore, the analysis of the content and ratio of photosynthetic pigments is of great importance in assessing the resistance of plants to various stress factors [48]. Chlorophyll *a* in *N. ovata* leaves proved to be the most sensitive to the adverse conditions in the fly ash substrate, showing significant reductions. This can be explained by a lower nitrogen content since it is one of the most important components of green pigments [49] and a high concentration of some toxic metals (perhaps, Cr and Pb). It is well known that an excess of toxic metals in plant cells can cause structural changes in chloroplasts, inhibit key enzymes in chlorophyll synthesis, and cause the destruction of pigment molecules [24,29,35].

The degree of photosynthetic apparatus activity and its resistance to unfavorable external stressors are often evaluated by the photosynthetic pigment ratio. A comparison of pigment ratios showed the decrease in the Chl *a/b* and Chl $(a + b)$/CAR ratios in plants from FAD. This fact is explained by a significantly lower Chl *a* concentration, while Chl *b* and carotenoid content were increased. The ratio of the sum of the auxiliary pigments (Chl *b* + CAR) to Chl *a*, characterizing the share of antenna forms, significantly increased in *N. ovata* plants growing in FAD. Obviously, the activation of the synthesis of auxiliary pigments is a compensatory reaction that contributes to better absorption of light for photosynthesis. However, carotenoids in chloroplasts perform not only antenna and photoprotective functions, but also an antioxidant one. Carotenoid molecules can interact with reactive oxygen species (ROS) due to double bonds [50]. Thus, an increase in the carotenoid content in plants growing on fly ash substrate is most likely a response to stress and is aimed at combating ROS.

The absence of significant changes in the assimilation index (mg CO_2 mg^{-1} chlorophyll h^{-1}) indicates that the unfavorable conditions of the fly ash substrate did not cause significant damage to the chlorophyll molecules, since its functional ability was preserved [49].

The absorption of CO_2 and the transpiration of water by plant leaves occurs through the stomata. Accordingly, the rate of CO_2 assimilation and transpiration per unit area showed similar changes. The level of moisture deficiency in plants can be associated with the intensity of water evaporation. The results obtained are consistent with the data of measuring transpiration for each plot. Therefore, transpiration was more intense in orchids from NFC, than from FAD, which caused a higher water deficit. In addition, a greater moisture deficit in plants from a natural habitat can be explained by the high value of a projective vegetation cover. Consequently, there was probably competition between the plants for resources, including water.

5. Conclusions

The present study revealed those adaptive structural and functional features of *Neottia ovata* that contribute to its survival strategies in a transformed habitat. Despite the adverse edaphic conditions of the fly ash dump, the population size of this species was noticeably higher than that in natural forest community. This is due to less phytocoenotic stress on the fly ash deposits while the population of *N. ovata* in its natural habitat experiences higher level of competition. Moreover, the natural colonization of *N. ovata* on the fly ash substrate was facilitated by adaptive changes in the mesostructure parameters of the leaves, such as an increase in epidermis thickness, the number of chloroplasts in the cell, and the internal assimilating surface. The *N. ovata* plants colonizing the fly ash dump were characterized by the higher chlorophyll *b* and carotenoids content compared to plants growing in the natural forest community that evidenced the compensatory response on the decrease in chlorophyll *a*. Furthermore, *N. ovata* growing on the fly ash substrate retained a relatively favorable water balance, which also contributed to its high resistance to adverse environmental conditions.

The content of most of the studied metals in the underground parts (rhizome + roots) of *N. ovata* was considerably higher than in the leaves, which diminished the harmful effect of toxic metals on the aboveground organs, including the generative ones. Despite the lower content of most of the macro- and micronutrients in the fly ash substrate and the higher concentration of some toxic heavy metals (lead and chromium), the plants from the transformed ecosystem showed high viability.

The study of the morphophysiological features of orchids in technologically disturbed habitats is necessary for developing measures to protect the gene pool of rare plant species and to solve applied problems associated with identifying optimal conditions for naturalization and introduction into a new environment.

Supplementary Materials: The following are available online at https://www.mdpi.com/article/10.3390/horticulturae7050109/s1, Table S1: Spearman's correlation between total metal content in the soil with metals in leaves and rhizome + roots of *N. ovata* from the natural forest community and the fly ash dump; Table S2: Spearman's correlation between available metal content in the soil with metals in leaves and rhizome + roots of *N. ovata* from the natural forest community and the fly ash dump; Table S3: Spearman's correlation between total and available metal content in the soil with physiological parameters of *N. ovata* from the natural forest community and the fly ash dump; Table S4: Spearman's correlation between metal content in the leaves and rhizome + roots with physiological parameters of *N. ovata* from the natural forest community and the fly ash dump; Table S5: The mesostructure parameters of the lower leaf of *N. ovata* from the natural forest community (P-1) and the fly ash dump (P-2).

Author Contributions: Conceptualization, M.M., G.B. and E.F.; methodology, M.M., G.B., N.C., O.S., E.F., N.L. and M.G.; software, M.M. and O.S.; validation, M.M., G.B. and E.F.; formal analysis, M.M., G.B., E.F., N.L. and M.G.; investigation, M.M., G.B., N.C., O.S., E.F., N.L. and M.G.; resources, M.M., E.F. and O.S.; data curation, M.M., G.B. and E.F.; writing—original draft preparation, M.M., G.B. and N.C.; writing—review and editing, M.M. and G.B.; visualization, M.M. and G.B.; supervision, M.M. and G.B.; project administration, M.M., G.B. and E.F.; funding acquisition, M.M., G.B., N.C., O.S., E.F., N.L. and M.G. All authors have read and agreed to the published version of the manuscript.

Funding: The reported study was partly funded by RFBR and the Government of Sverdlovsk region, project number 20-44-660011 and the Ministry of Science and Higher Education of the Russian Federation as part of state task of the Ural Federal University, FEUZ-2020-0057.

Informed Consent Statement: Informed consent was obtained from all subjects involved in the study.

Acknowledgments: The authors are very grateful to Tripti and Adarsh Kumar (UrFU, Ekaterinburg, Russia) for their constructive comments, and also Regional Center for Linguistic Support of Scientific and Publication Activity of Academic and Administrative Staff of UrFU. Two anonymous reviewers gave valuable comments for the improvement of this paper and we also gratefully acknowledged them.

Conflicts of Interest: The authors declare no conflict of interest.

References

1. Chase, M.W.; Cameron, K.M.; Freudenstein, J.V.; Pridgeon, A.M.; Salazar, G.A.; van den Berg, C.; Schuiteman, A. An updated classification of Orchidaceae. *Bot. J. Linn. Soc.* **2015**, *177*, 151–174. [CrossRef]
2. Christenhusz, M.J.M.; Byng, J.W. The number of known plants species in the world and its annual increase. *Phytotaxa* **2016**, *261*, 201–217. [CrossRef]
3. Vakhrameeva, M.G.; Tatarenko, I.V. Ecological characteristics of orchids of European Part of Russia. *Acta Univ. Wratislav.* **2001**, *79*, 49–54.
4. Vakhrameeva, M.G.; Tatarenko, I.V.; Varlygina, T.I.; Torosyan, G.K.; Zagulskii, M.N. *Orchids of Russia and Adjacent Countries (within the Borders of the Former USSR)*; A.R.G. Gantner Verlag: Ruggell, Liechtenstein, 2008.
5. Kindlmann, P.; Jersakova, J. Floral display, reproductive success, and conservation of terrestrial orchids. *Selbyana* **2005**, *26*, 136–144. [CrossRef]
6. Sasamori, M.H.; Endres, D.J.; Droste, A. Asymbiotic culture of *Cattleya intermedia* Graham (Orchidaceae): The influence of macronutrient salts and sucrose concentrations on survival and development of plantlets. *Acta Bot. Bras.* **2015**, *29*, 292–298. [CrossRef]
7. Einzmann, H.J.R.; Schickenberg, N.; Zotz, G. Variation in root morphology of epiphytic orchids along small–scale and large–scale moisture gradients. *Acta Bot. Bras.* **2020**, *34*, 66–73. [CrossRef]
8. Brodmann, J.; Twele, R.; Francke, W.; Hölzler, G.; Zhang, Q.H.; Ayasse, M. Orchids mimic green–leaf volatiles to attract prey–hunting wasps for pollination. *Curr. Biol.* **2008**, *18*, 740–744. [CrossRef] [PubMed]
9. Brys, R.; Jacquemyn, H.; Hermy, M. Pollination efficiency and reproductive patterns in relation to local plant density, population size, and floral display in the rewarding *Listera ovata* (Orchidaceae). *Bot. J. Linn. Soc.* **2008**, *157*, 713–721. [CrossRef]
10. Jakubska–Busse, A.; Kadej, M. The pollination of Epipactis Zinn, 1757 (Orchidaceae) species in central Europe—the significance of chemical attractants, floral morphology and concomitant insects. *Acta Soc. Bot. Pol.* **2011**, *80*, 49–57. [CrossRef]
11. Ehlers, B.K.; Pedersen, H. Genetic variation in three species of Epipactis (Orchidaceae): Geographic scale and evolutionary inferences. *Biol. J. Linn. Soc.* **2000**, *69*, 411–430. [CrossRef]
12. Hens, H.; Jäkäläniemi, A.; Tali, K.; Efimov, P.; Kravchenko, A.V.; Kvist, L. Genetic structure of a regionally endangered orchid, the dark red helleborine (*Epipactis atrorubens*) at the edge of its distribution. *Genetica* **2017**, *145*, 209–221. [CrossRef]
13. Jurkiewicz, A.; Turnau, K.; Mesjasz–Przybylowicz, J.; Przybylowicz, W.; Godzik, B. Heavy metal localisation in mycorrhizas of *Epipactis atrorubens* (Hoffm.) Besser (Orchidaceae) from zink mine tailings. *Protoplasma* **2001**, *218*, 117–124. [CrossRef]
14. Shefferson, R.; Kull, T.; Tali, K. Mycorrhizal interactions of orchids colonizing Estonian mine tailings hills. *Am. J. Bot.* **2008**, *95*, 156–164. [CrossRef]
15. Tešitelová, T.; Tešitel, J.; Jersáková, J.; Říhová, G.; Selosse, M.S. Symbiotic germination capability of four Epipactis species (Orchidaceae) is broader than expected from adult ecology. *Am. J. Bot.* **2012**, *99*, 1020–1032. [CrossRef]
16. Zhang, S.; Yang, Y.; Li, J.; Qin, J.; Zhang, W.; Hu, H. Physiological diversity of orchids. *Plant Divers.* **2018**, *40*, 196–208. [CrossRef]
17. Swarts, D.N.; Dixon, W.D. Terrestrial orchid conservation in the age of extinction. *Ann. Bot.* **2009**, *104*, 543–556. [CrossRef] [PubMed]
18. Adamowski, W. Expansion of native Orchids in anthropogenous habitats. *Pol. Bot. Stud.* **2006**, *22*, 35–44.
19. Jermakowicz, E.; Brzosko, E. Demographic responses of boreal–montane orchid *Malaxis monophyllos* (L.) Sw. populations to contrasting environmental conditions. *Acta Soc. Bot. Pol.* **2016**, *85*, 1–17. [CrossRef]
20. Rewicz, A.; Bomanowska, A.; Shevera, M.; Kurowski, J.; Krason, K.; Zielinska, K. Cities and disturbed areas as man–made shelters for orchid communities. *Not. Bot. Horti Agrobot. Cluj Napoca* **2017**, *45*, 126–139. [CrossRef]
21. Agostini, A.; Caltagirone, C.; Caredda, A.; Cicatelli, A.; Cogoni, A.; Farci, D.; Guarino, F.; Garau, A.; Labra, M.; Lussu, M.; et al. Heavy metal tolerance of orchid populations growing on abandoned mine tailings: A case study in Sardinia Island (Italy). *Ecotoxicol. Environ. Saf.* **2020**, *189*, 110018. [CrossRef] [PubMed]
22. Filimonova, E.I.; Lukina, N.V.; Glazyrina, M.A. Orchidaceae in the technogenic ecosystems of the Urals. *Ecosystems* **2014**, *11*, 68–75. (In Russian)
23. Filimonova, E.I.; Lukina, N.V.; Glazyrina, M.A.; Borisova, G.G.; Maleva, M.G.; Chukina, N.V. Endangered orchid plant *Epipactis atrorubens* on serpentine and granite outcrops of Middle Urals, Russia: A comparative morphophysiological study. *AIP Conf. Proc.* **2019**, 040016. [CrossRef]
24. Filimonova, E.; Lukina, N.; Glazyrina, M.; Borisova, G.; Tripti; Kumar, A.; Maleva, M. A comparative study of *Epipactis atrorubens* in two different forest communities of the Middle Urals, Russia. *J. For. Res.* **2020**, *31*, 2111–2120. [CrossRef]
25. Kotilínek, M.; Těšitelová, T.; Jersáková, J. Biological Flora of the British Isles: *Neottia ovata. J. Ecol.* **2015**, *103*, 1354–1366. [CrossRef]

26. Vladimir Nikolaevich, B.; Nikolai Sergeevich, K. *Red Book of the Sverdlovsk Region: Animals, Plants, Mushrooms*; Mir LLC: Yekaterinburg, Russia, 2018. (In Russian)

27. Wiegand, K.M.A. Revision of the Genus Listera. *J. Torrey Bot. Soc.* **1899**, *26*, 157–171. [CrossRef]

28. Brzosko, E. The dynamics of Listera ovata populations on mineral islands in the Biebrza National Park. *Acta Soc. Bot. Pol.* **2002**, *71*, 243–251. [CrossRef]

29. Gajic´, G.; Djurdjevic´, L.; Kostic, O.; Jaric´, S.; Mitrovic´, M.; Pavlovic, P. Ecological potential of plants for phytoremediation and ecorestoration of fly ash deposits and mine wastes. *Front. Environ. Sci.* **2018**, *6*, 124. [CrossRef]

30. Pandey, V.C.; Prakash, P.; Bajpai, O.; Kumar, A.; Singh, N. Phytodiversity on fly ash deposits: Evaluation of naturally colonized species for sustainable phytorestoration. *Environ. Sci. Pollut. Res.* **2015**, *22*, 2776–2787. [CrossRef]

31. Filimonova, E.I.; Lukina, N.V.; Glazyrina, M.A.; Veselkin, D.V.; Chibrik, T.S.; Stephanovich, G.S. Morphological features of juvenile *Pinus sylvestris* L. on the ash dumps in the Middle Urals. *AIP Conf. Proc.* **2019**, 030008. [CrossRef]

32. Mokronosov, A.T. Mesotructure and functional activity of the photosynthetic apparatus. In *Mesotructure and Functional Activity of the Photosynthetic Apparatus*; Mokronosov, A.T., Ed.; Ural. Gos. Univ.: Sverdlovsk, Russia, 1978; pp. 5–15. (In Russian)

33. Mokronosov, A.T.; Gavrilenko, V.F.; Zhigalova, T.V. *Photosynthesis. Physiological and Environmental and Biochemical Aspects*; Academia: Moscow, Russia, 2006. (In Russian)

34. Ivanova, L.A.; P'yankov, V.I. Structural adaptation of the leaf mesophyll to shading. *Russ. J. Plant Physiol.* **2002**, *49*, 419–431. [CrossRef]

35. Ivanova, L.A. Adaptive features of leaf structure in plants of different ecological groups. *Russ. J. Ecol.* **2014**, *45*, 107–115. [CrossRef]

36. Terashima, I.; Miyazawa, S.; Hanba, Y.T. Why are sun leaves thicker than shade leaves?—Consideration based on analyses of CO_2 diffusion in the leaf. *J. Plant Res.* **2001**, *114*, 93–105. [CrossRef]

37. Slaton, M.R.; Smith, W.K. Mesophyll architecture and cell exposure to intercellular air space in alpine, desert, and forest species. *Int. J. Plant Sci.* **2002**, *163*, 937–948. [CrossRef]

38. Nekrasova, O.; Radchenko, T.; Filimonova, E.; Lukina, N.; Glazyrina, M.; Dergacheva, M.; Uchaev, A.; Betekhtina, A. Natural forest colonisation and soil formation on ash dump in southern taiga. *Folia For. Pol. Ser. A For.* **2020**, *62*, 306–316. [CrossRef]

39. Polley, J.R. Colorimetric determination of nitrogen in biological materials. *Anal. Chem.* **1954**, *26*, 1523–1524. [CrossRef]

40. Fiske, C.H.; Subbarow, Y. The colorimetric determination of phosphorus. *J. Biol. Chem.* **1925**, *66*, 375–400. [CrossRef]

41. Hellmuth, E.O. Measurement of Leaf Water Deficit with Particular Reference to the Whole Leaf Method. *J. Ecol.* **1970**, *58*, 409–417. [CrossRef]

42. Lichtenthaler, H.K. [34] Chlorophylls and carotenoids: Pigments of photosynthetic biomembranes. *Methods Enzymol.* **1987**, *148*, 350–382. [CrossRef]

43. Pandey, V.C.; Abhilashb, P.C.; Upadhyaya, R.N.; Tewari, D.D. Application of fly ash on the growth performance and translocation of toxic heavy metals within *Cajanus cajan* L.: Implication for safe utilization of fly ash for agricultural production. *J. Hazard. Mater.* **2009**, *166*, 255–259. [CrossRef]

44. Kabata–Pendias, A.; Mukherjee, A.B. *Trace Elements from Soil to Human*; Springer: Berlin/Heidelberg, Germany, 2007.

45. Marschner, H. *Mineral Nutrition of Higher Plants*, 2nd ed.; Academic Press: San Diego, CA, USA, 1995.

46. Hejcman, M.; Schellberg, J.; Pavlu, V. Dactylorhiza maculata, Platanthera bifolia and Listera ovata survive N application under P limitation. *Acta Oecol.* **2010**, *36*, 684–688. [CrossRef]

47. Van Arendonk, J.J.C.M.; Poorter, H. The chemical composition and anatomical structure of leaves of grass species differing in relative growth rate. *Plant Cell Environ.* **1994**, *17*, 963–970. [CrossRef]

48. Ashraf, M.; Harris, P.J.C. Photosynthesis under stressful environments: An overview. *Photosynthetica* **2013**, *51*, 163–190. [CrossRef]

49. Golovko, T.; Tabalenkova, G. Pigments and productivity of the crop plants. In *Photosynthetic Pigments: Chemical Structure, Biological Function and Ecology*; Golovko, T.K., Gruszecki, W.I., Prasad, M.N.V., Strzalka, K.J., Eds.; Komi Scientific Centre of the Ural Branch of the Russian Academy of Sciences: Syktyvkar, Russia, 2014; pp. 207–220.

50. Gruszecki, W.; Szymanska, R.; Fiedor, L. Carotenoids as photoprotectors. In *Photosynthetic Pigments: Chemical Structure, Biological Function and Ecology*; Golovko, T.K., Gruszecki, W.I., Prasad, M.N.V., Strzalka, K.J., Eds.; Komi Scientific Centre of the Ural Branch of the Russian Academy of Sciences: Syktyvkar, Russia, 2014; pp. 161–170.

horticulturae

Article

Influence of Phosphite Supply in the MS Medium on Root Morphological Characteristics, Fresh Biomass and Enzymatic Behavior in Five Genotypes of Potato (*Solanum tuberosum* L.)

Richard Dormatey [1,2,†], Chao Sun [1,2,†], Kazim Ali [1,3], Tianyuan Qin [1,2], Derong Xu [1,2], Zhenzhen Bi [1,2] and Jiangping Bai [1,2,*]

[1] Department of Crop Genetics and Breeding, College of Agronomy, Gansu Agricultural University, Lanzhou 730070, China; rmddormatey@gmail.com (R.D.); sunc@gsau.edu.cn (C.S.); kazim76@gmail.com (K.A.); qty1637835362@sina.com (T.Q.); xudr1@sina.com (D.X.); bizz@gsau.edu.cn (Z.B.)

[2] Gansu Provincial Key Laboratory of Aridland Crop Science, Lanzhou 730070, China

[3] National Institute for Genomics and Advanced Biotechnology, National Agricultural Research Centre, Park Road, Islamabad 45500, Pakistan

* Correspondence: baijp@gsau.edu.cn

† Richard Dormatey and Chao Sun contributed equally to this work.

Citation: Dormatey, R.; Sun, C.; Ali, K.; Qin, T.; Xu, D.; Bi, Z.; Bai, J. Influence of Phosphite Supply in the MS Medium on Root Morphological Characteristics, Fresh Biomass and Enzymatic Behavior in Five Genotypes of Potato (*Solanum tuberosum* L.). *Horticulturae* 2021, 7, 265. https://doi.org/10.3390/horticulturae7090265

Academic Editors: Agnieszka Hanaka, Jolanta Jaroszuk-Ściseł and Małgorzata Majewska

Received: 11 July 2021
Accepted: 23 August 2021
Published: 26 August 2021

Abstract: Crop production is threatened by low phosphorus (P) availability and weed interference. Obtaining plant genotypes that can utilize Phosphite (Phi) as fertilizer can supplement phosphates (Pi) while providing an environmentally friendly means of weed control. The study was conducted to determine the tolerance and enzymatic behavior of five potato genotypes to PO_3. Explants were regenerated in vitro from two nodal cuttings and cultured on Murashige and Skoog (MS) medium under controlled conditions for 30 days. Matured plantlets were subcultured for 20 days in MS medium containing (0.25, 0.5 mM) Phi and Pi and No-P (-Phi + -Pi). The results showed significant genotypic variation in tolerance indices among the five genotypes. Atlantic showed greater tolerance to Phi, with highest total root length (50.84%), root projected area (75.09%), root surface area (68.94%), root volume (33.49%) and number of root forks (75.66%). Phi induced an increasing trend in the levels of hydrogen peroxide in the genotypes with the least effect in Atlantic. The comprehensive evaluation analysis confirmed the tolerance of Atlantic genotype with this ranking; Atlantic, Longshu3, Qingshu9, Longshu6 and Gannong2. Antioxidant enzyme activities and proline content also increased significantly under Phi and No-P treatments. The results suggested that potato genotypes with larger root systems may be more tolerant to Phi than genotypes with smaller root systems.

Keywords: phosphite stress; antioxidant enzyme; hydrogen peroxide; root morphology; potato; genotypes

1. Introduction

Phosphorus (P) is an important macronutrient required by all living organisms and a very important cellular component that plays a crucial role in biological activities [1,2]. Phosphorus is involved in the signaling of target proteins through phosphorylation and dephosphorylation that determine various cellular performances for optimal plant growth [3]. The major P forms include phosphate and phosphite, which are used in agriculture [4]. Phosphate anions ($H_2PO_4^-$, HPO_4^{2-} and PO_4^{3-}) are certainly the main forms of P used by plants for metabolic processes and development, while phosphite is a reduced form of Pi that can be readily taken up by plants through Pi transporters [3]. More than 90% of the Pi required by plants is supplied by the soil, which provides adequate storage [5]. However, an estimated 80% of Pi fertilizer applied to soil worldwide is lost through immobilization and conversion to inorganic forms that plants cannot utilize directly [6]. Since Pi is highly reactive and rapidly transformed by soil microbes, only 20–30% is effectively

Horticulturae 2021, 7, 265. https://doi.org/10.3390/horticulturae7090265 — https://www.mdpi.com/journal/horticulturae

utilized by plants [7]. According to Gianessi [8], in most soils, weeds and crops compete for the available Pi, resulting in Pi deficiency to meet plant growth requirements. As global demand for food increases, the overuse of PO_4 fertilizers and herbicides has become inevitable [9]. This can accelerate the depletion of non-renewable phosphorus reserves, and increase production costs and prices of agricultural products; it also has significant environmental impacts, such as runoff into water bodies, leading to algal blooms, eutrophication, etc. [10,11]. Overuse of herbicides in cultivation has led to the emergence of herbicide-resistant superweeds in recent years [12]. Thus, low soil phosphorus availability and herbicide-resistant weeds have been identified as major threats to the long-term sustainability of agriculture [7,12,13], for which an effective long-term solution is urgently needed.

Phosphite anions ($H_2PO_3^-$ and HPO_3^{2-}) have high solubility and low soil reactivity. Although plants and various microorganisms cannot utilize Phi, it can be used as a potential target to enhance germplasm for phosphorus utilization by plants [14,15]. Phosphite has an inhibitory effect on plant growth with similar properties to herbicides [16]. The phosphite salt does not pose any risk to human and animal health and is therefore massively used as an effective fungicide in crop production [2]. Thus, phosphite has a direct effect on phytopathogenic fungi by inhibiting mycelial proliferation and reducing conidiogenesis of *Fusarium* sp. isolated from the rhizosphere of plants [17]. In addition, Phi can act indirectly by stimulating the inherent defense mechanisms of plants to limit pathogen growth [18], and also activate host defense genes that help plants defend against disease [19] and directly suppress the growth of pathogens such as *Phytophthora* [20–22]. According to Mehta et al. [23] and Thao and Yamakawa [24], Phi anions cannot be utilized by plants as a phosphorus nutrient, although Phi is well taken up by plant leaves and roots. The supply of Phi to plants as a sole source of P fertilizer can hinder plant growth and a higher dose can completely destroy plants [25,26]. Moreover, McDonald, Grant and Plaxton [16] claimed that Phi-treated plants accumulate Phi rapidly in their cells. Phosphite is phloem-mobile and accumulates in sink tissues [27]. Since Phi is not metabolized by plants, it remains in tissues for a long time and consequently disrupts the signal transduction chain that allows plants to detect and respond to Pi deficiency at the molecular level, thereby amplifying the negative effects of Phi [23,28].

On the contrary, the stimulatory effect of Phi mediates structural and biochemical changes in potato periderm and rind [29]. Phi application improved fruit set and yield of *Persea americana* (avocado) and also restored optimal growth of Pi deficient *Citrus* species [20]. Again, several reports indicated impressive results of Phi on plant P nutrition, which ultimately increased crop yields [25,30]. When Phi is added to the soil, it comes into contact with microorganisms that help Phi to oxidize to Pi [2]. Thus, following microbial oxidation reactions, Phi may become available to the plant as a P nutrient through this indirect pathway. Interestingly, efforts to generate transgenic plants with microbial genes (*ptxD*) that allow plants to use Phi as a sole P source have opened new possibilities for the use of this P-containing compound for plant nutrition [30]. In contemporary agriculture, Phi is emerging as a unique biostimulant that improves crop productivity and quality, through direct antibiotic effects on microorganisms and inhibition via enhanced plant defense responses. In addition, Phi induces a variety of abiotic stress tolerance mechanisms, including heat tolerance [31,32]. Obtaining a potato genotype that is tolerant to Phi will enable us to conduct more advanced genetic studies to understand gene functions for subsequent molecular work. In this experiment, we investigated the effects of Phi concentrations, MS medium without P nutrient (No-P) and sufficient Pi under in vitro cultures to determine the tolerance of potato genotypes. Nevertheless, there is insufficient information to test the assumption that Phi can stimulate antioxidant enzyme activities and hydrogen peroxide levels. Therefore, the experiment aims to: (1) determine the tolerance of five potato genotypes to Phi stress using tolerance indices and (2) evaluate the responses of antioxidant enzymes in *Solanum tuberosum* L. plantlets grown in different concentrations of Phi.

2. Materials and Methods

2.1. Place of the Experiment and Materials

The experiment was conducted in Gansu Providential key laboratory of Aridland Crop Science, College of Agronomy, Gansu Agricultural University Lanzhou, China (36°03′ N; 103°40′ E). Potato genotypes; Qingshu9, Longshu6, Longshu3, Atlantic and Gannong2, were used in this experiment. The Atlantic genotype is reported to be drought-susceptible [33,34], the Longshu6 genotype is classified as moderately drought-tolerant [35], while Qingshu9 and Longshu3 were designated as drought-tolerant genotypes.

2.2. Source of Genotypes and Preparation of In Vitro Explants

The five potato genotypes were obtained from the laboratory of Crop Improvement and Germplasm Enhancement, Gansu Agricultural University, Lanzhou. Two years of field trials have yielded a wealth of germplasm traits with a large number of sterile tissue culture seedlings for genetic screening. Uniform explants from two nodal cuttings were cultured on the potato growth medium of Murashige and Skoog [36], which contained sucrose 30 gL^{-1} and agar 5 gL^{-1}. The pH was adjusted to 5.8 and the medium was autoclaved at 121 °C at 15 1b psi for 25 min. Cultures were maintained in a growth room at 25 ± 1 °C, 16 h photoperiod, active photosynthetic radiation of 45 μmol photons $m^{-2}s^{-1}$, and relative humidity of 55–66% for a 30-day growth period. Plantlets were harvested after 30 days and used for subsequent experiments. Uniform cuttings of each with two axillary buds were subcultured in an ethanol-sterilized chamber with laminar air flow and propagated in MS medium supplemented with final concentrations of (0.25, 0.5 mM) Phi and Pi and No-P supply, in sterilized glass vials (120 × 50 mm). Five cuttings of explants were cultured in each vial, which was tightly sealed with the lids and kept at 25 ± 1 °C in the growth room.

2.3. Experimental Design and Treatments

A 5 × 5 factorial trial in a completely randomized design with 3 replicates was conducted in a controlled growth room. Treatments included five potato genotypes, two concentrations (0.25 and 0.5 mM) each of phosphite and phosphate, and a medium without P fertilizer, representing (No-P). We used 0.5 mM Pi as a control. Fifty vials per genotype were cultured, with each vial containing five explants. After 20 days, the plantlets were examined for physio-morphological indices. The rest of the plantlets were plunged into liquid nitrogen and immediately preserved at −80 °C for biochemical analysis.

2.4. Measurements of Data

2.4.1. Physio-Morphological Parameters

Physiological parameters, such as shoot and root length (cm), were determined with a ruler by randomly selecting three plants from each replication and averaging these data. The number of roots and leaves were counted on each selected plant. Fresh stem weight, fresh root weight and total plant weight (g) were also measured using an electronic balance, with subsequent calculation of root to shoot ratio and tolerance biomass index for each genotype.

The roots of the sampled plantlets were carefully detached from the stems, washed in distilled water and scanned using a root scanner (STD) 4800, EPSON, Quebec City, QC, Canada and the root morphological indices such as total root length (TRL), root projected area (RPA), root surface area (RSA), root volume (RV), number of root tips (NRT), number of root forks (NRF) were calculated using root image analysis software Win RHIZO version 5.0 (Regent Instruments, Inc., Quebec City, QC, Canada).

2.4.2. Tolerance Indexes Determination

Based on the formula of Wilkins [37], the Phi tolerance indices (TIs) of the root system, which clearly indicate the tolerance of the root systems to Phi stress, were calculated according to the modifications of Dawuda et al. [38]. Since Phi uptake has direct effects on the different measured root morphological indices and plant tolerance to Phi stress, Phi TI was determined at the end of the experiment for each root index. Considering the score of TI, we classified the genotype with the largest TI for most of the calculated indices as the most tolerant genotype among the five potato genotypes studied. The formula of TI is as follows: TI = index under Phi stress/index without Phi stress $\times$ 100.

2.4.3. Determination of Hydrogen Peroxide and Malonaldehyde Contents in Samples

The content of hydrogen peroxide (H_2O_2) in the shoot samples was determined as described by Junglee et al. [39], with minor modifications. A 0.1 g fresh shoot sample was crushed using a mortar and pestle in liquid nitrogen. The homogenate was transferred to a 2 mL centrifuge tube and kept in an ice bath. An amount of 1.5 mL of 0.1% Trichloroacetic acid (TCA) was added and the uniform mixture was centrifuged at $12,000 \times g$ for 15 min at 4 °C. The supernatant 0.5 mL was carefully mixed with 0.5 mL Phosphate Buffer Saline (PBS) and 1 mL KI (1M) at 7.0 pH. The mixture was kept at 28 °C for one hour. The absorbance was measured using a spectrophotometer (model U-5100, Seya-Namioka, Hitachi High-technologies, Minato-ku, Tokyo, Japan) at 390 nm. The contents of H_2O_2 were determined with reference to the standard curve (0, 1, 2, 3, 4 and 5 mmol^{-1}). Malonaldehyde (MDA) Content Assessment Lipid peroxidation was measured by calculating the amount of MDA emitted using the technique for thiobarbituric acid (TBA) as presented in Hodges et al. [40]. Preserved fresh shoot samples of 0.15 g were crushed using a mortar and pestle and 4.5 mL of 10% TCA was added. Then, the homogenized substance was centrifuged at $500 \times g$ for 15 min at 4 °C. The supernatant was transferred to a centrifuge bottle and the volume (V) was recorded. Two mL of the supernatant was then mixed with 2 mL of 0.6% TBA. The homogenized mixture was warmed in boiling water for 20 min, the reaction stopped in an ice bath, and centrifuged at $5000 \times g$ for 10 min. The supernatant (2 mL) of V1 was transferred to a cuvette. The absorbance of the supernatant was measured at 450, 532, and 600 nm, respectively. The MDA content was estimated according to the following formula: MDA concentration (μmol/L) = 6.45 $\times$ (A532 − A600) − 0.56 $\times$ A450. MDA content (μmol/g FW) = C (μmol/L) $\times$ V (L) $\times$ V1 (mL)/2 mL $\times$ M (g FW).

2.4.4. Determination of Antioxidant Enzymes Activities and Proline Contents in Shoots

Enzyme samples were prepared from frozen tissue preserved at −80 °C. Each shoot sample (approximately 0.5 g) was crushed in liquid nitrogen using a mortar and pestle and homogenized in 5 mL of 0.1 M phosphate buffer (pH 7.8) containing 0.5 mM ethylenediamine tetraacetic acid (EDTA). Each homogenate was centrifuged at $12,000 \times g$ for 15 min at 4 °C. The supernatant was collected for determination of enzymatic activity. The activities of catalase (CAT), peroxidase (POD) and superoxide dismutase (SOD) in the shoot homogenate were determined using a reagent kit (Nanjing Jiancheng Bioengineering Institute, Nanjing, China) following the manufacturer's instructions. The principles of these kits are summarized as follows:

Catalase activity was determined by the spectrophotometric ammonium molybdate method, in which ammonium molybdate rapidly stops the H_2O_2 degradation reaction, catalyzed CAT, as the remaining H_2O_2 reaction produces a yellow compound that can be examined by absorbance at 405 nm. A catalase unit activity was classified as the amount of enzyme in 1 g of fresh tissue that reduces 1 μmol of H_2O_2 per minute at 37 °C.

SOD activity was calculated according to the method of Dhindsa et al. [41]. The method is based on photochemical reduction of SOD-motivated Nitrotetrazolium Blue Chloride (NBT) at 560 nm. A unit of SOD activity was well defined as the amount of enzyme that inhibits 50% of the oxidation.

Peroxidase activity was determined by catalysis of hydrogen peroxide by POD, observing absorbance changes at 420 nm and estimating the activity of POD. One unit of POD activity was defined as the amount of enzyme in 1 g of fresh plant tissue reducing 1 μg of H_2O_2 at 37 °C per min. Proline content was determined according to the method of Bates et al. [42]. Fresh shoot samples weighing 0.15 g were crushed with 4.5 mL of 3% (*w/v*) sulfosalicylic acid homogenization and the homogenate was heated in boiling water for 30 min. This was then filtered through 0.2 μm filter paper. The extract and the volume of extract were designated as Vt. The supernatant was used to determine the amount of proline. The reaction mixture consisted of 2 mL of plant extract and an appropriate amount of ninhydrin glacial acetic acid. The test tubes containing the substance were heated in boiling water for 30 min. The reaction was quenched with the addition of toluene in an ice bath. The substance was shaken vigorously on vortex mixer for 15–30 s and divided into two phases (upper and lower chromophase). The upper chromophase (toluene) was carefully aspirated with a pipette, and absorbance was taken at 520 nm. The amounts of proline were measured from the standard curve and expressed as $\mu g \cdot g^{-1} FW$. The amount of proline was calculated as: Proline content ($\mu g/gFW$) = C × Vt/(V × W).

2.5. Analyses of Data

All data collected were analyzed using SPSS software 22.0 version (IBM Corp., Chicago, IL, USA). Means of treatments were separated by Duncan's multiple range tests with a probability of 5%. The distribution of means was presented in the figures using standard deviations. All graphs were created using GraphPad Prism version 8.0 (GraphPad Software, Inc., San Diego, CA, USA). Principal component analysis (PCA) was performed using the software PAST-PAlaeontological Statistics, version 1.34. To confirm the tolerance status of potato genotypes to Phi, comprehensive evaluation analysis based on PCA was carried out using R software package Statistics, version 3.5.3, considering the following formula:

$$\mu(Xi) = (Xi - Xmin)/(Xmax - Xmin) \quad i = 1, 2, 3 \ldots n \tag{1}$$

In Formula (1), $\mu(Xi)$ refers to the membership function value of the i-th comprehensive index, and Xi refers to the i-th. Comprehensive index value, Xmax refers to the maximum value of the i-th, comprehensive index, Xmin refers to the i-th, the minimum value of a comprehensive index. Calculation of the weighting of each comprehensive index:

$$Wi = \frac{Pi}{\sum_{i=1}^{m} Pi} \quad i = 1, 2, 3 \ldots n \tag{2}$$

In Formula (2), Wi represents the importance of the i-th comprehensive index in all comprehensive indices. In terms of degree and weight, Pi is the contribution rate of the i-th comprehensive index of each genotype. Calculation of comprehensive evaluation value (D).

$$D = \sum_{i=1}^{m} [\mu(Xi) \times Wi] \quad i = 1, 2, 3 \ldots n \tag{3}$$

In Formula (3), D represents the phosphite tolerance of different potato genotypes. The D value is obtained by calculating the weighted membership function value. The larger the value of D, the more the Phi tolerance.

3. Results

3.1. Influence of Phosphite and Phosphate on Physiological Parameters of Five Potato Genotypes after 20 Days' Growth Period

The results showed significant ($p < 0.01$) genotype x phosphorus source and rates interaction effect on all physiological parameters (Figure 1a–d). Phosphite stress significantly reduced the growth of potato genotypes in this study. Reduction in growth was observed in both roots and shoots resulting in reduction in fresh biomass and fresh root to shoot ratio. Mostly all physiological indices across all genotypes were decreased by PO_3 (0.25 and 0.5 mM) and No-P, but Atlantic genotype was least affected. At Phi 0.5 mM, the decrease in leaf number was least (40.67%) in Longshu6 and greatest (57.16%) in Qingshu9. The least (61.93%) and greatest (81.73%) decrease in the number of roots occurred in Atlantic and Gannong2, respectively. The least (58.18, 6.22%) and greatest (69.95, 91.33%) decrease in shoot and root length was observed in Atlantic and Gannong2 genotypes, respectively. Moreover, similar results were observed among the five genotypes with respect to PO_3 (0.25 mM), but the decrease in physiological parameters due to Phi stress was more pronounced at 0.5 compared to 0.25 mM. In the treatment without P supply, physiological growth was also decreased in all the five potato genotypes except Atlantic and Longshu3, which recorded increase in root length as compared to their respective control. Compared to the control, the application of PO_3 at (0.25 and 0.5 mM) and No-P had a negative effect on the fresh biomass indices of the genotypes (Table 1). Application of PO_3 (0.25 mM) to genotype Gannong2 caused greater reductions in FRW (80.95%), FSW (82.48%) and TPW (82.00%). In addition, smaller reductions were observed in FSW (43.24%) and TPW (30.20%), and there was a slight increase in FRW (7.41%) of Atlantic genotype. The application of Phi (0.5 mM) gave similar results with a slight difference in severity compared to the results obtained with 0.25 mM. Thus, application of higher rates of Phi could be lethal to potato plantlets. Maximum decrease in FRW (87.21%), FSW (83.94%) and TPW (80.50%) and minimum decrease in FRW (14.81%), FSW (60.81%) and TPW (48.51%) were observed in potato genotypes Gannong2 and Atlantic, respectively.

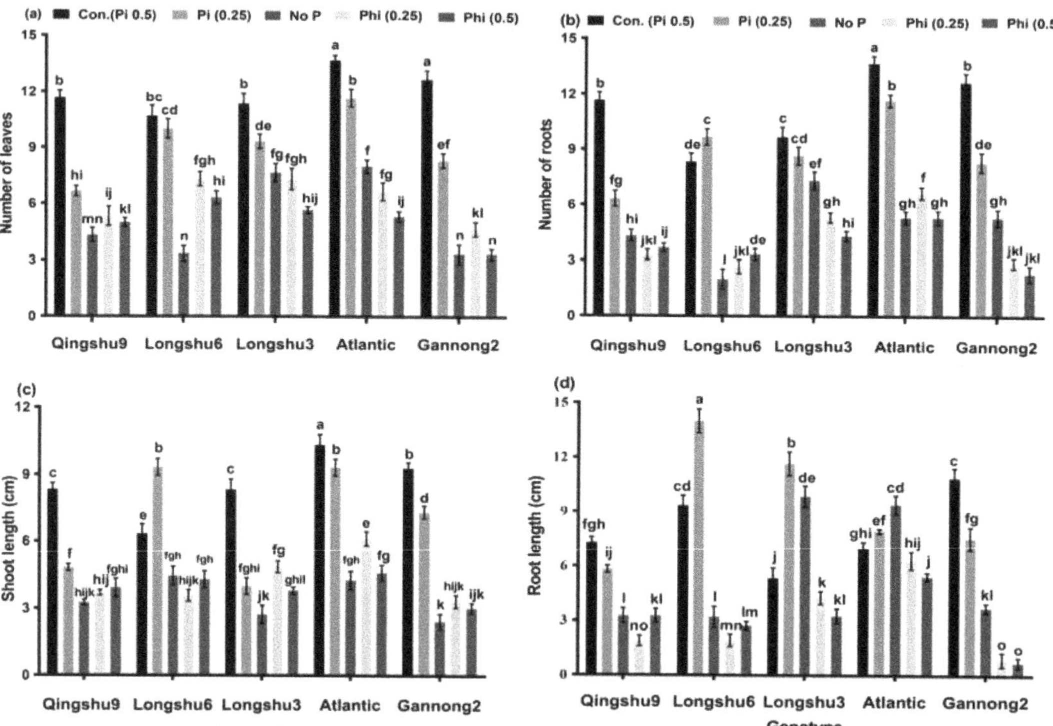

Figure 1. Effect of Phi, Pi and No-P on (**a**) number of leaves, (**b**) number of roots, (**c**) shoot length (cm), and (**d**) root length (cm) of five potato genotypes grown in modified MS media containing 0.25 and 0.5 mM Phi and Pi and No-P supply for 20 days. Values denote the mean of 3 replicates, ±standard deviation (SD). Means obtained with the same letter in the minuscule do not differ by Duncan Multiple range's test ($p \leq 0.05$).

Table 1. Effect of Phi, Pi and No-P supply on fresh biomass, root/shoot ratio and biomass tolerance index of five potato genotypes studied under in vitro conditions.

Genotype	Treatment	Biomass Accumulation (g)			Root/Shoot Ratio	* BTI (%)
		Fresh Root Weight	Fresh Shoot Weight	Fresh Plant Weight		
Qingshu9	Con.Pi 0.5	0.087 ± 0.002 [a]	0.156 ± 0.001 [a]	0.243 ± 0.003 [ab]	0.560 ± 0.006 [ef]	100
	Pi0.25 mM	0.067 ± 0.004 [b]	0.143 ± 0.0008 [a]	0.210 ± 0.004 [bc]	0.467 ± 0.031 [fgh]	96.42
	No-P	0.033 ± 0.001 [ef]	$0.\,054 \pm 0.0008$ [c]	$0.\,087 \pm 0.002$ [fg]	0.611 ± 0.019 [de]	35.80
	Phi0.25 mM	0.019 ± 0.004 [gh]	0.064 ± 0.001 [bc]	0.083 ± 0.003 [fg]	0.297 ± 0.072 [jk]	30.04
	Phi 0.5 mM	0.018 ± 0.006 [gh]	0.065 ± 0.001 [bc]	0.083 ± 0.006 [g]	0.277 ± 0.085 [jk]	30.04
Longshu6	Control	0.087 ± 0.0006 [a]	0.154 ± 0.002 [a]	0.241 ± 0.002 [ab]	0.565 ± 0.015 [ef]	100
	Pi0.25 mM	0.097 ± 0.0006 [a]	0.157 ± 0.001 [a]	0.254 ± 0.001 [a]	0.618 ± 0.007 [de]	105.39
	No-P	0.025 ± 0.001 [fg]	0.062 ± 0.003 [c]	0.087 ± 0.003 [fg]	0.403 ± 0.033 [ghi]	36.51
	Phi0.25 mM	0.014 ± 0.0006 [gh]	0.054 ± 0.0006 [c]	0.068 ± 0.0006 [g]	0.259 ± 0.015 [k]	27.80
	Phi 0.5 mM	0.026 ± 0.001 [fg]	0.057 ± 0.001 [c]	0.083 ± 0.001 [g]	0.456 ± 0.018 [fghi]	34.44
Longshu3	Control	0.048 ± 0.001 [cd]	0.147 ± 0.001 [c]	0.195 ± 0.002 [c]	0.327 ± 0.001 [jk]	100
	Pi0.25 mM	0.085 ± 0.0006 [a]	0.089 ± 0.0006 [b]	0.174 ± 0.000 [cd]	0.955 ± 0.013 [b]	89.23
	No-P	0.067 ± 0.002 [b]	0.055 ± 0.001 [c]	0.122 ± 0.003 [ef]	1.218 ± 0.007 [a]	63.08
	Phi0.25 mM	0.037 ± 0.0006 [def]	0.064 ± 0.0006 [bc]	0.101 ± 0.001 [fg]	0.578 ± 0.002 [ef]	57.95
	Phi 0.5 mM	0.024 ± 0.001 [fg]	0.048 ± 0.002 [c]	0.072 ± 0.001 [g]	0.500 ± 0.037 [efg]	36.92
Atlantic	Control	0.054 ± 0.0006 [bc]	0.148 ± 0.0006 [a]	0.202 ± 0.000 [c]	0.365 ± 0.005 [hij]	100
	Pi 0.25 mM	0.067 ± 0.001 [b]	0.135 ± 0.001 [a]	0.202 ± 0.0006 [c]	0.496 ± 0.012 [fg]	100
	No-P	0.057 ± 0.001 [bc]	0.058 ± 0.0006 [c]	0.115 ± 0.002 [ef]	0.983 ± 0.019 [b]	56.93
	Phi0.25 mM	0.058 ± 0.0006 [bc]	0.084 ± 0.0006 [b]	0.142 ± 0.0006 [de]	0.690 ± 0.010 [d]	69.80
	Phi 0.5 mM	0.046 ± 0.001 [cde]	0.058 ± 0.001 [c]	0.104 ± 0.002 [efg]	0.793 ± 0.005 [d]	51.48
Gannong2	Control	0.063 ± 0.001 [b]	0.137 ± 0.001 [a]	0.200 ± 0.001 [c]	0.460 ± 0.006 [fghi]	100
	Pi0.25 mM	0.067 ± 0.001 [b]	0.135 ± 0.001 [a]	0.202 ± 0.0006 [c]	0.496 ± 0.012 [fg]	101.00
	No-P	0.024 ± 0.001 [fg]	0.044 ± 0.001 [cd]	0.068 ± 0.001 [g]	0.545 ± 0.031 [ef]	34.00
	Phi0.25 mM	0.016 ± 0.0006 [fg]	0.052 ± 0.001 [c]	0.078 ± 0.001 [g]	0.500 ± 0.032 [efg]	18.00
	Phi0.5 mM	0.008 ± 0.001 [h]	0.022 ± 0.003 [d]	0.039 ± 0.003 [h]	0.350 ± 0.052 [ijk]	19.50

Data indicate the mean $\pm$ SD of 3 biological replications and were tested for significance using Duncan's multiple range tests. Individual columns marked with different lowercase letters indicate significant differences ($p < 0.01$). * The BTI was not subjected to an analysis of variance.

Fresh biomass indices of No-P treated potato genotypes were decreased due to the absence of Pi in the growth media. The effect of Pi deficiency was more pronounced in the shoots than in the roots of the five potato genotypes. The greatest decrease in fresh biomass indices (FRW, FSW and TPW) was recorded in genotype Gannong2, while the least occurred in genotype Longshu3. Moreover, the negative effects of Phi and No-P on growth of all genotypes were measured in biomass tolerance index (BTI). The result showed that Atlantic genotype had the highest (70.39 and 51.49%) BTIs, followed by LS3 (51.79 and 36.92%), while the lowest (34.16 and 15.00%) BTIs were recorded in Qinshu9 and Gannong2 genotypes at Phi (0.25 and 0.5 mM). In No-P treatment, genotype Longshu3 recorded the highest value (62.56%) followed by Atlantic (56.93%) and the lowest value (34.00%) was observed in Gannong2. Among Phi-treated genotypes in terms of root to shoot ratio (RSR), Atlantic recorded the highest value (0.79 and 0.69), and the lowest value (0.26 and 0.28) was recorded in Longshu6 and Qingshu9 at (0.25 and 0.5 mM). In No-P treatment, Longshu3 had the highest value (1.22), while Longshu6 had the lowest (0.40). However, in Pi-sufficient genotypes, the number of leaves, number of roots, shoot length and root length were increased compared to their respective controls. The maximum increase was recorded in the Atlantic genotype. Fresh biomass indices showed similar results in all the five genotypes as compared to the respective controls. The lowest root, shoot, and total plant weights were recorded in Qingshu9 and Longshu3, while maximum increase was recorded in Gannong2 and Longshu6 genotypes, respectively.

3.2. Effects of Phosphite, Phosphate and No-P Supply on Root Morphological Characteristics of Five Potato Genotypes for 20 Days Growth Period

There were significant ($p < 0.01$) genotype x phosphite interaction effects on total root length (TRL), root projected area (RPA), root surface area (RSA), root volume (RV), number of root tips (NRT), and number of root forks (NRF) (Figure 2). In general, root morphological indices were decreased in all five genotypes, with the Atlantic genotype being the least affected. The decrease in TRL due to Phi effects was least (32.86 and 43.17%) in Atlantic and greatest (80.85 and 80.93%) in Longshu6 and Gannong2 at Phi (0.25 and 0.5 mM), respectively. The least decrease in RPA (15.62 and 24.81%) was recorded in Atlantic while the greatest (68.57 and 75.85%) was observed in Gannong2 at (0.25 and 0.5 mM). Moreover, the least (16.86 and 30.85%) and greatest (85.57 and 84.50%) decrease in RSA was observed in Atlantic, Longshu3 and Longshu6 and Gannong2 genotypes, respectively. The least (53.85 and 66.51%) decrease in RV occurred in Atlantic, while the greatest (82.08 and 84.70%) was observed in Longshu6 and Gannong2. Moreover, the least decrease (29.49 and 24.34%) of NRT occurred in Atlantic and Longshu3 genotypes, while the greatest (85.02 and 89.86%) was observed in Longshu6 and Gannong2, at 0.25 and 0.5 mM, respectively. The least (16.57 and 31.06%) decrease in NRF was observed in Atlantic, while the greatest (83.94 and 80.01%) was measured in Longshu6 and Gannong2, at Phi 0.25 and 0.5 mM, respectively. The decrease in root morphological parameters observed in the five genotypes was due to Phi effects, which caused a decrease in the size of the root and shoot systems of the genotypes. The root morphological indices of the treatment without P supply were also decreased in all five genotypes. Mainly due to the absence of Pi in the No-P treatment, this reduced the root growth of the susceptible genotypes, while the tolerant genotypes developed longer root systems. The least decrease in morphological indices was observed in Longshu3 and Atlantic, while the greatest decrease was observed in Gannong2 and Longshu6 genotypes. However, Pi-sufficient potato genotypes exhibited much higher root morphological indices than genotypes grown under Phi and No-P. Root morphological indices were slightly increased in Pi-sufficient plants of the five potato genotypes. Among the genotypes: Longshu3 and Longshu6 recorded the highest increase, while Atlantic and Qingshu9 had the least increase in all root morphological indices.

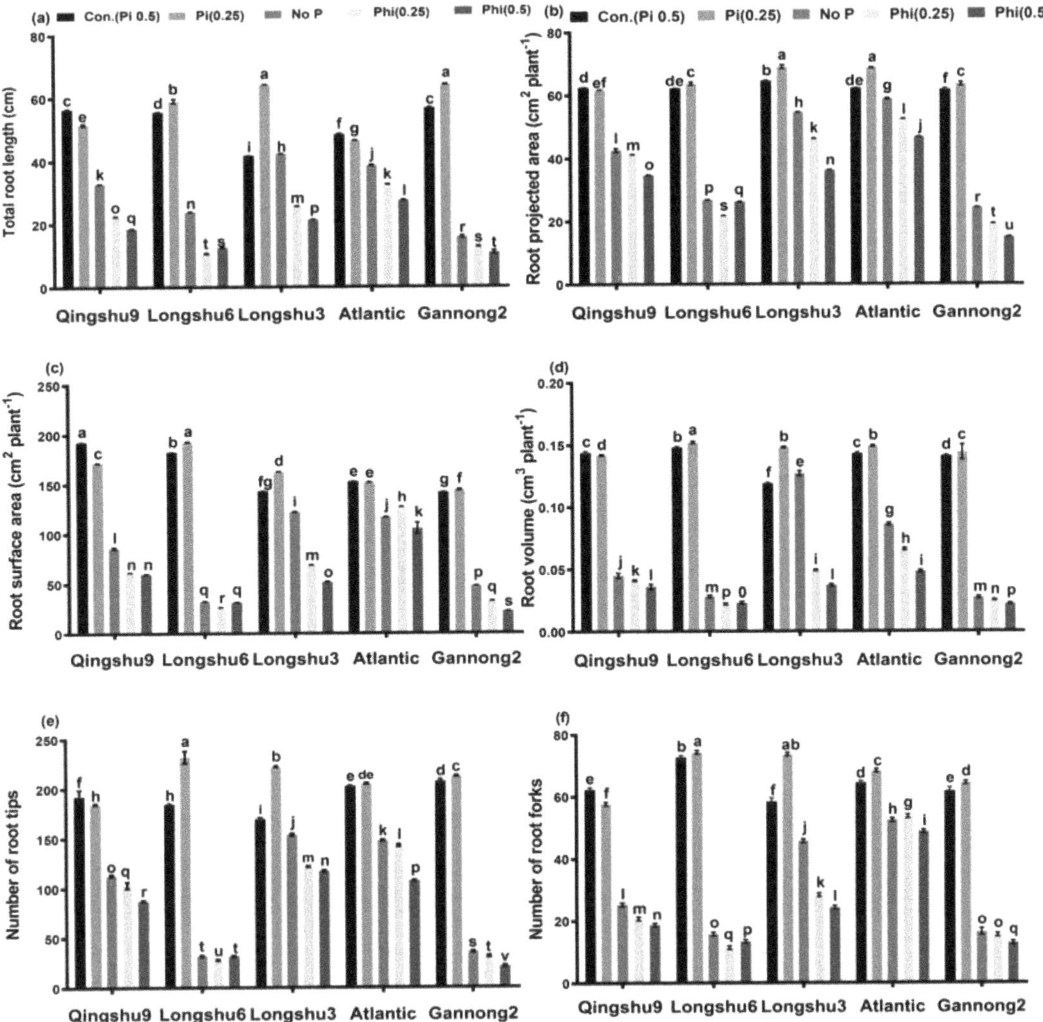

Figure 2. Effect of PO$_3$, PO$_4$ and No-P on (**a**) Total root length (**b**) Root projected area (**c**) Root surface area (**d**) Root volume (**e**) Number of root tips and (**f**) Number of root forks of five potato genotypes grown for 20 days in medium containing concentrations of 0.25, 0.5 mM PO$_3$ and PO$_4$ and No-P. Values symbolize the mean of 3 replicates ± standard deviation (SD). Means followed by the same lowercase letter do not differ among treatments by Duncan Multiple range's test ($p \leq 0.05$).

3.3. Tolerance of the Five Potato Genotypes to Phosphite Stress

There were significant differences ($p < 0.01$) in the tolerance indices (TIs) among the five potato genotypes at 0.25, 0.5 mM and No-P (Figure 3). The result showed that after the plantlets were grown for 20 days in the Phi media, Atlantic had the highest TI for majority of the measured indices such as TRL (56.84%), RPA (75.09%), RSA (68.94%), RV (33.49%) and NRF (75.66%) at Phi (0.5 mM). Considering this result, among the five potato genotypes tested, the Atlantic genotype proved to be the most tolerant, except TI for NRT, which was highest in Longshu3 (69.16%). However, compared to Atlantic, the other genotypes obtained lower TI values and were found to be susceptible to PO$_3$ stress.

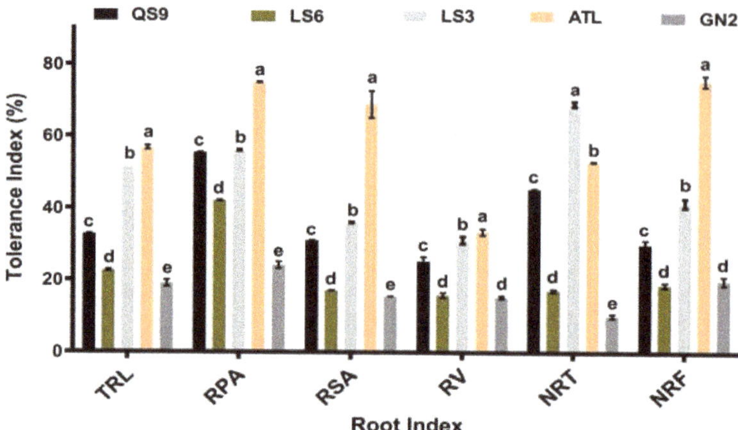

Figure 3. Root PO_3 tolerance indexes (%) of five potato genotypes under 0.5 mM Phi treatment. Qingshu9 = QS9; Longshu6 = LS6; Longshu3 = LS3; Atlantic = ATL and Gannong2 = GN2. Total root length = TRL; Root projected area = RPA; Root surface area = RSA; Root volume = RV; Number of root tips = NRT and Number of root forks = NRF. Values represent the mean of 3 replicates ± standard deviation (SD). Bars associated with different lowercase letters show significant differences by Duncan Multiple Range's test ($p \leq 0.05$).

3.4. Content of H_2O_2 and MDA in the Shoots

There was a significant ($p < 0.001$) genotype x phosphite x rate effect on H_2O_2 and MDA content in shoots of potato genotypes (Figure 4). Compared to the control plants, Phi increased H_2O_2 and MDA content in the shoots of the five genotypes by 50.96 to 68.84% and 43.63 to 70.17% at 0.25 mM, 56.01 to 71.64% and 48.32 to 71.48% at 0.5 mM, respectively. The highest H_2O_2 and MDA values were observed in GN2, followed by genotypes Longshu6, Qingshu9 and Longshu3. The lowest H_2O_2 and MDA levels were observed in Atlantic genotypes at Phi 0.25 and 0.5 mM, respectively. Moreover, similar results were obtained with respect to treatment without P supply in the five potato genotypes. However, Pi-sufficient plants in all five potato genotypes exhibited low H_2O_2 and MDA contents, compared to genotypes under Phi stress.

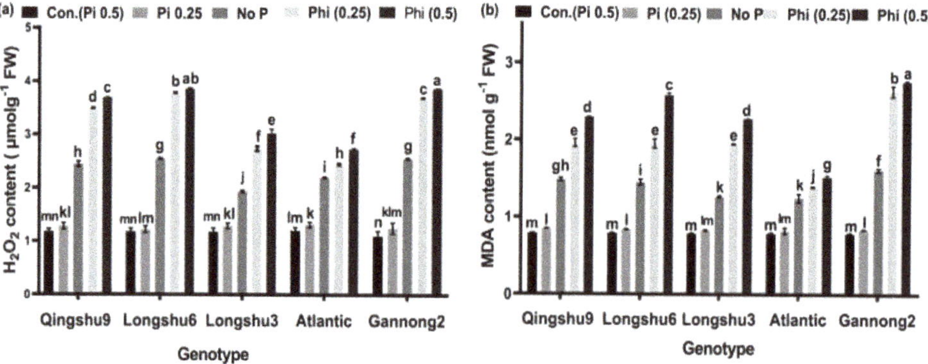

Figure 4. Effects of Phi, Pi and No-P on H_2O_2 and MDA content; (**a**) Shows H_2O_2 content and (**b**) shows MDA content in the shoots of potato genotypes. Values denote the mean of 3 replicates ± standard deviation (SD). Duncan multiple range test ($p \leq 0.05$) shows large differences between bars assigned to different lowercase letters.

3.5. Antioxidant Enzymes Activities and Content of Proline

The effect of genotype x Phi interaction on antioxidant enzyme activities as well as the levels of proline was also significant ($p < 0.01$) in the shoots of the five potato genotypes (Figure 5). Compared with the individual control plants, Phi stress increased the proline contents in the shoots of all genotypes by 80.11 to 81.09% and 76.79 to 79.21% at Phi 0.5 and 0.25 mM, respectively. Moreover, the activities of CAT, POD and SOD increased by 46.07 to 55.54%, 50.77 to 53.14% and 54.39 to 63.09%, respectively, in all genotypes at 0.5 mM. The greatest increase in CAT activity was observed in the ATL genotype, while the LS6 genotype had the greatest POD and SOD activities. Similar increases in CAT, POD and SOD activities were observed in all genotypes at Phi (0.25 mM) and in No-P supplied potato genotypes.

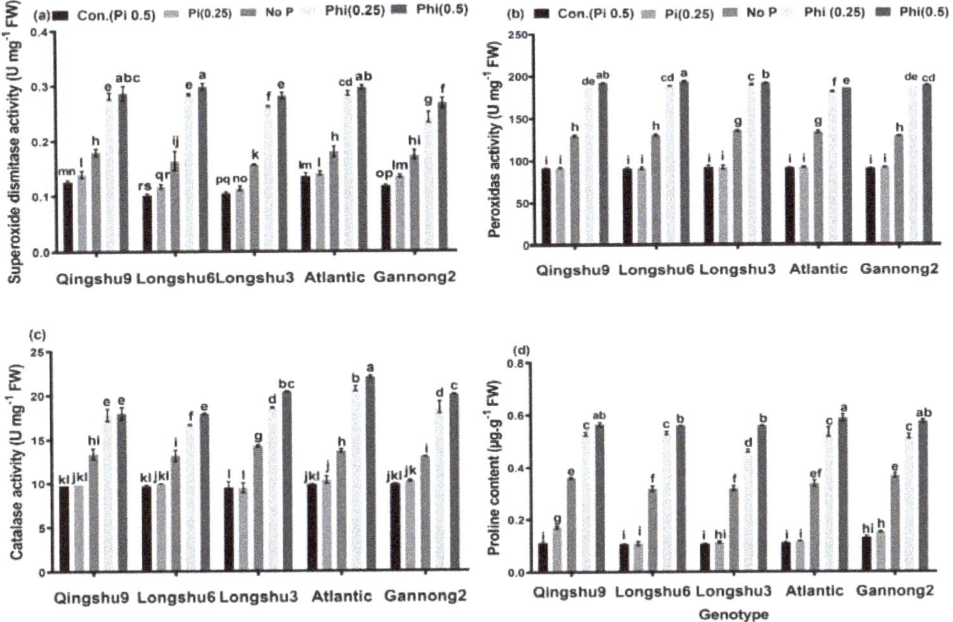

Figure 5. Phi and Pi effects on enzyme activities and contents of proline; (**a**) Shows the activities of SOD; (**b**) Shows the activities of POD; (**c**) Shows the activities of CAT; (**d**) Contents of proline. Values denote the mean of 3 replications ± standard deviation (SD). Bars associated with different lowercase letters show significant differences by Duncan Multiple Range's test ($p \leq 0.05$).

3.6. Relationships between Root Morphological Characteristics, Fresh Biomass and Biochemical Responses of the Five Potato Genotypes under Phi Stress

The correlation matrix between root morphological indices, fresh biomass, antioxidant enzyme activities, MDA, H_2O_2 and proline of the five potato genotypes under Phi stress were significantly negative (Table 2). All six root morphological indices were negatively correlated with MDA, H_2O_2, CAT, SOD, POD and Pro., e.g., TRL showed a significant negative correlation with MDA, H_2O_2, CAT, SOD, POD and Pro ($r = -0.87$ **, $r = -0.94$ **, $r = -0.80$ **, $r = -0.83$ **, $r = -0.87$ ** and $r = -0.89$ **). The fresh biomass, i.e., FRW, FSW and TPW were also negatively correlated with MDA, H_2O_2, CAT, SOD, POD and Pro, e.g., FRW was negatively correlated with MDA, H_2O_2, CAT, SOD, POD and Pro ($r = -0.67$ *, $r = -0.75$ **, $r = -0.62$ *, $r = -0.69$ *, $r = -0.69$ * and $r = -0.72$ **). Principal component analysis (PCA) was used to determine the effects of Phi and No-P on root morphological indices, fresh biomass, antioxidant enzymes, MDA, H_2O_2 and proline (Table 3). The cumulative contribution percentage of the two principal components associated with the response is 90.16%, the eigenvalue of PC1 is 8.818 and the contribution percentage is 62.98%. The eigenvectors include TRL, FRW, RPA, RSA, FRSR, NRF, RV and NRT. The eigenvalue of PC2 is 3.971, which corresponds to 27.18%, where the higher charges are CAT, SOD, MDA, POD and Pro show significant separations between the treatments. Genotype LS3 had the highest score followed by ATL, QS9, LS6 and GN2 at No-P, along PC1. On the other hand, ATL recorded the highest score followed by LS3, while GN2 had the lowest score at Phi (0.25 and 0.5 mM), along PC2. The treatments No-P and Phi (0.25 and 0.5) were far from the origin, implying that Phi and No-P strongly affected the root morphological characteristics and fresh biomass of potato genotypes (Figure 6). The comprehensive evaluation according to the principal component analysis of Phi tolerance was calculated using Formula (1) to calculate different products based on the two independent comprehensive indicators to obtain the membership function value $\mu(Xi)$ of each comprehensive index. Table 4 shows that, using the higher concentration of Phi (0.5 mM) treatment, in the same comprehensive index CI(1), Atlantic $\mu(X1)$ is the largest, with 0.937, indicating that Atlantic has the highest Phi tolerance on the CI(1) comprehensive index. According to the contribution rate of each comprehensive index, the weights of the two comprehensive indicators in terms of Phi tolerance were calculated using Formula (2) as follows: 0.699, 0.301. In addition, Formula (3) was used to calculate the Phi tolerance value according to the D value, and the Phi tolerance among the five genotypes was ranked as: Atlantic > Longshu3 > Qingshu9 > Longshu6 > Ganannong 2, respectively.

Table 2. Correlation matrix describing the relationship between root morphological characteristics, fresh biomass and activities of antioxidant enzymes, MDA, H_2O_2 and proline in potato plants under Phi stress at 20 days after treatments.

Index	TRL	RPA	RSA	RV	NRT	NRF	FRW	FSW	TPW
MDA	−0.87 **	−0.77 **	−0.81 **	−0.74 **	−0.78 **	−0.78 **	−0.67 *	−0.75 **	−0.84 **
H_2O_2	−0.94 **	−0.90 **	−0.92 **	−0.86 **	−0.90 **	−0.91 **	−0.75 **	−0.79 **	−0.92 **
CAT	−0.80 **	−0.71 **	−0.74 **	−0.68 **	−0.70 **	−0.70 **	−0.62 *	−0.74 **	−0.81 **
SOD	−0.83 **	−0.71 **	−0.75 **	−0.71 **	−0.71 **	−0.73 **	−0.69 *	−0.72 **	−0.78 **
POD	−0.87 **	−0.77 **	−0.82 **	−0.75 **	−0.77 **	−0.80 **	−0.69 *	−0.76 **	−0.85 **
Pro	−0.89 **	−0.80 **	−0.84 **	−0.77 **	−0.80 **	−0.82 **	−0.72 **	−0.81 **	−0.89 **

MDA = Malonaldehyde; H_2O_2 = Hydrogen peroxide; CAT = Catalase; SOD = Superoxide dismutase; POD = Peroxidase; Pro = Proline; TRL = Total root length; RPA = Root projected area; RSA = Root surface area; RV = root volume; NRT = Number of root tips; NRF = Number of root forks; FRW = Fresh root weight; FSW = Fresh shoot weight and TPW = Total plant weight. * = significant difference at 5% probability level; ** = significant difference at 1% probability level.

Table 3. Principal Component, loadings related to Phi treatments variable and explained variance for final sampling.

Principal Component	PC 1	PC 2
Total root length	0.323	0.098
Root projected area	0.284	0.234
Root surface area	0.290	0.235
Root volume	0.260	0.224
Number of root tips	0.259	0.266
Number of root forks	0.270	0.279
Fresh root weight	0.298	0.044
Fresh root/shoot ratio	0.295	0.095
Malonaldehyde	−0.232	0.353
Hydrogen peroxide	−0.319	0.044
Catalase	−0.162	0.409
Superoxide dismutase	−0.229	0.374
Peroxidases	−0.238	0.348
Proline	−0.245	0.339
Eigenvalue	8.818	3.805
Contribution rate %	62.98	27.18
Cumulative contribution rate %	62.98	90.16

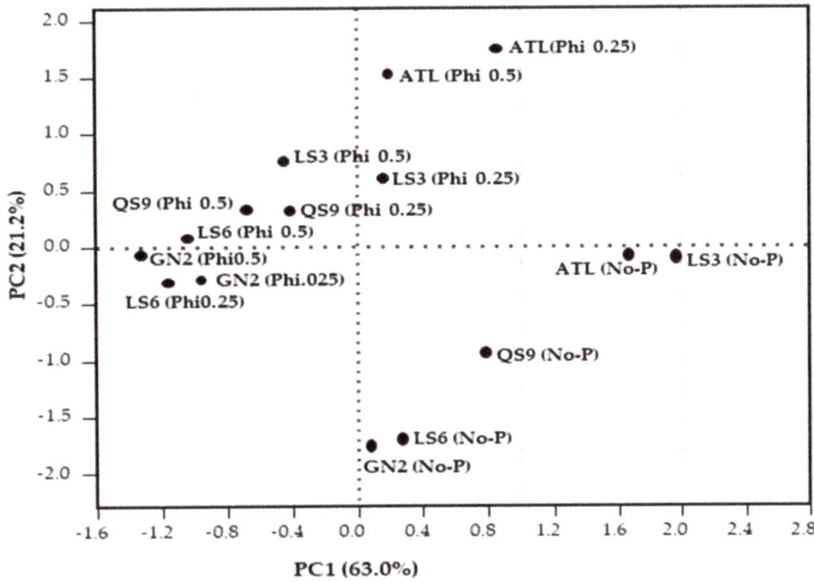

Figure 6. Scatter diagrams of arranged first two main components; PC1/PC2.

Table 4. Comprehensive index (C1), index weight μ(X), and comprehensive evaluation (D) of five potato genotypes under Phi and No-P stresses.

Genotype	Treatment	CI(1)	CI(2)	$\mu(X_1)$	$\mu(X_2)$	Comprehensive Assessment Value (D)	Ranking
Qingshu9	No-P	2.394	−1.940	0.639	0.235	0.517	
	Phi 0.25	−1.252	0.648	0.277	0.597	0.373	
	Phi 0.5	−2.079	0.689	0.195	0.603	0.318	3
Longshu6	No-P	0.832	−3.531	0.484	0.013	0.342	
	Phi 0.25	−3.567	−0.622	0.047	0.419	0.159	
	Phi 0.5	−3.172	0.169	0.086	0.539	0.220	4
Longshu3	No-P	6.034	−0.266	1.000	0.469	0.849	
	Phi 0.25	0.474	1.228	0.448	0.678	0.517	
	Phi 0.5	−1.384	1.539	0.264	0.721	0.402	2
Atlantic	No-P	5.160	−0.213	0.913	0.477	0.782	
	Phi 0.25	2.649	3.536	0.663	1.000	0.765	
	Phi 0.5	0.589	3.088	0.469	0.937	0.604	1
Gannong2	No-P	0.249	−3.627	0.426	0.000	0.297	
	Phi 0.25	−2.876	−0.582	0.116	0.425	0.209	
	Phi 0.5	−4.043	−0.117	0.000	0.499	0.148	5
Weight		0.699	0.301				

4. Discussion

In vitro regeneration studies are considered an efficient technique for defining the regulatory mechanisms of stress tolerance in potato explants. Several studies have shown the importance of in vitro shoot regeneration in plant breeding programs for economic drives [43]. To select more resistant cultivars for production, plant breeders need knowledge on how to improve regeneration of explants grown under stress conditions [44,45]. Five genotypes with different drought stress tolerance were selected based on sound data on their drought tolerance availability. These genotypes are widely grown and consumed in China. Therefore, it is important for farmers, consumers and breeders to understand their tolerance to PO_3. In addition, we need to investigate whether the mechanism of drought tolerance in potato differs from the mechanism that may support PO_3 tolerance. In this study, we investigated the effects of Phi, Pi deficiency and Pi sufficiency on potato plant growth. Our data showed that Phi at concentrations of 0.25 and 0.5 mM severely inhibited plant growth, whereas Pi adequate plants grew normally across the five genotypes. We found that the number of leaves, roots, shoot length and root length of five potato genotypes cultured in Phi media were significantly reduced compared to those propagated in Pi-adequate media. Similar negative growth effects of Phi on *B. nigra* seedlings and *B. napus* cell suspension cultures have been documented previously [46]. Plants cultured in Phi media are very sensitive to Phi and show signs of toxicity such as leaf chlorosis and stunted growth [16,24,30]. These findings coincided with our observation in the present study. The observed decrease in plant growth in the presence of Phi could be related to two important morphological changes in the plants. Reduced internode length was an identifying feature of Phi-treated plants. Plant height was significantly reduced in Phi-treated plants in all five genotypes, whereas P-sufficient plants appeared normal in the presence of Pi concentrations (0.25 and 0.5 mM). The second important cause was a significant decrease in root development in Pi deficiency-treated plants and Phi treated plants. However, potato genotypes Atlantic and Longshu3 showed a modest increase in root development rate compared with the other genotypes studied at 0.25 and 0.5 mM Phi.

We further found that Phi at 0.25 and 0.5 mM altered the fresh biomass parameters such as fresh root weight, fresh shoot weight, total plant weight and root to shoot ratio in all genotypes as compared to Pi-sufficient plants. The reduction in fresh biomass of the five genotypes caused by Phi toxicity was more pronounced in fresh root weight than fresh shoot weight at both concentrations. This reduction had a large effect on total plant weight, resulting in reduced plant growth and development in Phi-treated plants, as previously documented for *B. nigra* plants [47]. The reduced root to shoot ratio observed in the majority of genotypes is consistent with the deleterious effect of Phi on root hair production, anthocyanin accumulation in the shoot, and stimulation of enzymes induced by Pi deficiency [48]. In summary, our experiment in potato is consistent with previous reports on the biological effects of Phi in *Brassica* sp. [49] and in tomato [50]. Under Phi supply conditions, even a low dose (0.25 mM) was sufficient to induce a decrease in fresh biomass. The majority of the potato genotypes showed a decrease in root/shoot ratio compared to Pi-sufficient plants, except Atlantic and Longshu3, which showed a marginal increase. Furthermore, the effect of Pi deficiency on potato genotypes was studied using a medium containing -Phi + -Pi, which corresponds to No-P. All genotypes under this treatment showed a significant decrease in fresh biomass except Atlantic and Longshu3 which showed an increase in root weight compared to the other genotypes. The considerable increase in root length and root weight of these genotypes resulted in a slight increase in root to shoot ratio compared to the respective controls. This observation corroborates the findings of Lambers and Plaxton [51], who observed that the absence of Pi in the growth media altered root phenotypes such as increased root hair density and root length as well as metabolism (e.g., release of carboxylates and Pi scavenging enzymes into the rhizosphere). The mechanism for overcoming the P deficit in growth media containing No-P plants is explained by an increase in root length [4,46,52]. Moreover, several studies have shown that plants under water stress increase root length in deeper profiles and that the main difference between shallow and deeper rooting genotypes is manifested in the stress conditions imposed in each case [53,54]. Biomass tolerance index (BTI) was calculated for each genotype in relation to the stress variables (0.25, 0.5 mM and No-P) to clearly define the tolerance of potato genotypes to Phi. The result showed that in No-P treatment, Longshu3 genotype had the highest BTI, followed by Atlantic, and Gannong2 had the lowest BTI. At Phi 0.25 and 0.5 mM, the Atlantic genotype had the highest BTI followed by Longshu3 genotype, while Gannong2 genotype had the lowest BTI.

Root growth and shoot growth are correlated with each other. According to Polania et al. [55], shoot growth supplies carbon and some hormones to the roots, while root growth supplies water, nutrients and hormones to the shoot. Despite the fact that no previous study has explicitly investigated the utility of root morphological traits for plant performance under Phi stress, the results of the current study suggest that Phi interference reduces root morphological traits in all genotypes except the Atlantic genotype. It has been suggested that the ability of Phi to limit *Arabidopsis* development is due to the competitive inhibition of Pi uptake and the inability of plants to readily utilize Phi through oxidation to Pi [46]. Phi cannot enter P metabolic pathways unless it is converted to Pi [16,56]. Moreover, the growth of plants cultivated in the Phi treatments was comparable to that of plants grown under the No-P supply treatment in terms of root morphological indices. These results confirm the findings of Lee et al. [57] for *Ulva lactuca*, Schroetter et al. [58] for *Zea mays*, Thao et al. [59] for *Brassica rapa*, Avila et al. [60] for *Zea mays*, Zambrosi et al. [61] for *Citrus* spp., and Hirosse et al. [62] for *Ipomoea batatas*. These researchers observed that Phi anion does not replace Pi anion in P nutrition of plants. They added that the use of Phi as the sole source of P resulted in a significant reduction in plant development compared to treatments with insufficient Pi fertilization. Root morphological traits such as architecture, branching, root volume, root hair length and density were found to have reflective effects on nutrient uptake from nutrient sources, which could be used to determine plant tolerance under stress conditions. According to Dawuda et al. [38], the size of root system of lettuce plants influenced their tolerance to cadmium stress in nutrient solution. The results of this study showed that the addition of Phi (0.25 and 0.5 mM) to MS media decreased the TRL, RPA, RSA, RV, NRT and NRF of all potato genotypes studied. This consequently reduced the size of root systems of most genotypes except Atlantic. Other researchers have found that plant uptake of P and other nutrients depends on root surface area, root system length and lateral roots to capture a large volume of nutrients in the soil/growth medium [63–65]. In addition, our results show that the Atlantic genotype, which had the larger root system, had root tolerance indices to Phi stress at both concentrations compared with the rest of the genotypes, which had the smallest root systems. The tolerance of the Atlantic genotype to Phi stress was confirmed by the largest tolerance indices for TRL, RPA, RSA, RV, and NRF. The larger root system possessed by Atlantic probably enhanced the uptake of other trace nutrients in the growth media, which contributed to its tolerance. The results of this study are in agreement with those of Wang et al. [66], who postulated that soybean genotypes with larger root systems are more tolerant to cadmium stress. Pandey et al. [67] also reported that genotype PDM-139, a green Gram genotype, with larger root surface area and root volume was the more tolerant genotype to low P compared to those with smaller root surface area and root volume. Several other reports suggest that larger root surface area, root volume, and root hair length of plants under P-stressed growth conditions are characteristics of tolerant genotypes [68,69].

In addition, plant tolerance to Phi-stress was also determined by the hydrogen peroxide (H_2O_2) content formed during stress exposure. Increased formation of reactive oxygen species (ROS) such as H_2O_2 induces oxidative stress due to the toxicity of Phi [70]. Zhang et al. [71] postulated that *Vicia sativa*, which had the lowest H_2O_2 content was more tolerant than *Phaseolus aureus* which had higher H_2O_2 content when both were exposed to cadmium stress. According to Oyarburo et al. [72], H_2O_2 accumulation in leaves is reduced, antioxidant enzyme activities are increased, gene expression is upregulated and accumulation of glucanases and chitinases is induced, which correlates positively with stress tolerance of the plant. In our current study, Phi interference was observed to increase the hydrogen peroxide content in the shoots of the five potato genotypes. Nevertheless, the Atlantic genotype had the lowest increase in H_2O_2 content, indicating that Atlantic is more tolerant to Phi stress than the other genotypes tested. These results indicate that potato genotypes with larger root systems are more tolerant to Phi stress than genotypes with smaller root systems. Antioxidant enzymes play a critical role in plant cell defense against stress-induced cell damage caused by the formation of free radicals, mainly in the form of

ROS. As a result, it has been suggested that increasing antioxidant enzyme activity may improve plant growth and yield. This result is in agreement with Ramos et al. [73], who indicated that increased SOD and CAT activity induced by low selenium concentrations increased leaf yield of *Lactuca sativa* L. Avila et al. [4] noted that CAT activity was low in Pi-sufficient plants but high by 71% in Phi-treated plants. Our current experiment provided similar results: the activities of antioxidant enzymes (e.g., SOD, POD and CAT) increased in all five genotypes in the presence of Phi and No-P treated plants, but decreased in Pi-sufficient plants. These antioxidant enzymes are involved in the detoxification of reactive oxygen species. Recent research has shown that Phi-anion can induce molecular changes that promote stress tolerance, such as activation of guaiacol peroxidase activity and lignin biosynthesis in maize [60], and structural and biochemical changes in periderm and cortex of potato tubers [20]. On the other hand, studies on the effects of Phi on antioxidant enzymes are still rare. All morphological variables of roots were significantly negatively correlated with antioxidant enzyme activities, MDA, H_2O_2 and Pro. Fresh biomass indices also showed significant negative correlation with antioxidant enzyme activity, MDA, H_2O_2 and Pro. The negative correlations observed under Phi stress indicated that Phi had a deleterious effect on potato root morphology and fresh biomass indices. To further confirm the tolerant potato genotype to Phi stress, a comprehensive evaluation analysis was conducted based on principal component analysis. The results evaluated the genotypes in this order: Atlantic, Longshu3, Qingshu9, Longshu6 and Gannong2.

5. Conclusions

Roots are responsible for the uptake of water and inorganic nutrients and are the primary organs affected by phosphite stress. Therefore, adaptation of roots to phosphite stress affects shoot response, physiological functions and plant growth. In the present study, the responses of five potato genotypes to 0.25, 0.5 mM Phi and Pi, and No-P were investigated. The results showed that Phi stress and No-P supply significantly reduced the size of root and shoot systems of the five potato genotypes tested. Nevertheless, the Atlantic genotype with the largest root system showed the highest tolerance to Phi stress by exhibiting the highest tolerance index values for total root length, root projected area, root surface area, root volume, number of root forks, and fresh biomass tolerance. H_2O_2 and MDA levels increased in shoots of all genotypes, but Atlantic genotype showed the least increase, indicating greater tolerance to Phi stress. Major antioxidant enzymes such as CAT, POD and SOD activities and proline content increased under stress conditions. Greater tolerance parameters and lower H_2O_2 contents were obtained from the Atlantic potato genotype under Phi stress, suggesting that potato genotypes with larger root systems may be more tolerant to Phi stress. The tolerance character of the Atlantic genotype was confirmed by comprehensive evaluation analysis using principal component analysis. The obtained results may be very useful for the selection of the genetically modified potato plants using the *ptxD* selection marker gene. However, the concentration and accumulation of P in shoot and root of Pi starved plants were not determined in the present study. The determination of this index will contribute to a better understanding of the mechanism of the negative effect of Phi anion on the physiological and morphological growth of potato genotypes. Therefore, future research should focus on the concentration and accumulation of P in shoot and root of plants grown under Phi stress and also determine the details of the molecular and genetic mechanisms of Phi tolerance in potato genotypes, especially in genotypes with relatively large root systems.

Author Contributions: Conceptualization, R.D. and C.S.; methodology, R.D. and C.S.; soft-ware, T.Q. and D.X.; validation, K.A. and Z.B.; formal analysis, R.D. and C.S.; investigation, R.D. and C.S.; resources, J.B.; data curation, D.X.; writing—original draft preparation, R.D., K.A. and C.S.; writing—review and editing, R.D., K.A. and C.S.; visualization, Z.B.; supervision, J.B.; project administration, J.B.; funding acquisition, J.B. All authors have read and agreed to the published version of the manuscript.

Funding: This work was supported by the National Natural Science Foundation of China (Grant No. 32060502 and 31960442), the Special Fund for Discipline Construction of Gansu Agricultural University (GAU-XKJS-2018-085, GAU-XKJS -2018-084), the special Fund for Talents of Gansu agricultural University (Grant No. 2017RCZX-44) and the Gansu Provincial Department of Education Fund (2019B-073).

Data Availability Statement: Not applicable.

Conflicts of Interest: The authors declare no conflict of interest.

References

1. Kalayu, G. Phosphate solubilizing microorganisms: Promising approach as biofertilizers. *Int. J. Agron.* **2019**, *2019*, 1–7. [CrossRef]
2. Achary, V.M.M.; Ram, B.; Manna, M.; Datta, D.; Bhatt, A.; Reddy, M.K.; Agrawal, P.K. Phosphite: A novel P fertilizer for weed management and pathogen control. *Plant Biotech. J.* **2017**, *15*, 1493–1508. [CrossRef] [PubMed]
3. Manna, M.; Achary, V.M.M.; Islam, T.; Agrawal, P.K.; Reddy, M.K. The development of a phosphite-mediated fertilization and weed control system for rice. *Sci. Rep.* **2016**, *6*, 24941. [CrossRef] [PubMed]
4. Ávila, F.W.; Faquin, V.; da Silva Lobato, A.K.; Ávila, P.A.; Marques, D.J.; Silva Guedes, E.M.; Tan, D.K.Y. Effect of phosphite supply in nutrient solution on yield, phosphorus nutrition and enzymatic behavior in common bean ('*Phaseolus vulgaris*' L.) Plants. *Aust. J. Crop Sci.* **2013**, *7*, 713.
5. Heuer, S.; Gaxiola, R.; Schilling, R.; Herrera-Estrella, L.; López-Arredondo, D.; Wissuwa, M.; Delhaize, E.; Rouached, H. Improving phosphorus use efficiency: A complex trait with emerging opportunities. *Plant J.* **2017**, *90*, 868–885. [CrossRef] [PubMed]
6. Yang, L.; Yang, Z.; Zhong, X.; Xu, C.; Lin, Y.; Fan, Y.; Wang, M.; Chen, G.; Yang, Y. Decreases in soil P availability are associated with soil organic P declines following forest conversion in subtropical China. *CATENA* **2021**, *205*, 105459. [CrossRef]
7. Brödlin, D.; Kaiser, K.; Kessler, A.; Hagedorn, F. Drying and rewetting foster phosphorus depletion of forest soils. *Soil Biol. Biochem.* **2019**, *128*, 22–34. [CrossRef]
8. Gianessi, L.P. The increasing importance of herbicides in worldwide crop production. *Pest Manag. Sci.* **2013**, *69*, 1099–1105. [CrossRef] [PubMed]
9. Parry, M.A.; Hawkesford, M.J. Food security: Increasing yield and improving resource use efficiency. *Proc. Nutr. Soc.* **2010**, *69*, 592–600. [CrossRef]
10. Haque, S.E. How Effective Are Existing Phosphorus Management Strategies in Mitigating Surface Water Quality Problems in the US? *Sustainability* **2021**, *13*, 6565. [CrossRef]
11. Peng, L.; Dai, H.; Wu, Y.; Peng, Y.; Lu, X. A comprehensive review of the available media and approaches for phosphorus recovery from wastewater. *Water Air Soil Pollut.* **2018**, *229*, 1–28. [CrossRef]
12. Powles, S.B.; Yu, Q. Evolution in action: Plants resistant to herbicides. *Annu. Rev. Plant Biol.* **2010**, *61*, 317–347. [CrossRef] [PubMed]
13. Asaduzzaman, M.; Pratley, J.E.; Luckett, D.; Lemerle, D.; Wu, H. Weed management in canola (*Brassica napus* L): A review of current constraints and future strategies for Australia. *Archives Agron. Soil Sci.* **2020**, *66*, 427–444. [CrossRef]
14. Guo, M.; Li, B.; Xiang, Q.; Wang, R.; Liu, P.; Chen, Q. Phosphite translocation in soybean and mechanisms of Phytophthora sojae inhibition. *Pest. Biochem. Physiol.* **2021**, *172*, 104757. [CrossRef] [PubMed]
15. García-Gaytán, V.; Hernández-Mendoza, F.; Coria-Téllez, A.V.; García-Morales, S.; Sánchez-Rodríguez, E.; Rojas-Abarca, L.; Daneshvar, H. Fertigation: Nutrition, stimulation and bioprotection of the root in high performance. *Plants* **2018**, *7*, 88. [CrossRef]
16. McDonald, A.E.; Grant, B.R.; Plaxton, W.C. Phosphite (phosphorous acid): Its relevance in the environment and agriculture and influence on plant phosphate starvation response. *J. Plant Nutr.* **2001**, *24*, 1505–1519. [CrossRef]
17. Solis-Palacios, R.; Hernández-Ramírez, G.; Salinas-Ruiz, J.; Hidalgo-Contreras, J.V.; Gómez-Merino, F.C. Effect and Compatibility of Phosphite with Trichoderma sp. Isolates in the Control of the Fusarium Species Complex Causing Pokkah Boeng in Sugarcane. *Agronomy* **2021**, *11*, 1099. [CrossRef]
18. Cerqueira, A.; Alves, A.; Berenguer, H.; Correia, B.; Gómez-Cadenas, A.; Diez, J.J.; Monteiro, P.; Pinto, G. Phosphite shifts physiological and hormonal profile of Monterey pine and delays Fusarium circinatum progression. *Plant Physiol. Biochem.* **2017**, *114*, 88–99. [CrossRef]
19. Rampersad, S.N. Pathogenomics and management of Fusarium diseases in plants. *Pathogens* **2020**, *9*, 340. [CrossRef]
20. Olivieri, F.P.; Feldman, M.L.; Machinandiarena, M.F.; Lobato, M.C.; Caldiz, D.O.; Daleo, G.R.; Andreu, A.B. Phosphite applications induce molecular modifications in potato tuber periderm and cortex that enhance resistance to pathogens. *Crop Protect.* **2012**, *32*, 1–6. [CrossRef]
21. Silva, O.; Santos, H.; Dalla Pria, M.; May-De Mio, L. Potassium phosphite for control of downy mildew of soybean. *Crop Protect.* **2011**, *30*, 598–604. [CrossRef]
22. Huang, Z.; Carter, N.; Lu, H.; Zhang, Z.; Wang-Pruski, G. Translocation of phosphite encourages the protection against Phytophthora infestans in potato: The efficiency and efficacy. *Pest. Biochem. Physiol.* **2018**, *152*, 122–130. [CrossRef] [PubMed]
23. Mehta, D.; Ghahremani, M.; Pérez-Fernández, M.; Tan, M.; Schläpfer, P.; Plaxton, W.C.; Uhrig, R.G. Phosphate and phosphite have a differential impact on the proteome and phosphoproteome of Arabidopsis suspension cell cultures. *Plant J.* **2021**, *105*, 924–941. [CrossRef] [PubMed]

24. Thao, H.T.B.; Yamakawa, T. Phosphite (phosphorous acid): Fungicide, fertilizer or bio-stimulator? *Soil Sci. Plant Nutr.* **2009**, *55*, 228–234. [CrossRef]
25. César Bachiega Zambrosi, F.; Mattos, D.J.; Syvertsen, J.P. Plant growth, leaf photosynthesis, and nutrient-use efficiency of citrus rootstocks decrease with phosphite supply. *J. Plant Nutr. Soil Sci.* **2011**, *174*, 487–495. [CrossRef]
26. Xi, Y.; Han, X.; Zhang, Z.; Joshi, J.; Borza, T.; Aqa, M.M.; Zhang, B.; Yuan, H.; Wang-Pruski, G. Exogenous phosphite application alleviates the adverse effects of heat stress and improves thermotolerance of potato (*Solanum tuberosum* L.) seedlings. *Ecotoxicol. Environ. Saf.* **2020**, *190*, 110048. [CrossRef]
27. Leong, S.J.; Lu, W.-C.; Chiou, T.-J. Phosphite-mediated suppression of anthocyanin accumulation regulated by mitochondrial ATP synthesis and sugars in Arabidopsis. *Plant Cell Physiol.* **2018**, *59*, 1158–1169. [CrossRef]
28. Trejo-Téllez, L.I.; Estrada-Ortiz, E.; Gómez-Merino, F.C.; Becker, C.; Krumbein, A.; Schwarz, D. Flavonoid, nitrate and glucosinolate concentrations in Brassica species are differentially affected by photosynthetically active radiation, phosphate and phosphite. *Front. Plant Sci.* **2019**, *10*, 371. [CrossRef] [PubMed]
29. Burra, D.D. Defence Related Molecular Signalling in Potato. *Acta Univ. Agric. Sueciae* **2016**, *7*, 1652–6880.
30. Gómez-Merino, F.C.; Trejo-Téllez, L.I. Biostimulant activity of phosphite in horticulture. *Sci. Hort.* **2015**, *196*, 82–90. [CrossRef]
31. Han, X.; Xi, Y.; Zhang, Z.; Mohammadi, M.A.; Joshi, J.; Borza, T.; Wang-Pruski, G. Effects of phosphite as a plant biostimulant on metabolism and stress response for better plant performance in Solanum tuberosum. *Ecotoxicol. Environ. Saf.* **2021**, *210*, 111873. [CrossRef]
32. Gómez-Merino, F.C.; Trejo-Téllez, L.I. Conventional and novel uses of phosphite in horticulture: Potentialities and challenges. *Italus Hortus* **2016**, *23*, 1–13.
33. Quandahor, P.; Lin, C.; Gou, Y.; Coulter, J.A.; Liu, C. Leaf morphological and biochemical responses of three potato (*Solanum tuberosum* L.) cultivars to drought stress and aphid (Myzus persicae Sulzer) infestation. *Insects* **2019**, *10*, 435. [CrossRef]
34. Jia, H.; Wang, Q.; Tao, Q.; Xue, W. Analysis and circuit implementation for the fractional-order Chen system. In Proceedings of the 8th CHAOS Conference Proceedings, Paris, France, 26–29 May 2015.
35. Cabello, R.; De Mendiburu, F.; Bonierbale, M.; Monneveux, P.; Roca, W.; Chujoy, E. Large-scale evaluation of potato improved varieties, genetic stocks and landraces for drought tolerance. *Am. J. Potato Res.* **2012**, *89*, 400–410. [CrossRef]
36. Murashige, T.; Skoog, F. A revised medium for rapid growth and bio assays with tobacco tissue cultures. *Physiol. Plant.* **1962**, *15*, 473–497. [CrossRef]
37. Wilkins, D. The measurement of tolerance to edaphic factors by means of root growth. *New Phytol.* **1978**, *80*, 623–633. [CrossRef]
38. Dawuda, M.M.; Liao, W.; Hu, L.; Yu, J.; Xie, J.; Calderón-Urrea, A.; Jin, X.; Wu, Y. Root tolerance and biochemical response of Chinese lettuce (Lactuca sativa L.) genotypes to cadmium stress. *PeerJ* **2019**, *7*, e7530. [CrossRef]
39. Junglee, S.; Urban, L.; Sallanon, H.; Lopez-Lauri, F. Optimized assay for hydrogen peroxide determination in plant tissue using potassium iodide. *Am. J. Analyt. Chem.* **2014**, *5*, 730. [CrossRef]
40. Hodges, D.M.; DeLong, J.M.; Forney, C.F.; Prange, R.K. Improving the thiobarbituric acid-reactive-substances assay for estimating lipid peroxidation in plant tissues containing anthocyanin and other interfering compounds. *Planta* **1999**, *207*, 604–611. [CrossRef]
41. Dhindsa, R.S.; Plumb-Dhindsa, P.; Thorpe, T.A. Leaf senescence: Correlated with increased levels of membrane permeability and lipid peroxidation, and decreased levels of superoxide dismutase and catalase. *J. Exp. Bot.* **1981**, *32*, 93–101. [CrossRef]
42. Bates, L.S.; Waldren, R.P.; Teare, I. Rapid determination of free proline for water-stress studies. *Plant Soil* **1973**, *39*, 205–207. [CrossRef]
43. Pocketbook, F.S. *World Food and Agriculture*; Food and Agriculture Organization: Rome, Italy, 2015.
44. Vinocur, B.; Altman, A. Recent advances in engineering plant tolerance to abiotic stress: Achievements and limitations. *Curr. Opin. Biotech.* **2005**, *16*, 123–132. [CrossRef]
45. Zhu, J.-K. Plant salt tolerance. *Trends Plant Sci.* **2001**, *6*, 66–71. [CrossRef]
46. Ticconi, C.A.; Delatorre, C.A.; Abel, S. Attenuation of phosphate starvation responses by phosphite in Arabidopsis. *Plant Physiol.* **2001**, *127*, 963–972. [CrossRef] [PubMed]
47. Ram, B.; Fartyal, D.; Sheri, V.; Varakumar, P.; Borphukan, B.; James, D.; Yadav, R.; Bhatt, A.; Agrawal, P.K.; Achary, V.M.M. Characterization of phoA, a Bacterial Alkaline Phosphatase for Phi Use Efficiency in Rice Plant. *Front. Plant Sci.* **2019**, *10*, 37. [CrossRef] [PubMed]
48. Plaxton, W.C.; Carswell, M.C. Metabolic aspects of the phosphate starvation response in plants. In *Plant Responses to Environmental Stresses*; Routledge: Oxfordshire, UK, 2018; pp. 349–372.
49. Carswell, M.C.; Grant, B.R.; Plaxton, W.C. Disruption of the phosphate-starvation response of oilseed rape suspension cells by the fungicide phosphonate. *Planta* **1997**, *203*, 67–74. [CrossRef] [PubMed]
50. Vinas, M.; Mendez, J.C.; Jiménez, V.M. Effect of foliar applications of phosphites on growth, nutritional status and defense responses in tomato plants. *Sci. Hort.* **2020**, *265*, 109200. [CrossRef]
51. Lambers, H.; Plaxton, W.C. Phosphorus: Back to the roots. *Annu. Plant Rev.* **2015**, *48*, 3–22.
52. Devaiah, B.N.; Karthikeyan, A.S.; Raghothama, K.G. WRKY75 transcription factor is a modulator of phosphate acquisition and root development in Arabidopsis. *Plant Physiol.* **2007**, *143*, 1789–1801. [CrossRef]
53. Correa, J.; Postma, J.A.; Watt, M.; Wojciechowski, T. Soil compaction and the architectural plasticity of root systems. *J. Exp. Bot.* **2019**, *70*, 6019–6034. [CrossRef]

54. Vanhees, D.J.; Loades, K.W.; Bengough, A.G.; Mooney, S.J.; Lynch, J.P. The ability of maize roots to grow through compacted soil is not dependent on the amount of roots formed. *Field Crop. Res.* **2021**, *264*, 108013. [CrossRef]

55. Polania, J.; Rao, I.M.; Cajiao, C.; Grajales, M.; Rivera, M.; Velasquez, F.; Raatz, B.; Beebe, S.E. Shoot and root traits contribute to drought resistance in recombinant inbred lines of MD 23–24× SEA 5 of common bean. *Front. Plant Sci.* **2017**, *8*, 296. [CrossRef]

56. Varadarajan, D.K.; Karthikeyan, A.S.; Matilda, P.D.; Raghothama, K.G. Phosphite, an analog of phosphate, suppresses the coordinated expression of genes under phosphate starvation. *Plant Physiol.* **2002**, *129*, 1232–1240. [CrossRef] [PubMed]

57. Lee, T.M.; Tsai, P.F.; Shyu, Y.T.; Sheu, F. The Effects of Phosphite on Phosphate Starvation Responses of *Ulva Lactuca* (Ulvales, Chlorophyta) 1. *J. Phycol.* **2005**, *41*, 975–982. [CrossRef]

58. Schroetter, S.; Angeles-Wedler, D.; Kreuzig, R.; Schnug, E. Effects of phosphite on phosphorus supply and growth of corn (*Zea mays*). *Landbauforsch. Volkenrode* **2006**, *56*, 87.

59. Thao, H.T.B.; Yamakawa, T.; Myint, A.K.; Sarr, P.S. Effects of phosphite, a reduced form of phosphate, on the growth and phosphorus nutrition of spinach (*Spinacia oleracea* L.). *Soil Sci. Plant Nutrit.* **2008**, *54*, 761–768. [CrossRef]

60. Avila, F.W.; Faquin, V.; Araujo, J.L.; Marques, D.J.; Júnior, P.M.R.; da Silva Lobato, A.K.; Ramos, S.J.; Baliza, D.P. Phosphite supply affects phosphorus nutrition and biochemical responses in maize plants. *Aust. J. Crop Sci.* **2011**, *5*, 646–653.

61. Zambrosi, F.C.B.; Mattos, D., Jr.; Quaggio, J.A.; Cantarella, H.; Boaretto, R.M. Phosphorus uptake by young citrus trees in low-P soil depends on rootstock varieties and nutrient management. *Commun. Soil Sci. Plant Anal.* **2013**, *44*, 2107–2117. [CrossRef]

62. Hirosse, E.H.; Creste, J.E.; Custódio, C.C.; Machado-Neto, N.B. In vitro growth of sweet potato fed with potassium phosphite. *Acta Scient. Agron.* **2012**, *34*, 85–91. [CrossRef]

63. Balemi, T.; Negisho, K. Management of soil phosphorus and plant adaptation mechanisms to phosphorus stress for sustainable crop production: A review. *J. Soil Sci. Plant Nutr.* **2012**, *12*, 547–562. [CrossRef]

64. Richardson, A.E.; Barea, J.-M.; McNeill, A.M.; Prigent-Combaret, C. Acquisition of phosphorus and nitrogen in the rhizosphere and plant growth promotion by microorganisms. *Plant Soil* **2009**, *321*, 305–339. [CrossRef]

65. Lambers, H.; Clode, P.L.; Hawkins, H.J.; Laliberté, E.; Oliveira, R.S.; Reddell, P.; Shane, M.W.; Stitt, M.; Weston, P. Metabolic adaptations of the non-mycotrophic Proteaceae to soils with low phosphorus availability. *Annu. Plant Rev. Phosphorus Metab. Plants* **2015**, *48*, 289.

66. Wang, Y.; Yang, R.; Zheng, J.; Shen, Z.; Xu, X. Exogenous foliar application of fulvic acid alleviate cadmium toxicity in lettuce (*Lactuca sativa* L.). *Ecotoxico. Environ. Saf.* **2019**, *167*, 10–19. [CrossRef] [PubMed]

67. Pandey, R.; Meena, S.K.; Krishnapriya, V.; Ahmad, A.; Kishora, N. Root carboxylate exudation capacity under phosphorus stress does not improve grain yield in green gram. *Plant Cell Rep.* **2014**, *33*, 919–928. [CrossRef]

68. Gahoonia, T.S.; Nielsen, N.E. Root traits as tools for creating phosphorus efficient crop varieties. *Plant Soil* **2004**, *260*, 47–57. [CrossRef]

69. Wu, P.; Ma, X.; Tigabu, M.; Wang, C.; Liu, A.; Oden, P.C. Root morphological plasticity and biomass production of two Chinese fir clones with high phosphorus efficiency under low phosphorus stress. *Can. J. For. Res.* **2011**, *41*, 228–234. [CrossRef]

70. Tripathi, D.K.; Singh, S.; Singh, V.P.; Prasad, S.M.; Dubey, N.K.; Chauhan, D.K. Silicon nanoparticles more effectively alleviated UV-B stress than silicon in wheat (Triticum aestivum) seedlings. *Plant Physiol. Biochem.* **2017**, *110*, 70–81. [CrossRef] [PubMed]

71. Zhang, F.; Zhang, H.; Wang, G.; Xu, L.; Shen, Z. Cadmium-induced accumulation of hydrogen peroxide in the leaf apoplast of Phaseolus aureus and Vicia sativa and the roles of different antioxidant enzymes. *J. Haz. Mat.* **2009**, *168*, 76–84. [CrossRef] [PubMed]

72. Oyarburo, N.S.; Machinandiarena, M.F.; Feldman, M.L.; Daleo, G.R.; Andreu, A.B.; Olivieri, F.P. Potassium phosphite increases tolerance to UV-B in potato. *Plant Physiol. Biochem.* **2015**, *88*, 1–8. [CrossRef]

73. Ramos, S.; Faquin, V.; Guilherme, L.; Castro, E.; Ávila, F.; Carvalho, G.; Bastos, C.; Oliveira, C. Selenium biofortification and antioxidant activity in lettuce plants fed with selenate and selenite. *Plant Soil Environ.* **2010**, *56*, 584–588. [CrossRef]

horticulturae

Article

Twenty-Years of Hop Irrigation by Flooding the Inter-Row Did Not Cause a Gradient along the Row in Soil Properties, Plant Elemental Composition and Dry Matter Yield

Sandra Afonso [1,2], **Margarida Arrobas** [1] **and Manuel Ângelo Rodrigues** [1,*]

[1] Centro de Investigação de Montanha, Instituto Politécnico de Bragança, Campus de Santa Apolónia, 5300-253 Bragança, Portugal; sandraafonso@ipb.pt (S.A.); marrobas@ipb.pt (M.A.)

[2] Faculdade de Ciências, Universidade do Porto, Rua do Campo Alegre, s/n, 4169-007 Porto, Portugal

* Correspondence: angelor@ipb.pt; Tel.: +351-273303260

Abstract: In hops (*Humulus lupulus* L.), irrigation by flooding the inter-row can carry away suspended particles and minerals, causing gradients in soil fertility. The effect of more than 20 years of flooding irrigation on soil and plants was evaluated in two hop fields by measuring soil and plant variables in multiple points along the rows. In a second experiment 1000 kg ha^{-1} of lime was applied and incorporated into the soil to assess whether liming could moderate any gradient created by the irrigation. At different sampling points along the rows, significant differences were recorded in soil properties, plant elemental composition and dry matter yield, but this was not found to exist over a continuous gradient. The variations in cone yield were over 50% when different sampling points were compared. However, this difference cannot be attributed to the effect of irrigation, but rather to an erratic spatial variation in some of the soil constituents, such as sand, silt and clay. Flooding irrigation and frequent soil tillage resulted in lower porosity and higher soil bulk density in the 0.0–0.10 m soil layer in comparison to the 0.10–0.20 m layer. In turn, porosity and bulk density were respectively positively and negatively associated with crop productivity. Thus, irrigation and soil tillage may have damaged the soil condition but did not create any gradient along the row. The ridge appeared to provide an important pool of nutrients, probably caused by mass flow due to the evaporation from it and a regular supply of irrigation water to the inter-row. Liming raised the soil pH slightly, but had a relevant effect on neither soil nor plants, perhaps because of the small amounts of lime applied.

Keywords: *Humulus lupulus* L.; soil porosity; soil bulk density; liming; hop ridges

Citation: Afonso, S.; Arrobas, M.; Rodrigues, M.Â. Twenty-Years of Hop Irrigation by Flooding the Inter-Row Did Not Cause a Gradient along the Row in Soil Properties, Plant Elemental Composition and Dry Matter Yield. *Horticulturae* **2021**, *7*, 194. https://doi.org/10.3390/horticulturae7070194

Academic Editors: Jolanta Jaroszuk-Ściseł, Małgorzata Majewska and Agnieszka Hanaka

Received: 18 June 2021
Accepted: 13 July 2021
Published: 15 July 2021

Publisher's Note: MDPI stays neutral with regard to jurisdictional claims in published maps and institutional affiliations.

1. Introduction

Hop plants (*Humulus lupulus* L.) require an adequate supply of water during the growing season to sustain their huge canopy [1]. In most of the hop producing regions of the world, the crop needs to be irrigated, particularly in lower latitudes of reduced precipitation in summer. Although hop fields have started to be drip irrigated all over the world, there is a long tradition of surface watering of this crop, by flooding the space between rows [1,2]. In this kind of surface or furrow irrigation system, water is applied at the top end of each furrow (in hops to the inter-row space) and flows down the field under the influence of gravity [3]. This is still the most commonly used irrigation method for hops in northern Portugal [4]. The water use efficiency with this irrigation technique is highly dependent on the field gradient and water infiltration rate, which can vary considerably, inducing spatial and temporal variability in the main soil properties [5]. In addition, flood irrigation can affect the spatial distribution of soil physicochemical properties which may exacerbate the spatial variability in crop growth and yield [6].

Flood irrigation can have a major impact on soil properties by varying salinity, redox potential, compaction and/or porosity [7–10]. Furthermore, hop fields which are flood-irrigated need to be frequently tilled to control summer weeds and to reduce soil compaction and superficial crusts in the short term. This allows a better infiltration of water, but that can also have a negative impact on the soil in the long term [11,12]. Soil compaction, increased by furrow irrigation, may also reduce soil drainage and aeration, contributing to the reduction of soil redox potential which influences soil chemistry and plant nutrient availability [10,13]. The degree of compaction of a soil can be assessed by measuring some physical properties, such as bulk density and porosity [12,13]. As the soil becomes more compact, bulk density increases and soil porosity decreases, which reduces water and air diffusion into the soil [11,14]. In some hop fields in northern Portugal it was found that the decrease in soil redox potential, associated with an excess of water and/or poor drainage, was the main cause of the spatial variability found in crop growth and yield [4].

Soil pH is another relevant issue in hop production. The range of pH most suited for growing hops is considered to be between 5.7 and 7.5 [15,16]. The application of lime is recommended for acidic soils, and a positive relationship has been found between the increase in soil pH and hop yield [15,17]. However, the effect of liming on crops can also vary with the irrigation system. Some researchers have studied the influence of liming in rice under flooding conditions, since great interactions between flooding, soil acidity and nutritional disorders are usually found [18–20]. In hops, these interactions are less well known, or the response to liming, but it is believed that it may be relevant enough to be studied, since the crop continues to be irrigated by flooding in several parts of the world.

This study evaluated the variation in soil properties and nutritional status and the productivity of hop plants created along the rows by flooding irrigation. As a second line of study, the effect of the application of lime on soil properties and on hop nutritional status, growth and yield was evaluated, to ascertain if the application of lime could compensate for the variability created by the irrigation system. Both lines of study were carried out in commercial hop fields which had been flood-irrigated for over 20 years.

2. Materials and Methods

2.1. General Experimental Conditions

The field experiments were carried out during two growing seasons (2017 and 2018) on a commercial farm located in Pinela (41°40′33.6″ N; 6°44′32.7″ W), Bragança, north-eastern Portugal. A detailed location of field experiments is shown in Figure 1.

Figure 1. Map of Portugal indicating Pinela (**left**) and the two hop fields identified in this study as field 1 and field 2 (**right**). Images from https://www.google.com/maps/place/Pinela (accessed on 7 July 2021).

The region benefits from a Mediterranean-type climate, with an annual average temperature and accumulated precipitation of 12.7 °C and 772.8 mm, respectively. The average monthly temperatures and precipitation recorded during the experimental period are shown in Figure 2.

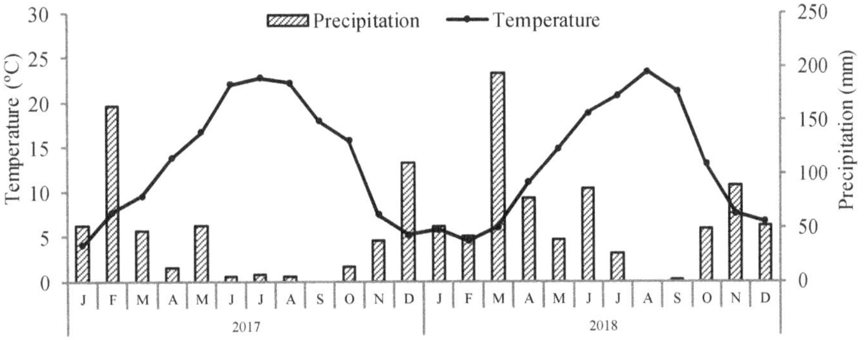

Figure 2. Average monthly temperatures and precipitation recorded during the experimental period at a weather station located in Bragança, north-eastern Portugal.

The hop plots where the study was undertaken are ~2 ha in size each, with the rows having a length ranging from 150 to 180 m, and established with the cultivar Nugget. The fields are arranged in a 7 m conventional high trellis system, with concrete poles connected with steel cables, in a "V" design system. The farmer has managed the fields by flooding irrigation since the hop crop was installed more than 20 years ago. Several tillage passes (3 to 4) are performed every year to remove the crusts and facilitate water infiltration. The fertilization programme includes the application of a compound nitrogen (N): phosphorus (P): potassium (K) fertilizer (7:14:14, 7% N, 14% P_2O_5, 14% K_2O) early in the spring, followed by two applications of N fertilizer (ammonium nitrate, 27% N) as a side-dressing, totalling ~150, 44 and 83 kg ha^{-1} of N, P and K, respectively. The farmer also follows a phytosanitary programme for crop protection against pests and diseases.

2.2. Field Experiments and Soil and Plant Sampling

The first experiment (Experiment 1) was carried out during the growing season of 2017 in two hop fields. It consisted of the evaluation of soil properties, plant nutritional status and crop yield, searching for any gradient along the rows created by the irrigation system. The rows used in this experiment were divided into nine segments of equivalent length, creating nine positions (P1, P2, … , P9) for soil and plant sampling. The soil was sampled between rows and on the ridges to a depth of 0.0 to 0.2 m. Three rows and inter-rows of hops were used to create three replicates for each position. Each soil sample for analysis resulted from six sampling points (composite samples). The soil was sampled by using an open-face auger.

For the determination of soil bulk density and porosity, a different approach to soil sampling was followed. It was found unnecessary to sample in the ridges since no compaction was expected in this part. Instead, the soil was sampled at two different depths, 0.0 to 0.10 m and 0.10 to 0.20 m. Due to the increased difficulty of sampling, particularly in the 0.10 to 0.20 m layer, only five positions were considered (P1, P3, P5, P7 and P9) and sampled in three replicates. For these analyses, undisturbed soil cores were taken by using appropriate cylinders of 100 cm^3. Soil samplings were carried out on 10 March 2017.

The plants used in this experiment for the evaluation of their nutritional status and crop productivity were randomly selected and marked when plant height was close to 3 m (to avoid using very atypical plants) and close to each of the positions used for soil sampling. Leaf sampling for crop nutritional status assessment was done at ~2 m in height, on 17 July 2017. At harvest (1 September 2017), plant biomass was cut at ground level. Subsequently, the aboveground biomass was separated into leaves, stems and cones and weighed fresh. Subsamples of each plant part were weighed fresh again and then oven-dried at 70 °C and weighed dry for determination of dry matter yield.

The second experiment (Experiment 2) consisted of the application of 1000 kg ha^{-1} of lime (55% CaCO$_3$, 28% CaO and 20% MgO) in February 2017, to assess the liming effects on soil properties and plants in comparison to the untreated control. This experiment was also carried out in two hop plots. The general methodology for soil and plant sampling was similar to that reported for Experiment 1, consisting of marking nine positions along the rows. The soil was sampled on 4 January 2019, only between the rows, at 0.0–0.20 m soil depth, using an open-face auger. Leaf samples were taken at ~2 m in height, on 17 July 2017 and 18 July 2018. At harvest (1 September 2017 and 31 August 2018), plant biomass was cut at ground level and treated as reported for Experiment 1.

2.3. Laboratory Analyses

The undisturbed soil samples from Experiment 1 were oven-dried at 105 °C and weighed. Soil bulk density was estimated from the weight of dry soil divided by the volume of the cylinder. Soil porosity was determined as the ratio of nonsolid volume (soil particle density—bulk density) to the total volume of soil (soil particle density) [21]. The other soil samples from Experiments 1 and 2 were oven-dried at 40 °C and sieved in a mesh of 2 mm. The samples were analysed for pH (H$_2$O and KCl), electrical conductivity (soil:solution, 1:2.5), exchangeable complex (ammonium acetate, pH 7.0) and organic carbon (C) (Walkley–Black method). Extractable P and K were determined by a combination of ammonium lactate and acetic acid buffered at pH 3.7. Soil boron (B) was extracted by hot water and the extracts analysed by the azomethine-H method. More details of these analytical procedures are given in Van Reeuwijk [22]. Other micronutrients [copper (Cu), iron (Fe), zinc (Zn), and manganese (Mn)] were determined by atomic absorption spectrometry after extraction with ammonium acetate and EDTA, following the methodology reported by Lakanen and Erviö [23].

Tissue samples (leaves, stems and cones) from both experiments were oven-dried at 70 °C and ground. Elemental tissue analyses were performed by Kjeldahl (N), colorimetry (B and P), flame emission spectrometry (K) and atomic absorption spectrophotometry (calcium (Ca), magnesium (Mg), Cu, Fe, Zn and Mn) methods after nitric digestion of the samples [24].

2.4. Data Analysis

Data was subjected to analysis of variance, according to the experimental designs, using SPSS program version 25. When significant differences were found between the experimental treatments, the means were separated by the Tukey HSD (sampling position) and Student's-t (field, sampling site, lime treatment) tests (α = 0.05). Linear regression analysis was performed to understand the effects of gradient on soil properties and plant nutritional status and productivity in Experiment 1 and the relationship between soil pH and plant variables in Experiment 2. The relation between the variables was obtained through correlation analysis with the Pearson coefficient, when the assumption of normality and linearity was accomplished; when this was not the case, the Spearman coefficient was used.

3. Results

3.1. Gradients in Soil and Plants along the Rows

3.1.1. Soil Properties

The silt and sand contents varied significantly between the sampling positions (Table 1). The two fields also differed significantly in clay and sand content. The soil bulk density and soil porosity varied significantly between the sampling positions and fields but in the opposite way. The interaction between sampling position and field was significant for soil porosity, which means that the effect of the irrigation on this variable depended on the field.

Table 1. Soil separates and soil bulk density and porosity from samples collected at 0.0–0.20 m depth, in March 2017, as a function of sampling position (1, … , 9), and field. Means followed by the same letter are not significantly different by Tukey HSD (sampling position) or Student's *t* (field) tests ($\alpha = 0.05$).

	Clay	Silt (%)	Sand	Bulk Density (kg dm^{-3})	Porosity (%)
Sampling position (P)					
Lowest value	15.6 a	34.5 a	59.7 a	1.26 a	52.1 a
Highest value	11.8 a	28.5 b	49.9 b	1.18 b	48.0 b
Field (F)					
Field 1	16.0 a	33.2 a	50.8 b	1.25 a	49.1 b
Field 2	11.5 b	32.1 a	56.4 a	1.21 b	50.8 a
Prob (P)	0.2770	0.0386	0.0307	0.0143	0.0020
Prob (F)	0.0005	0.3741	0.0072	0.0260	0.0259
Prob (P × F)	0.8998	0.0432	0.1221	0.0874	0.0256

The soil bulk density was higher in the soil surface (0.0–0.1 m) when compared to the deeper (0.1–0.2 m) layer (Figure 3). The soil bulk density did not vary significantly along the rows for both soil depths. The soil porosity, in turn, was lower in the surface layer, and the gradient found along the rows was not significant for any of the soil layers. The soil bulk density and porosity varied significantly between the two fields, but the gradients found along the rows were not statistically significant.

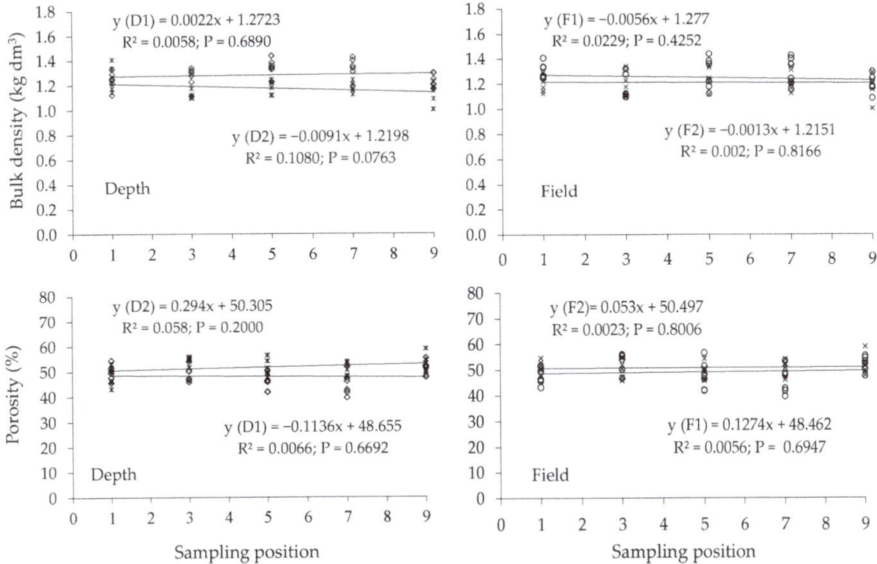

Figure 3. Soil bulk density and porosity from soil samples taken at different sampling positions along the gradient of irrigation (1, … , 9), as a function of depth (D1, 0.0–0.10 m; D2, 0.10–0.20 m) and field (F1, field 1; F2, field 2).

Some other soil properties determined from the samples collected at 0.0–0.20 m depth varied significantly between sampling sites, sampling positions and fields (Table 2). Extractable P and K, conductivity, organic C, CEC and extractable Zn and B showed significantly higher values in the samples collected in the ridges. However, soil pH (H$_2$O and KCl), base saturation and extractable Mn were significantly higher in the samples collected in the inter-rows. Most of the soil properties varied significantly between the sampling positions, the exceptions being soil pH, conductivity and extractable K. Soil properties also differed significantly between fields, except for soil conductivity. Significant interaction between the sampling site and the field was found for extractable P, conductivity, exchangeable Ca and extractable Fe. Significant interaction between the sampling position and the field was found for organic C and extractable Fe, Mn, Zn, Cu and B. No significant interaction was found between the three factors of this experiment.

3.1.2. Hop Dry Mater Yield and Leaf Nutrient Concentration

Aboveground dry biomass (stems, leaves, cones and total) in Field 1 showed a clear tendency for a decrease along the rows (Figure 4). However, the decrease was only statistically significant for stem dry matter yield (DMY). For all plant parts, the coefficients of determination (R^2) were not particularly high, which helps to explain the lack of significant correlation between the two variables. In Field 2, no clear tendency was found in aboveground DMY.

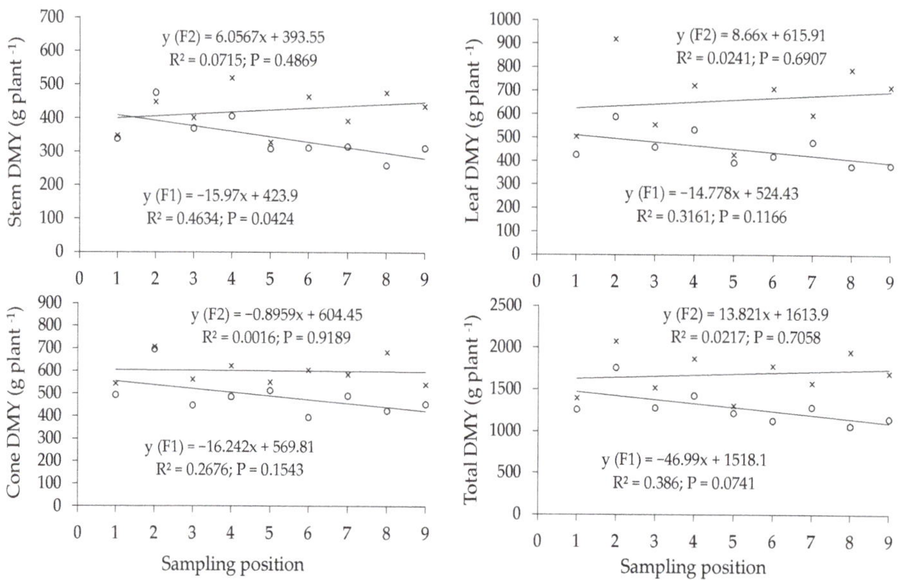

Figure 4. Dry matter yield (DMY) from plants collected at harvest in September 2017, at different sampling positions along the gradient of irrigation (1, . . . , 9), and as a function of field (F1, field 1; F2, field 2).

N concentration in the leaves taken at 2 m height did not vary significantly along the rows in any of the fields (Figure 5). Leaf P also did not vary significantly along the rows but the values in Field 1 were lower than in Field 2. Leaf K levels did not vary significantly along the rows in Field 1 but increased significantly in Field 2. Leaf Ca and Mg levels showed a slight tendency to increase in both fields but without statistical significance. In general, the micronutrients showed even more erratic tendencies when the values of the two fields were compared and only the values of leaf Cu showed a significant decrease along the rows in Field 1.

Table 2. Selected soil properties from samples collected at 0.0–0.20 m depth, in March 2017, as a function of sampling site, sampling position (1, … , 9), and field. Means followed by the same letter are not statistically different by Tukey HSD (Sampling position) or Student's t (Sampling site and Field) tests (α = 0.05).

	Extract. K	Extract. P	Conductivity	pH_{H_2O}	pH_{KCl}	Organic C	Exchan. Ca	CEC	Base Saturation	Extract. Fe	Extract. Mn	Extract. Zn	Extract. Cu	Extract. B
	(mg K_2O kg^{-1})[1]	(mg P_2O_5 kg^{-1})[1]	(μs/m)			(g kg^{-1})	(cmolc kg^{-1})[2]		(%)		(mg kg^{-1})[3]			(mg kg^{-1})[4]
Sampling site (S)														
Ridge	310.4 a	349.7 a	78.9 a	5.42 b	4.42 b	20.9 a	4.94 a	7.34 a	87.7 b	213.4 a	166.8 b	5.03 a	7.86 a	1.16 a
Inter-row	246.5 b	292.7 b	54.7 b	5.53 a	4.52 a	18.3 b	4.77 a	6.63 b	89.8 a	222.1 a	194.2 a	4.28 b	7.84 a	0.79 b
Sampling position (P)														
Lowest value	228.3 a	195.0 c	60.2 a	5.42 a	4.35 a	17.8 b	3.80 c	6.21 b	86.0 b	178.1 c	136.0 d	3.47 d	6.64 d	0.79 c
Highest value	313.6 a	399.7 a	70.0 a	5.60 a	4.64 a	21.3 a	5.73 a	7.68 a	92.1 a	262.2 a	217.3 a	5.64 a	9.22 a	1.17 a
Field (F)														
Field 1	361.8 a	286.3 b	65.9 a	5.75 a	4.67 a	18.4 b	5.22 a	7.31 a	93.6 a	207.2 b	134.1 b	4.96 a	10.35 a	0.91 b
Field 2	195.1 b	356.2 a	67.6 a	5.21 b	4.27 b	20.8 a	4.49 b	6.66 b	84.0 b	228.4 a	226.8 a	4.35 b	5.36 b	1.04 a
Prob (S)	<0.0001	0.0022	<0.0001	0.0028	0.0236	<0.0001	0.4028	0.0069	0.0179	0.2836	0.0003	0.0009	0.9386	<0.0001
Prob (P)	0.0758	<0.0001	0.9345	0.3221	0.0710	0.0058	<0.0001	0.0268	0.0012	<0.0001	<0.0001	<0.0001	0.0003	0.0007
Prob (F)	<0.0001	0.0002	0.5991	<0.0001	<0.0001	<0.0001	0.0004	0.0130	<0.0001	0.0100	<0.0001	0.0070	<0.0001	0.0075
Prob (S × P)	0.6500	0.9648	0.9341	0.9859	0.7781	0.8954	0.7956	0.7826	0.0938	0.9836	0.0644	0.8658	0.2274	0.5881
Prob (S × F)	0.8057	0.0001	0.0013	0.8860	0.7597	0.7527	0.0287	0.1220	0.0901	0.0053	0.6578	0.0738	0.1982	0.0810
Prob (P × F)	0.0904	0.0374	0.8663	0.1846	0.2467	0.0199	0.8269	0.6486	0.1179	<0.0001	<0.0001	0.0018	<0.0001	0.0497
Prob (S × P × F)	0.4096	0.9991	0.9791	0.5470	0.9433	0.5569	0.9503	0.9365	0.2005	0.9005	0.5541	0.6590	0.6793	0.4975

[1] Egner–Rhiem; [2] ammonium acetate, pH 7; [3] ammonium acetate and EDTA; [4] azomethine-H.

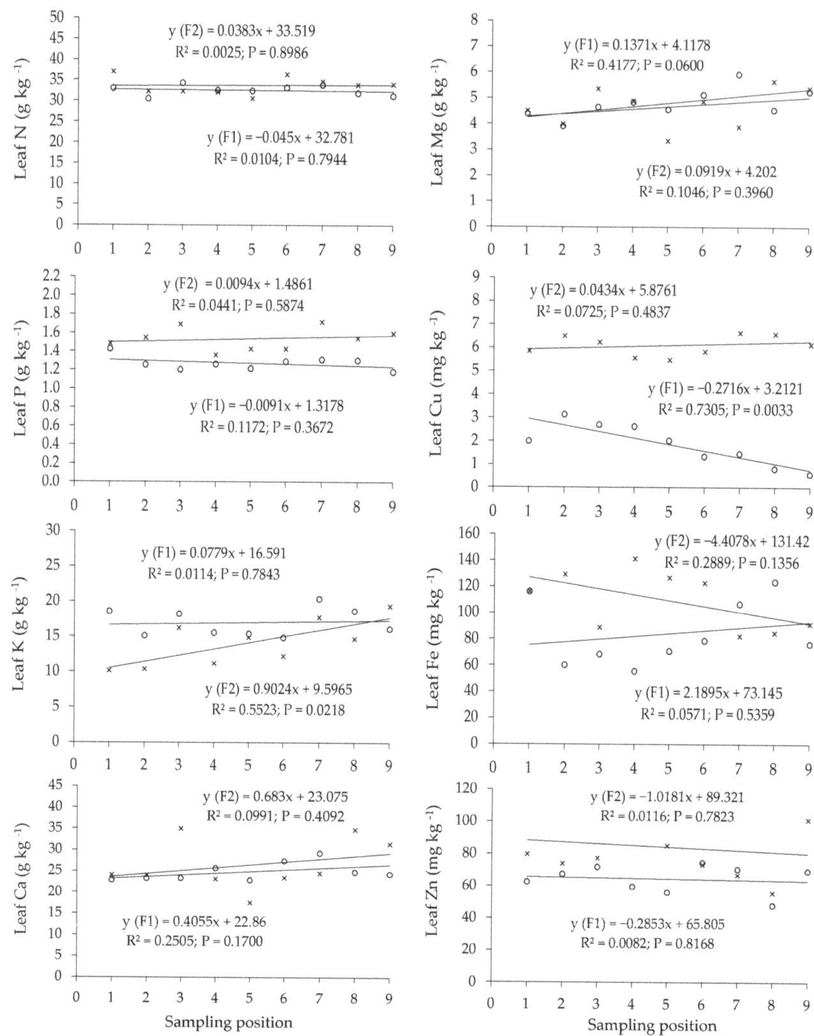

Figure 5. Leaf nutrient concentration from samples taken at 2 m height and at different sampling positions along the gradient of irrigation (1, . . . , 9), as a function of field (F1, field 1; F2, field 2).

3.1.3. Correlation Analysis between Soil Properties and Plant Dry Matter Yield

Soil bulk density and porosity correlated in a different way with soil pH (H_2O and KCl), leaf P and total DMY (Table 3). That is, the correlations of soil pH were positive for soil bulk density and negative for soil porosity at 0.0–0.10 m depth. Leaf P concentration was significantly and negatively correlated with soil bulk density at 0.10–0.20 m depth, in contrast to the positive correlation found with soil porosity. Leaf Fe concentration was found significant and negatively correlated only with soil porosity at 0.10–0.20 m depth. The strongest correlations were found for total DMY with soil bulk density (r = −0.706) and soil porosity (r = 0.714), both at 0.10–0.20 m depth.

Table 3. Correlation coefficients of soil bulk density and soil porosity of samples collected in the inter-rows, at different depths (D1, 0–0.10 m; and D2, 0.10–0.20 m), with soil pH (H_2O and KCl) and leaf nutrient concentrations from samples taken at 2 m height in July 2017, and total and cone dry matter yield (DMY) from plants collected in September 2017.

	Soil [†]							Leaf Nutrient [†]						DMY	
	pH_{H_2O}	pH_{KCl}	N	P	K	Ca	Mg	Fe	Mn	Cu	Zn	B	Total [‡]	Cone [†]
					(g kg^{-1})					(mg kg^{-1})			(g Plant^{-1})	
Soil bulk density														
D1 (0.0–0.10 m depth)	0.422 *	0.440 *	−0.442	−0.190	0.043	−0.209	−0.130	0.128	−0.067	−0.322	−0.515	−0.333	−0.243	0.139
D2 (0.10–0.20 m depth)	0.087	0.062	−0.239	−0.690 *	−0.046	−0.512	−0.249	0.626	−0.220	−0.525	−0.312	−0.459	−0.706 *	−0.128
Soil porosity														
D1 (0.0–0.10 cm depth)	−0.396 *	−0.400 *	0.418	−0.038	−0.110	0.055	0.075	−0.055	0.139	0.097	0.370	0.285	0.168	−0.261
D2 (0.10–0.20 m depth)	−0.020	0.015	0.248	0.646 *	<0.0001	0.406	0.185	−0.632 *	0.273	0.535	0.309	0.418	0.714 *	0.127
Soil separates														
Clay	0.806 **	0.542 *	−0.241	−0.563 *	0.639 **	0.038	0.057	−0.389	−0.197	−0.773 **	−0.459	−0.707 **	−0.676 **	−0.666 **
Silt	0.387	0.129	−0.220	0.066	0.323	−0.049	−0.084	0.042	−0.292	−0.179	−0.042	−0.503 *	−0.117	−0.005
Sand	−0.703 **	−0.391	0.276	0.339	−0.562 *	0.034	0.046	0.247	0.300	0.571 *	0.307	0.639 **	0.427	0.410

Significant correlations at the correspondent levels of * 0.05 and ** 0.01; [†] Spearman and [‡] Pearson correlation coefficients.

Significant correlations were found for soil clay content, positive for soil pH and leaf K, and negative for leaf P, leaf Cu, total DMY and cone DMY. In contrast, soil sand content correlated significantly and negatively with soil pH (H_2O) and leaf K, and positively with leaf Cu and B. Soil silt content correlated significantly and negatively with leaf B.

3.2. Liming Experiment

3.2.1. Soil Properties

Most soil properties, such as extractable K, P, Mn, Zn, Cu, B, conductivity and pH presented significantly higher values in the limed plot in comparison to the untreated control (Table 4). Exchangeable Ca and CEC showed higher values in the limed plot but not significantly different to those observed in the control. Significant differences between the two fields used in this experiment were also found for most of the soil properties, the values of extractable K, P, Zn, Cu and B, conductivity, pH, exchangeable Ca, CEC and base saturation being significantly higher in Field 1. Only extractable Fe was significantly higher in Field 2. The interaction between liming and field was significant for extractable K, conductivity, and pH.

3.2.2. Plant Response to Liming

The concentration of nutrients in the leaves taken at 2 m height showed significant differences between treatments for leaf P in 2017 and for leaf Fe and B in 2018 (Table 5). The values reported for P and Fe were significantly higher in the limed plots, and those reported for B were significantly higher in the control. Total and cone DMY were significantly lower in the limed plots with the exception of total DMY in 2017, whose differences between treatments were not statistically significant. When comparing fields, significant differences were found for some nutrients and total and cone DMY. However, only leaf concentrations of K, Cu and B, and total and cone DMY, maintained the same trend in both years and fields. In 2017, significant interaction between the liming treatment and the field was only found for leaf N and Mn and in 2018 for leaf P and total DMY.

3.2.3. Correlation Analysis between Soil pH and Plant Variables

Significant correlations between the soil pH (H_2O and KCl) and leaf nutrient concentration were found for several nutrients, but a similar trend over the two years was found only for leaf Cu and B, both presenting negative correlations with soil pH (Table 6). Soil pH and leaf P, for instance, showed a negative correlation in 2017 and a positive correlation in 2018. Significant and negative relations between soil pH and total and cone DMY were also found for the first year of plant sampling.

Table 4. Soil properties from samples collected at 0.0–0.2 m depth, in January 2019, in the inter-rows, as a function of liming treatment and field. Means followed by the same letter are not statistically different by Student's t test (α = 0.05).

	Extract. K	Extract. P	Conductivity	pH_{H_2O}	pH_{KCl}	Organic C	Exchan. Ca	CEC	Base Saturation	Extract. Fe	Extract. Mn	Extract. Zn	Extract. Cu	Extract. B
	(mg K_2O kg^{-1})[1]	(mg P_2O_5 kg^{-1})[1]	(μs/m)			(g kg^{-1})	(cmolc kg^{-1})[2]		(%)		(mg kg^{-1})[3]			(mg kg^{-1})[4]
Treatment (T)														
Control	82.5 b	162.2 b	69.8 b	5.20 b	4.09 b	14.3 a	3.76 a	7.38 a	82.7 a	160.4 a	96.1 b	2.54 b	4.74 b	0.82 b
Lime	100.8 a	216.7 a	85.5 a	5.40 a	4.35 a	14.6 a	3.90 a	7.64 a	79.5 a	165.8 a	123.7 a	3.49 a	5.74 a	1.11 a
Field (F)														
Field 1	126.5 a	244.6 a	82.6 a	5.54 a	4.53 a	14.7 a	4.47 a	8.21 a	91.2 a	153.2 b	105.8 a	3.32 a	7.23 a	1.11 a
Field 2	56.8 b	134.3 b	72.6 b	5.06 b	3.91 b	14.2 a	3.19 b	6.81 b	71.1 b	173.0 a	114.0 a	2.71 b	3.25 b	0.82 b
Prob (T)	<0.0001	<0.0001	<0.0001	<0.0001	<0.0001	0.6084	0.4582	0.1793	0.0638	0.5629	0.0009	<0.0001	0.0001	0.0002
Prob (F)	<0.0001	<0.0001	0.0002	<0.0001	<0.0001	0.4112	<0.0001	<0.0001	<0.0001	0.0357	0.3074	0.0052	<0.0001	0.0003
Prob (T × F)	0.0009	0.1647	<0.0001	<0.0001	<0.0001	0.8899	0.1608	0.4658	0.0098	0.8719	0.4696	0.4311	0.0661	0.0079

[1] Egner–Rhiem; [2] ammonium acetate, pH 7; [3] ammonium acetate and EDTA; [4] azomethine-H.

Table 5. Leaf concentration of macro and micronutrients in July 2017 and 2018, from samples collected at 2 m height, and total and cone dry matter yield (DMY) from plants collected in August 2017 and September 2018, as a function of liming treatment and field. Means followed by the same letter are not statistically different by Student's t test (α = 0.05).

		N	P	K	Ca	Mg	Fe	Mn	Cu	Zn	B	Total DMY	Cone DMY
		(g kg^{-1})					(mg kg^{-1})					(g $Plant^{-1}$)	
	Treatment (T)												
	Control	3.31 a	0.14 b	1.56 a	2.57 a	0.47 a	96.7 a	374.1 a	3.97 a	74.3 a	71.7 a	1483 a	544.3 a
	Lime	3.39 a	0.15 a	1.66 a	2.72 a	0.52 a	95.3 a	316.9 a	4.59 a	82.9 a	69.3 a	1379 a	441.2 b
	Field (F)												
2017	Field 1	3.39 a	0.13 b	1.76 a	2.56 a	0.49 a	87.9 b	355.9 a	2.54 b	64.35 b	63.29 b	1271 b	446.2 b
	Field 2	3.31 a	0.16 a	1.46 b	2.73 a	0.50 a	104.2 a	335.1 a	6.03 a	92.88 a	77.70 a	1591 a	539.3 a
	Prob. (T)	0.2043	0.0440	0.3130	0.3597	0.1445	0.8597	0.0703	0.2111	0.2636	0.2542	0.2139	0.0033
	Prob. (F)	0.2180	<0.0001	0.0049	0.2909	0.7214	0.0447	0.5024	<0.0001	0.0007	<0.0001	0.0005	0.0073
	Prob. (T × F)	0.0024	0.7290	0.8820	0.9293	0.4216	0.2588	0.0358	0.1327	0.2608	0.0918	0.3402	0.5728

Table 5. *Cont.*

		N	P	K	Ca	Mg	Fe	Mn	Cu	Zn	B	Total DMY	Cone DMY
				(g kg^{-1})					(mg kg^{-1})			(g Plant^{-1})	
	Treatment (T)												
	Control	3.48 a	0.19 a	3.26 a	1.53 a	0.56 a	94.6 b	513.0 a	6.93 a	21.0 a	57.5 a	1681 a	475.2 a
	Lime	3.56 a	0.19 a	3.19 a	1.57 a	0.61 a	109.3 a	495.2 a	7.07 a	19.5 a	51.6 b	1407 b	380.2 b
2018	Field (F)												
	Field 1	3.55 a	0.20 a	3.83 a	1.43 b	0.62 a	91.0 b	408.9 b	6.26 b	20.51 a	52.37 b	1421 b	428.9 a
	Field 2	3.49 a	0.17 b	2.63 b	1.67 a	0.55 b	112.9 a	599.4 a	7.73 a	19.96 a	56.70 a	1666 a	426.4 a
	Prob. (T)	0.2155	0.8374	0.6215	0.4655	0.0550	0.0035	0.7162	0.5532	0.1170	0.0024	0.0016	0.0021
	Prob. (F)	0.3461	<0.0001	<0.0001	0.0001	0.0279	<0.0001	0.0004	<0.0001	0.5648	0.0218	0.0043	0.9304
	Prob. (T × F)	0.6601	0.0185	0.2051	0.0983	0.2205	0.9394	0.8796	0.6927	0.1228	0.4463	0.0057	0.3068

Table 6. Spearman correlation coefficients for soil pH (H_2O and KCl) from samples collected in January 2019, at 0.0–0.20 m depth, in the inter-rows, with leaf nutrient concentration from samples collected at 2 m height in July (2017 and 2018), and total and cone dry matter yield (DMY) from plant samples collected in August 2017 and September 2018.

					Leaf Nutrient Concentration							DMY	
Soil	N	P	K	Ca	Mg	Fe	Mn	Cu	Zn	B		Total	Cone
			(g kg^{-1})					(mg kg^{-1})				(g Plant^{-1})	
						2017							
pH$_{H_2O}$	0.492 *	−0.648 **	0.323	−0.131	−0.088	−0.157	0.324	−0.531 **	−0.503 *	−0.788 **		−0.635 **	−0.657 **
pH$_{KCl}$	0.519 **	−0.651 **	0.318	−0.068	−0.076	−0.137	0.315	−0.524 **	−0.538 **	−0.787 **		−0.563 **	−0.590 **
						2018							
pH$_{H_2O}$	0.315	0.544 **	0.606 **	−0.289	0.492 *	−0.290	−0.714 **	−0.611 **	−0.144	−0.477 *		0.093	−0.104
pH$_{KCl}$	0.269	0.526 **	0.585 **	−0.321	0.436 *	−0.317	−0.717 **	−0.670 **	−0.127	−0.477 *		0.188	−0.066

Significant correlations at the correspondent levels of * 0.05 and ** 0.01.

4. Discussion

The results of Experiment 1 showed significant differences in some soil properties at different positions along the rows, but not over a continuous gradient. Thus, the results cannot be attributed to the flooding irrigation, but they were probably caused by heterogeneity in spatial variability of important soil constituents such as clay, sand and silt, since it is well-known that soil texture determines many other soil physical and chemical properties [25]. Variations in soil properties were also found when comparing different soil layers. The soil bulk density was higher in the soil surface layer (0.0–0.1 m), and porosity was found to be higher in the deeper (0.1–0.2 m), layer. The soil bulk density and porosity in agricultural fields are influenced not only by soil texture but also by external loads which cause soil compaction [13,26,27]. In this particular case, it seems that the effects of frequent irrigation and soil tillage prevailed, which may have prevented a proper soil aggregation, leading to an increase in soil bulk density and a reduction in soil porosity on the surface layer which was directly impacted by the cultivator. The variation in soil properties was also significant when comparing fields. The field higher in clay and lower in sand presented significantly higher soil bulk density. Usually, clayey soils tend to have a lower bulk density and higher porosity than sandy soils [28]. However, these results indicate an opposite trend, probably because of the negative effect on soil aggregation and compaction caused by frequent soil tillage. Other studies have also found spatial variability in bulk density and water infiltration on flooded fields caused mainly by tillage practices, particularly when heavy machinery is used [8,29].

The soil samples collected from the ridges showed significantly higher values of extractable P and K. In the ridges, the conditions for nutrient uptake were poor since they are created every year by soil pushed from the inter-rows, which means that they contain nutrients barely taken up by the plant due to the limited expansion of roots in this position. In addition, in this irrigation system, the water flows from the inter-row to the ridge due to the gradient of water potential caused by the evapotranspiration from the latter and the continuous water supply to the inter-row. This means that nutrients tend to accumulate in the ridge, carried by mass flow, in contrast to what happens in the inter-row, from which nutrients tend to be leached out. Mass flow is the main driving force causing the movement of most nutrients in the soil [30–32]. Thus, soil conductivity was higher in the ridge, due to the increased presence of salts as demonstrated by the increase in CEC. Organic C also appeared higher in the ridge, probably because this zone is not tilled so frequently, which reduces the exposure of organic matter to the heterotrophic microorganisms that cause its oxidation [33]. This zone also contains the remaining bines (those that do not climb) and weeds, which are incorporated into the soil when the ridge is created, which usually represent more debris than that incorporated in the inter-row. B also increased in the ridge, perhaps due to higher levels of organic C, which have the ability of retaining B in the soil [34,35].

Soil pH (H_2O and KCl), base saturation and extractable Mn were significantly higher in the samples collected in the inter-rows. These results are probably related to the decrease in the potential redox, which may have increased the pH of the soil [36]. The increase in soil pH in the inter-rows was probably also related to the increase in the concentration of cation ions, such as Ca and Fe [37]. Base saturation increased in the inter-rows probably due to the presence of the divalent cations, less available to move into the ridge by mass flow. The higher concentration of Mn in the inter-rows might have also been due to the reduction of Mn that occurred at the beginning of the reduction process. This can occur when the redox potential is still positive [38].

A clear gradient along the rows was not observed for total and cone DMY. These results did not corroborate the hypothesis that flood irrigation is creating a spatial variation in plant performance along the rows. The differences detected in the plants seem to be due to spatial variability in the soil constituents, namely the soil separates which, in turn, influence soil bulk density and porosity. The results from the correlation analysis showed significant and negative relations between total DMY and soil bulk density (r = −0.706) and between total DMY and clay content (r = −0.676). In contrast, total DMY and soil porosity at 0.10–0.20 m correlated significantly and positively (r = 0.714). The soil surface layer presented a higher bulk density, which has already been explained by the effect of irrigation and frequent soil tillage, which reduces the stability of soil aggregates, increasing bulk density and decreasing porosity [10,26]. On the other hand, it seems that the higher porosity in the 0.10–0.20 m layer was an important factor affecting DMY, likely because in the surface layer the diffusion of oxygen to ensure the biological processes of the soil is always easier. Soils with a higher clay content tend to retain more water, decreasing soil aeration which negatively affects the function of root and plant metabolism [13,39]. Under the conditions of this experiment, the clay content in the soil seemed to be negatively associated with hop DMY, mainly because clay is a determinant factor of soil bulk density and porosity, which were identified in this study as determinant factors in crop productivity.

Irrigation also did not cause any relevant gradient in tissue nutrient concentration as detected by the analysis of variance. However, correlation analysis provided some data that deserves to be commented on. Leaf P was significantly and positively correlated with soil porosity at 0.10–0.20 m, but was negatively correlated with soil bulk density at 0.10–0.20 m and clay content. Leaf P did not show any consistent gradient along the rows, but was lower in the field presenting a higher soil bulk density and clay content. This reveals that P uptake was enhanced by the increased porosity of the soil at the deeper layer and by the lower clay content. Similarly, on barley (*Hordeum vulgare* L.) there was reported a reduction in P uptake and yield associated with heavy soil compaction [13]. The higher porosity of soil may have facilitated P root uptake from the deeper layer, which is richer in P, probably due to the increase in the vertical movement of P as the result of fertilization and flooding as reported by [40]. In turn, the higher clay content may have resulted in higher P adsorption and lower P availability. In contrast, leaf Fe was significantly and negatively correlated with soil porosity at 0.10–0.20 m depth. Leaf Fe also presented an opposite tendency between fields, decreasing along the rows in the field with a lower clay content and higher soil porosity. This result is probably related to soil reduction conditions, as the availability of Fe decreases when soil oxygen and redox potential increases [37]. Leaf K showed a significant and positive correlation with soil clay content and a negative one with sand content. The availability of K in the soil is not directly affected by redox potential, but its fixation in 2:1 clay minerals is facilitated by the increase in soil pH [36]. There has also been reported an antagonistic effect between Fe and K in paddy fields [41,42], an aspect that may also have influenced these results.

In Experiment 2, the application of lime increased several variables of soil fertility, including pH, but did not significantly increase exchangeable Ca and CEC. In fact, the rate of lime applied in this experiment was too low to cause important changes to soil properties, as is usually achieved when using high rates of lime [32]. In a previous study, Čeh and Čremožnik [17] applied 2.3 t lime ha^{-1} and reported similar results, that is, a reduced effect on soil properties due to the application of lime.

The main effect on the elemental composition of the leaves resulting from the application of lime would have been the significant increase of leaf P in the first growing season after the lime application. This raised the soil pH contributing to a reduction in P fixation, which in acidic soils is due to reactions with Al and Fe oxides, which precipitate P as $AlPO_4$ and $FePO_4$ [43].

Total and cone DMY did not increase with the application of lime, but rather showed a decreasing trend. It is generally considered that the optimal pH for hop growth is between 5.7 and 7.5 [15,44]. In this study, soil pH was below the lowest value of the reported range, which would have favoured a positive effect on the vegetation. However, the lime application influenced some soil properties, but not enough to have a high impact on the elemental composition of the leaves. In general, the nutrient content of the leaves was found to be within the sufficiency ranges established for hops [45], both in the limed and in the control treatments. Regarding total DMY, a significant interaction between lime treatment and field was recorded, which may also have contributed to difficulties in the interpretation of these results.

Correlation analysis, in turn, also did not show coherent trends over the two years of the study. Perhaps the most relevant result was the negative correlation between soil pH and biomass production in the first year, which again refers to diverse interactions which may have occurred between environmental variables (year) and factors under study (field and liming). The effect of environmental variables on the performance of the hop plant is well known [46–48], although in this study it was not possible to clarify the isolated effects of any of them.

5. Conclusions

Irrigation by flooding the space between rows over more than 20 years was not responsible for any gradient in soil properties, plant elemental composition and plant performance, although variations in those variables were found at different positions in the row caused by erratic spatial variability of some constituents of the soil, such as sand, silt and clay. However, irrigation followed by soil tillage on repeated occasions during the growing season seems to have reduced soil porosity and increased soil bulk density in the surface 0.0–0.1 m soil layer. These variables were found to be related to crop productivity in positive and negative ways, respectively.

This study also showed that the ridge is a point of nutrient accumulation, particularly for those that move more easily in the soil by mass flow, thereby showing also higher conductivity and CEC. The reduced water potential in the ridge created by evapotranspiration is the driving force causing the water flow from the inter-row. Organic C was also higher in the ridge in comparison with the inter-row, probably due to the annual incorporation of weeds and weaker hop bines (those that did not climb) when the ridge is created in early spring.

Although the original soil was acidic, and the application of 1000 kg ha^{-1} of lime caused a small increase in pH, this did not lead to other relevant changes in soil properties, nor in plant nutrition status or total and cone DMY. The liming effect might not have been enough to nullify the effects of the interaction between factors that always occur in field experiments.

Author Contributions: Conceptualization, M.Â.R.; methodology, M.A. and S.A.; formal analysis, S.A.; investigation, S.A.; resources, M.Â.R. and M.A.; data curation, S.A.; writing—original draft preparation, S.A.; writing—review and editing, M.Â.R.; supervision, M.Â.R. and M.A.; project administration, M.Â.R.; funding acquisition, M.Â.R. and M.A. All authors have read and agreed to the published version of the manuscript.

Funding: The authors are grateful to the Foundation for Science and Technology (FCT, Portugal) for financial support from national funds FCT/MCTES, to CIMO (UIDB/AGR/00690/2020) and for Sandra Afonso's doctoral scholarship (BD/116593/2016).

Institutional Review Board Statement: Not applicable.

Informed Consent Statement: Not applicable.

Data Availability Statement: No new data were created or analyzed in this study. Data sharing is not applicable to this article.

Conflicts of Interest: The authors declare no conflict of interest.

References

1. Turner, S.F.; Benedict, C.A.; Darby, H.; Hoagland, L.A.; Simonson, P.; Sirrine, J.R.; Murphy, K.M. Challenges and Opportunities for Organic Hop Production in the United States. *Agron J.* **2011**, *103*, 1645–1654. [CrossRef]
2. Rossini, F.; Virga, G.; Loreti, P.; Iacuzzi, N.; Ruggeri, R.; Provenzano, M.E. Hops (*Humulus lupulus* L.) as a Novel Multipurpose Crop for the Mediterranean Region of Europe: Challenges and Opportunities of Their Cultivation. *Agriculture* **2021**, *11*, 484. [CrossRef]
3. Hoque, M. The Way Ahead. In *Biotechnology for Sustainable Agriculture*; Singh, R.L., Mondal, S., Eds.; Woodhead Publishing: Cambridge, UK, 2018; pp. 375–397. [CrossRef]

4. Afonso, S.; Arrobas, M.; Rodrigues, M.Â. Soil and Plant Analyses to Diagnose Hop Fields Irregular Growth. *J. Soil Sci. Plant Nutr.* **2020**, *20*, 1999–2013. [CrossRef]
5. Hedley, C.B.; Knox, J.W.; Raine, S.R.; Smith, R. Water: Advanced Irrigation Technologies. In *Encyclopedia of Agriculture and Food Systems*; Van Alfen, N.K., Ed.; Academic Press: Oxford, UK, 2014; pp. 378–406. [CrossRef]
6. Simmonds, M.B.; Plant, R.E.; Peña-Barragán, J.M.; van Kessel, C.; Hill, J.; Linquist, B.A. Underlying Causes of Yield Spatial Variability and Potential for Precision Management in Rice Systems. *Precis. Agric.* **2013**, *14*, 512–540. [CrossRef]
7. Cox, C.; Jin, L.; Ganjegunte, G.; Borrok, D.; Lougheed, V.; Ma, L. Soil Quality Changes Due to Flood Irrigation in Agricultural Fields Along The Rio Grande in Western Texas. *Appl. Geochem.* **2018**, *90*, 87–100. [CrossRef]
8. Cerdà, A.; Daliakopoulos, I.N.; Terol, E.; Novara, A.; Fatahi, Y.; Moradi, E.; Salvati, L.; Pulido, M. Long-Term Monitoring of Soil Bulk Density and Erosion Rates in Two *Prunus persica* (L) Plantations Under Flood Irrigation and Glyphosate Herbicide Treatment in La Ribera District, Spain. *J. Environ. Manag.* **2021**, *282*, 111965. [CrossRef] [PubMed]
9. González-Méndez, B.; Webster, R.; Fiedler, S.; Siebe, C. Changes in Soil Redox Potential in Response to Flood Irrigation with Waste Water in Central Mexico. *Eur. J. Soil Sci.* **2017**, *68*, 886–896. [CrossRef]
10. Batey, T. Soil Bompaction and Soil Management—A Review. *Soil Use Manag.* **2009**, *25*, 335–345. [CrossRef]
11. Shapiro, C.A.; Elmore, R.W. Agricultural Crops. In *Encyclopedia of Applied Plant Sciences*, 2nd ed.; Thomas, B., Murray, B.G., Murphy, D.J., Eds.; Academic Press: Oxford, UK, 2017; pp. 1–8. [CrossRef]
12. Arriaga, F.J.; Guzman, J.; Lowery, B. Conventional Agricultural Production Systems and Soil Functions. In *Soil Health and Intensification of Agroecosytems*; Al-Kaisi, M.M., Lowery, B., Eds.; Academic Press: Cambridge, MA, USA, 2017; pp. 109–125. [CrossRef]
13. Nawaz, M.F.; Bourrié, G.; Trolard, F. Soil Compaction Impact and Modelling. A Review. *Agron. Sustain. Dev.* **2013**, *33*, 291–309. [CrossRef]
14. Indoria, A.K.; Sharma, K.L.; Reddy, K.S. Hydraulic properties of soil under warming climate. In *Climate Change and Soil Interactions*; Prasad, M.N.V., Pietrzykowski, M., Eds.; Elsevier: Amsterdam, The Netherlands, 2020; pp. 473–508. [CrossRef]
15. Sirrine, J.R.; Rothwell, N.; Lizotte, E.; Goldy, R.; Marquie, S.; Brown-Rytlewski, D. Sustainable Hop Production in the Great Lakes Region. *Ext. Bull. E-3083* **2010**. Available online: https://www.uvm.edu/sites/default/files/media/Sirrine-Sustainable-Hop-Production-in-the-Great-Lakes-Region.pdf (accessed on 13 April 2019).
16. Gent, D.H.; Sirrine, J.R.; Darby, H.M. Nutrient Management and Imbalances. In *Field Guide for Integrated Pest Management in Hops*; Washington Hop Commission: Moxee, WA, USA, 2015; pp. 98–100.
17. Čeh, B.; Čremožnik, B. Soil pH and Hop (*Humulus lupulus*) Yield Related to Liming Material Rate. *Hmelj. Bilt.* **2015**, *22*, 49–57.
18. Seng, V.; Bell, R.W.; Willett, I.R. Effect of Lime and Flooding on Phosphorus Availability and Rice Growth on Two Acidic Lowland Soils. *Commun. Soil Sci Plant Anal.* **2006**, *37*, 313–336. [CrossRef]
19. Sadiq, A.A.; Babagana, U. Influence of Lime Materials to Ameliorate Acidity on Irrigated Paddy Fields: A Review. *Acad. Res. Int.* **2012**, *3*, 413.
20. Shi, L.; Guo, Z.; Liang, F.; Xiao, X.; Peng, C.; Zeng, P.; Feng, W.; Ran, H. Effect of Liming with Various Water Regimes on Both Immobilization of Cadmium and Improvement of Bacterial Communities in Contaminated Paddy: A Field Experiment. *Int. J. Environ. Res. Public Health* **2019**, *16*, 498. [CrossRef] [PubMed]
21. Rowell, D.L. *Soil Science: Methods & Applications*, 1st ed.; Longman Group UK Ltd.: Harlow, UK, 1994; p. 368. [CrossRef]
22. Van Reeuwijk, L. Procedures for soil analysis (International Soil Reference and Information Centre). *Tech. Pap.* **2002**, *9*, 120.
23. Lakanen, E.; Erviö, R. A Comparison of Eight Extractants for the Determination of Plant Available Micronutrients in Soils. *Acta Agric. Fenn.* **1971**, *123*, 223–232.
24. Walinga, I.; Van Vark, W.; Houba, V.; Van der Lee, J. *Soil and Plant Analysis, Part 7: Plant Analysis Procedures*; Wageningen Agricultural University: Wageningen, The Netherlands, 1989.
25. Delgado, A.; Gómez, J.A. The Soil. Physical, Chemical and Biological Properties. In *Principles of Agronomy for Sustainable Agriculture*; Villalobos, F.J., Fereres, E., Eds.; Springer International Publishing: Cham, Switzerland, 2016; pp. 15–26. [CrossRef]
26. Hamza, M.A.; Al-Adawi, S.S.; Al-Hinai, K.A. Effect of Combined Soil Water and External Load on Soil Compaction. *Soil Res.* **2011**, *49*, 135–142. [CrossRef]
27. Alaoui, A.; Rogger, M.; Peth, S.; Blöschl, G. Does Soil Compaction Increase Floods? A Review. *J. Hydrol.* **2018**, *557*, 631–642. [CrossRef]
28. Chaudhari, P.R.; Ahire, D.V.; Ahire, V.D.; Chkravarty, M.; Maity, S. Soil Bulk Density as Related to Soil Texture, Organic Matter Content and Available Total Nutrients of Coimbatore Soil. *Int. J. Sci. Res. Publ.* **2013**, *3*, 1–8.
29. Green, T.R.; Ahuja, L.R.; Benjamin, J.G. Advances and Challenges in Predicting Agricultural Management Effects on Soil Hydraulic Properties. *Geoderma* **2003**, *116*, 3–27. [CrossRef]
30. Lambers, H.; Chapin, F.S.; Pons, T.L. Mineral Nutrition. In *Plant Physiological Ecology*; Lambers, H., Chapin, F.S., Pons, T.L., Eds.; Springer: New York, NY, USA, 2008; pp. 255–320. [CrossRef]
31. Comerford, N.B. Soil Factors Affecting Nutrient Bioavailability. In *Nutrient Acquisition by Plants: An Ecological Perspective*; BassiriRad, H., Ed.; Springer: Berlin/Heidelberg, Germany, 2005; pp. 1–14. [CrossRef]
32. Weil, R.R.; Brady, N.C. *The Nature and Properties of Soils*, 15th ed.; Global Edition: London, UK, 2017.
33. Liu, X.; Herbert, S.; Hashemi, A.; Zhang, X.; Ding, G. Effects of Agricultural Management on Soil Organic Matter and Carbon Transformation—A Review. *Plant. Soil Environ.* **2006**, *52*, 531. [CrossRef]

34. Goldberg, S. Reactions of Boron with Soils. *Plant Soil* **1997**, *193*, 35–48. [CrossRef]

35. Das, A.K.; Purkait, A. Boron Dynamics in Soil: Classification, Sources, Factors, Fractions, and Kinetics. *Commun. Soil Sci. Plant Anal.* **2020**, *51*, 2778–2790. [CrossRef]

36. Husson, O. Redox Potential (Eh) and pH as Drivers of Soil/Plant/Microorganism Systems: A Transdisciplinary Overview Pointing to Integrative Opportunities for Agronomy. *Plant Soil* **2013**, *362*, 389–417. [CrossRef]

37. Osman, K.T. *Soils: Principles, Properties and Management*, 1st ed.; Springer: Dordrecht, The Netherlands, 2013.

38. George, E.; Horst, W.J.; Neumann, E. Adaptation of Plants to Adverse Chemical Soil Conditions. In *Marschner's Mineral Nutrition of Higher Plants*; Marschner, P., Ed.; Elsevier: Amsterdam, The Netherlands, 2012; pp. 409–472.

39. Pezeshki, S.R.; DeLaune, R.D. Soil Oxidation-Reduction in Wetlands and its Impact on Plant Functioning. *Biology* **2012**, *1*, 196–221. [CrossRef]

40. Tian, J.; Dong, G.; Karthikeyan, R.; Li, L.; Harmel, R. Phosphorus Dynamics in Long-Term Flooded, Drained, and Reflooded Soils. *Water* **2017**, *9*, 531. [CrossRef]

41. Chen, J.; Xuan, J.; Du, C.; Xie, J. Effect of Potassium Nutrition of Rice on Rhizosphere Redox Status. *Plant Soil* **1997**, *188*, 131–137. [CrossRef]

42. Kundu, D.; Neue, H.; Singh, R. Iron and Potassium Availability to Rice in Tropudalf and Sulfaquept as Influenced by Water Regime. *J. Indian Soc. Soil Sci.* **2001**, *49*, 130–134.

43. Havlin, J.L.; Tisdale, S.L.; Nelson, W.L.; Beaton, J.D. *Soil Fertility and Fertilizers: An Introduction to Nutrient Management*, 8th ed.; Pearson: Boston, MA, USA, 2014.

44. Gingrich, C.; Hart, J.; Christensen, N. *Fertilizer Guide 79*; Oregon State University, Extension Service: Corvallis, OR, USA, 1994.

45. Bryson, G.; Mills, H.; Sasseville, D.; Jones, J.B., Jr.; Barker, A. *Plant. Analysis Handbook III: A Guide to Sampling, Preparation, Analysis and Interpretation for Agronomic and Horticultural Crops*; Micro-Macro Publishing: Athens, GA, USA, 2014.

46. Marceddu, R.; Carrubba, A.; Sarno, M. Cultivation Trials of Hop (*Humulus lupulus* L.) in Semi-Arid Environments. *Heliyon* **2020**, *6*, e05114. [CrossRef]

47. MacKinnon, D.; Viljem, P.; Čeh, B.; Naglič, B.; Pavlovic, M. The Impact of Weather Conditions on Alpha-Acid Content in Hop (*Humulus lupulus* L.) cv. Aurora. *Plant. Soil Environ.* **2020**, *66*, 519–525. [CrossRef]

48. Rossini, F.; Virga, G.; Loreti, P.; Provenzano, M.E.; Danieli, P.P.; Ruggeri, R. Beyond Beer: Hop Shoot Production and Nutritional Composition under Mediterranean Climatic Conditions. *Agronomy* **2020**, *10*, 1547. [CrossRef]

Article

Signal Intensity of Stem Diameter Variation for the Diagnosis of Drip Irrigation Water Deficit in Grapevine

Chen Ru, Xiaotao Hu *, Wene Wang, Hui Ran, Tianyuan Song and Yinyin Guo

Key Laboratory of Agricultural Soil and Water Engineering in Arid and Semiarid Areas, Ministry of Education, Northwest A&F University, Yangling 712100, China; chenru1024@nwafu.edu.cn (C.R.); wangwene@nwsuaf.edu.cn (W.W.); huiran@nwsuaf.edu.cn (H.R.); tianyuansong1014@nwafu.edu.cn (T.S.); guoyinyin0513@nwafu.edu.cn (Y.G.)
* Correspondence: huxiaotao11@nwsuaf.edu.cn; Tel.: +86-138-9281-6133; Fax: +86-29-8708-2117

Citation: Ru, C.; Hu, X.; Wang, W.; Ran, H.; Song, T.; Guo, Y. Signal Intensity of Stem Diameter Variation for the Diagnosis of Drip Irrigation Water Deficit in Grapevine. *Horticulturae* 2021, 7, 154. https://doi.org/10.3390/horticulturae7060154

Academic Editors: Agnieszka Hanaka, Jolanta Jaroszuk-Ściseł and Małgorzata Majewska

Received: 16 May 2021
Accepted: 9 June 2021
Published: 15 June 2021

Publisher's Note: MDPI stays neutral with regard to jurisdictional claims in published maps and institutional affiliations.

Abstract: Precise irrigation management of grapevines in greenhouses requires a reliable method to easily quantify and monitor the grapevine water status to enable effective manipulation of the water stress of the plants. This study describes a study on stem diameter variations of grapevine planted in a greenhouse in the semi-arid area of Northwest China. In order to determine the applicability of signal intensity of stem diameter variation to evaluate the water status of grapevine and soil. The results showed that the relative variation curve of the grapevine stem diameter from the vegetative stage to the fruit expansion stage showed an overall increasing trend. The correlations of MDS (maximum daily shrinkage) and DI (daily increase) with meteorological factors were significant ($p < 0.05$), and the correlations with SWP, RWC and soil moisture were weak. Although MDS and DI can diagnose grapevine water status in time, SI_{MDS} and SI_{DI} have the advantages of sensitivity and signal intensity compared with other indicators. Compared with MDS and DI, the R^2 values of the regression equations of SI_{MDS} and SI_{DI} with SWP and RWC were high, and the correlation reached a very significant level ($p < 0.01$). Thus, SI_{MDS} and SI_{DI} are more suitable for the diagnosis of grapevine water status. The SI_{MDS} peaked at the fruit expansion stage, reaching 0.957–1.384. The signal-to-noise ratio of SI_{DI} was higher than that of MDS across the three treatments at the vegetative stage. The value and signal-to-noise ratio of SI_{DI} at the flowering stage were similar to those of SI_{MDS}, while the correlation between SI_{DI} and the soil moisture content was higher than that of SI_{MDS}. It can be concluded that that SI_{DI} is suitable as an indicator of water status of grapevine and soil during the vegetative and flowering stages. In addition, the signal-to-noise ratio of SI_{MDS} during the fruit expansion and mature stages was significantly higher than that of SI_{DI}. Therefore, SI_{MDS} is suitable as an indicator of the moisture status of grapevine and soil during the fruit expansion and mature stages. In general, SI_{MDS} and SI_{DI} were very good predictors of the plant water status during the growth stage and their continuous recording offers the promising possibility of their use in automatic irrigation scheduling in grapevine.

Keywords: grapevine; maximum daily shrinkage; daily increase; stem water potential; leaf relative water content; signal intensity

1. Introduction

Fruit tree orchards are common in arid and semi-arid areas where water for irrigation is scarce. This, together with an increasing world population that has to be fed and with other water-using sectors competing for the limited water resources, makes the use of precise irrigation techniques in those orchards unavoidable. The response of the scientific community to this challenge has been to invest a substantial amount of research in the development of deficit irrigation approaches [1,2] and of new irrigation technologies based on more-precise, user-friendly water-stress indicators. Some can be continuously and automatically recorded, having a great potential for irrigation scheduling [3].

In recent years, it has become very popular to study the relationship between plant and water based plant water status indicators [4]. There are many ways to monitor and diagnose crop water status. From the perspective of plant physiology, the short-term microchange dynamics of plant organs (stems, leaves, fruits, etc.) are closely related to the water status of plants and have been widely focused on by many scholars [2–4]. The most widely used approach for evaluating plant water status has been to determine leaf [5,6], stem water potential [6] and leaf water content [7–10]. Plant water potential is a significant and reliable indicator of plant water status for scheduling the irrigation of plants. Argyrokastritis et al. [5] established the relationship between leaf water potential and water stress index, which can well characterize the water deficit status of plants. Wang et al. [6] used leaf and stem water potential to characterize the water status of grapevine, which can well judge whether the grapevine is in deficit state and determine when to need irrigation. However, the main disadvantage of plant water potential is the relatively cumbersome measurement procedure, including the necessity of frequent trips to the field and a considerable input of labor. The robustness of the sensors used to measure stem diameter fluctuations have renewed interest in using these parameters as plant water status indicators [1,6,11,12]. Apart from being capable of an early detection of water stress, even if this is mild, these techniques permit continuous and automated recordings of the plant water status and an immediate, consistent and reliable response to water deficit [13–15].

Monitoring crop moisture conditions using the stem diameter microchange method has been popular since the mid-1980s. The microchange of plant stem diameter is closely related to its water status, when the root system absorbs enough water, the stem expands gradually, and when the water is deficient, the stem shrinks gradually. Therefore, it is possible to diagnose the water status of the plant with the microchange of stem diameter, which can provide an index for the real-time prediction of crop water shortage and precision irrigation [12,16,17]. Today, in fact, fruit tree orchards and vineyards are being irrigated based on changes in the stem diameter [18,19]. Among them, MDS and DI indexes are commonly studied today as the indicators of plant water content [20–22]. According to some previous studies, the ability of an index to be suitable for use as a water diagnosis indicator was mainly evaluated in terms of three qualities: sensitivity, signal intensity and variability. An appropriate indicator should have better sensitivity and signal intensity to water stress and exhibit less variability [4,14,23]. However, the disadvantage of this method is that it cannot determine whether the critical value is independent of crop species or growth stage. In addition, meteorological factors have a great influence on MDS under the same water conditions [24]. The maximum stem diameter over time can be used to diagnose water deficit, but the growth rate of the crop is different at different growth stages. Under high evaporation intensity at the mature stage, crop stems may also shrink even if the crop does not lack water, and the variation in daily MXSD (maximum stem diameter) has no more significance [25]. Therefore, it may be difficult to apply the MDS of stem diameter as a crop moisture stress signal in practice.

The observation of stem diameter and its dynamic changes is beneficial to the study of plant moisture changes under interlaced internal and external factors. However, the observation of stem diameter is easily influenced by meteorological factors [26–28]. How to eliminate the interference of external environmental factors on the variation in stem diameter is always the difficulty in determining the most appropriate indicator. Due to the influence of other factors, it was difficult to diagnose plant water content. Only the observed values of stem diameter variation and the prediction values of stem diameter variation under no water stress, should be calculated to assist in the diagnosis. The comparison of MDS and DI with reference factors (relative values) can be used to directly reflect crop water status. The signal intensity is obtained by standardizing the reference value of the stem diameter indicator under fully irrigated conditions and the measured value of actual growth conditions, which can effectively eliminate the influence of meteorological factors [29]. Thus, the accuracy of these equations is very important to future studies, as these equations are the foundation for diagnosing plant water content and making irrigation schemes based on plant stem diameter variations. Currently, existent studies concentrate mostly on the grapevine stem diameter variations in outdoor conditions and seldom did it to grapevines planted in a greenhouse [22,30,31]. In addition, the past research on the variation in stem diameter has mainly focused on the feasibility of moisture diagnosis, maximum daily shrinkage, daily increase, and other stem diameter indicators, which have been verified for use in the moisture status diagnosis of different crops [11,24,32–34]. The variation in stem diameter is influenced by environmental factors and the crop growth characteristics. The difference in crop growth at different growth stages may significantly affect the potential of stem diameter variation indicators for use in determining irrigation regimes. Therefore, different indicators should be applied in different growth stages [26]. There have been few reports on this topic, and further research is needed.

For these reasons above mentioned, this research selected the greenhouse grapevine as the study object. Because of its rich nutrition and delicious taste, grapevine has become a kind of world major fruit. In China, grapevines have been widely cultivated as well, and a considerable part among them are planted in sunlight greenhouses. Due to the present water shortage in the Northwest China, it is important to acquire accurate crop water content information and timely plan water-saving irrigation schemes, which benefit the sustainable development of local agriculture. Thus, the main aims of this study were as follows: (1) to explore the relative variation in the changes in MDS and DI in stem diameter during different stages; (2) to clarify the correlation of microchanges in stem diameter with stem water potential, leaf relative water content, and soil water content; (3) to evaluate whether SI_{MDS} and SI_{DI} can be applied to diagnose grapevine moisture and soil moisture status; (4) to analyze the sensitivity of signal intensity indicators and to determine the suitability of SI_{MDS} and SI_{DI} under different stages.

2. Materials and Methods

2.1. Study Area

The experiment was carried out in greenhouse of Yuhe Farm, Shaanxi Province, from March to July 2018 (108.58° E, 37.49° N). The annual average rainfall in this area is 365.7 mm, the annual average temperature is 8.3 °C, the annual relative humidity is 69.37%, and the annual average duration of sunshine is 2893.5 h, which is representative of the typical continental marginal monsoon climate of the area. Table 1 shows the meteorological data (cultivation stage averages) recorded over the experimental year. The test soil was an aeolian sandy soil. The chemical properties of the soil were as follows: the soil ammonium nitrogen was 7.48 mg·kg^{-1}, the nitrate nitrogen was 22.91 mg·kg^{-1}, the available phosphorus was 4.07 mg·kg^{-1}, and the available potassium was 163.47 mg·kg^{-1}. The physical properties of the soil are shown in Table 2.

2.2. Experimental Design

Six-year-old grapevine (early-maturing variety 6–12, which was selected from the scarlet bud transformation in 1998) were planted in greenhouse, and grapevines with good growth and similar shapes were selected for the experiments. The entire growth period of grapevines can be divided into four main growth stages: the vegetative stage, the flowering stage, the fruit expansion stage, and the coloring mature stage, the cultivation period was 121 days during the growth season. The greenhouse was oriented east-west and was 70 m long and 9 m wide. The grapevine row width and row spacing were 0.8 m and 1.5 m, respectively, and the plant spacing was 0.6 m, with 14 grapevines per row. Artificial warming was carried out in greenhouse to ensure the growth temperature of grapevines on 11 March 2018. Drip irrigation was used in the experiment. A single-wing labyrinth drip irrigation belt (produced by Xinjiang Dayu Water Saving Company, Xinjiang, China) was adopted. The inner diameter and wall thickness of drip irrigation belt was 0.02 m and 0.018 m, respectively. the distance between the drippers was 0.3 m, the design flow of the dripper was 4.0 L·h^{-1}, and the laying mode of the drip belt was one row of two pipes.

Table 1. Meteorological data of different cultivation stages in greenhouse.

Cultivation Stage	T_a	R_a	RH	VPD
vegetative stage	18.8	301.0	53.8	0.33
flowering stage	19.7	327.9	50.4	0.35
fruit expansion stage	22.3	421.1	51.2	0.34
coloring mature stage	25.1	362.3	55.7	0.32

Note: T_a (°C): air temperature; R_a (w·m^{-2}): solar radiation; RH (%): relative humidity; VPD (kpa): vapor pressure deficit.

Table 2. Physical properties of the soil.

Depth (cm)	Textural Analysis				FC ($g·g^{-1}$)	PWP ($g·g^{-1}$)	Bulk Density ($g·cm^{-3}$)
	Sand (%)	Clay (%)	Silt (%)	Texture Class			
0–40	87.54	5.27	7.19	Aeolian soil	13.18	2.31	1.64
40–80	70.23	19.53	10.24	Sandy loam	17.45	6.38	1.46

Note: FC: field capacity; PWP: permanently wilting point.

The experiment was conducted with drip-irrigated grapevines under three irrigation treatments: a full irrigation treatment (T_1:100% M) and two regulated deficit irrigation treatment (T_2: 80% M; T_3: 60% M), M represents the irrigation quota. There were three treatments in total and three plots per treatment (each plot had a length of 8 m, a width of 4.5 m, and an area of 36 m^2), with a random block arrangement. The irrigation dates and irrigation amount is shown in Table 3, the grapevines were irrigated 12 times during the entire growth period. The irrigation quota was calculated by Equation (1). The irrigation time was determined according to whether or not T_1 reached the lower limit of the water quantity, which was 65% of β_1 at the vegetative and coloring mature stages and 70% of β_1 at the flowering and fruit expansion stage. The predicted wet layer depth of the soil was 0.8 m. The total amount of fertilization during the entire growth period was 0.84 t·ha^{-1}, and the proportion of N:P:K was 1.0:0.6:1.2. Fertilization was carried out over three periods: the germination stage accounted for 20% of the total amount of fertilization, the flowering and fruit expansion stages accounted for 60%, and the coloring mature stage accounted for 20%. The drip irrigation and fertilization were controlled by integrated irrigation and fertilization equipment.

$$M = 0.1\gamma_s HP(\beta_1 - \beta_2) \tag{1}$$

where M represents the irrigation quota, mm; γ_s represents the soil dry bulk density, 1.64 g·cm^{-3} in 0–40 cm soil depth, 1.46 g·cm^{-3} in 40–80 cm soil depth; H is the predicted wet layer depth of soil, 0.8 m; P is the designed wet soil ratio, 0.8; β_1 is the field water holding capacity, 13.18% in 0–40 cm soil depth, 17.45% in 40–80 cm soil depth; and β_2 is the lower limit of the soil moisture content, 65% of β_1 at the vegetative and coloring mature stages and 70% of β_1 at the flowering and fruit expansion stage.

2.3. Observation Indicators

2.3.1. Meteorological Factors

All meteorological data are automatically measured and recorded every 30 min using a Watchdog micro series (Spectrum Technologies Inc., Chicago, IL, USA) meteorological station in the middle of the greenhouse. The monitoring indicators includes air temperature (T_a), relative humidity (RH), solar radiation (R_a) and other meteorological parameters. The vapor pressure deficit (VPD) was estimated by the RH and T_a and was calculated by the modified Penman formula. The formula [35] was as follows (2):

$$VPD = 0.6108 \times EXP(\frac{17.27 \times T_a}{T_a + 237.3}) \times (1 - \frac{RH}{100}) \tag{2}$$

Table 3. Irrigation amount of grapevine under different cultivation stages.

Cultivation Stage	Irrigation Date	Irrigation Amount ($m^3 \cdot ha^{-1}$)		
		T_1	T_2	T_3
vegetative stage	3/11	293.61	234.80	176.02
	3/19	293.61	234.65	175.55
	3/27	293.60	234.60	175.34
	4/06	293.62	234.65	175.62
Flowering stage	4/15	341.40	273.05	204.60
	4/25	341.39	273.04	204.15
Fruit expansion stage	5/05	341.39	273.00	204.20
	5/13	341.37	272.58	204.14
	5/20	341.42	272.69	204.47
	5/27	341.41	273.10	204.34
Coloring mature stage	6/05	293.60	234.50	175.59
	6/15	293.61	234.55	176.30
Total irrigation amount		3810	3045	2280

2.3.2. Soil Moisture Content

The soil moisture automatic monitoring system consisted of an EM50 data recorder (Environmental Logging System, Decision Devices, Inc., Pullman, WA, USA) and four ECH_2O_5TE sensors (Decision Devices, Pullman, WA, USA). The soil moisture automatic monitoring system was installed 30 cm from the base of the grapevines and perpendicular to the planting row. Three representative grapevines were selected, and three monitoring systems were installed for each treatment. A sensor was installed every 20 cm, the buried depth was 80 cm, the soil volume moisture content was recorded every 30 min. Before the beginning of the growing season, in order to ensure the accuracy of the ECH_2O_5TE sensor, soil samples were taken every 20 cm with a soil drill until 80 cm, and the soil moisture content was calculated by drying method. At the same time, the data recorded by the ECH_2O_5TE sensor in different soil layers were recorded. Three days of soil moisture data were used to calibrate ECH_2O_5TE by drying method. The regression equation was established by regression analysis between the soil water content calculated by drying method and the soil water content monitored by ECH_2O_5TE. In addition, the same method was used to calibration ECH_2O_5TE every 10 days during grape growth.

2.3.3. Stem Diameter Microchanges

The stem diameter microchanges were automatically monitored continuously using a DEX20 (Dynamax, USA, 0.050 mm) instrument. The instrument was installed at the stem 10 cm above the ground, with a maximum displacement of 5 mm and a recording interval of 30 min. The relative variation (RV) in stem diameter was defined as the change value at the time of probe installation, and the grapevine stem diameters were inconsistent when the sensor was installed due to differences among different plants. To explain the difference in plant growth caused by different water treatments, the initial value of stem diameter at the time of sensor installation in each growth period was set to 1 mm, so that the three treatments could be compared easily. The maximum daily shrinkage (MDS) was calculated by subtracting the minimum stem diameter (MNSD) from the maximum stem diameter (MXSD). Periodic changes in stem diameter were observed daily; MXSD usually occurred in the early morning and MNSD occurred at noon. The daily increase (DI) of stem diameter was obtained by subtracting the daily MXSD from that of the day before.

2.3.4. Stem Water Potential and Relative Water Content of Leaves

The pressure chamber (TP-PW-II, Top Cloud-agri Technology Company, Zhejiang, China) was used to measure the stem water potential (φ_s) every 5–7 days, and the φ_s is measured at 9:00 to 10:00 BJS. Three grapevines were selected for each treatment, and one branch under good growth conditions was selected as the sample on the sunny side outside the crown. The sample was put into a plastic bag containing moist gauze and quickly brought into the laboratory. The sample was clamped in a pressure chamber and pressurized by gas (compressed nitrogen), the pressure used for exudation of tissue fluid was observed. At this time, the pressure value was the stem water potential.

The relative water content (RWC) of leaves was determined by the drying method. The selection and determination of leaves were the same as those used for leaf water potential measurement. After weighing the fresh weight, the leaves were immersed in water for 12 h and then taken out. The water on the surface of the leaves was wiped with absorbent paper and weighed. Then, the leaves were immersed in water for 1 h, taken out, wiped dry and weighed until reaching a consistent weight. After 0.5 h of dehumidification at 105 °C for 0.5 h, the leaves were dried to constant weight at 80 °C. Leaf relative water content (RWC) = (initial fresh weight − dry weight)/(saturated fresh weight − dry weight) × 100%.

2.3.5. Signal Intensity Calculation of Stem Diameter Indicator

The reference value is usually calculated by the stem diameter indicator under non-water stressed conditions or by substituting the meteorological indicator into the reference equation [21]. The calculation formulas of SI_{MDS} and SI_{DI} are as follows:

$$SI_{MDS} = \text{Measured MDS/Reference MD} \qquad (3)$$

$$SI_{DI} = \text{Measured DI/Reference DI} \qquad (4)$$

In this study, the regression equation between the MDS and DI of stem diameter and meteorological factors showed that the correlation between the MDS and DI of three treatments and T_a was the best. The soil moisture of the W1 treatment always remained above 65% of the field capacity at the vegetative and mature stages and 70% of the field capacity at the flowering stage and fruit expansion stage. The reference equation for stem diameter was established between the MDS and DI values of the W1 and T_a to calculate the reference value of MDS and DI for each growth stage.

2.3.6. Flexible Evaluation of Signal Intensity

High variability indicators need to be measured many times to reduce error, which increases the costs of such methods. Therefore, the intensity and variability (coefficient of variance, CV) of indicators should also be considered. When soil moisture changes, the ratio of signal intensity (SI) to noise is greater in the short term, indicating that the indicator is more suitable for moisture status diagnosis [36]. The formula for calculating the signal-to-noise ratio is shown in formula (5).

$$\text{Signal-to-noise ratio} = \text{signal intensity/coefficient of variation} \qquad (5)$$

2.4. Data Analysis

The correlation and regression analysis were carried out using SPSS 21.0 software (SPSS Inc., Chicago, IL, USA). Multiple comparisons were performed by least significant difference tests, with a significance level of 0.05. Microsoft Excel 2010 Software was used for processing data. The graphs were created by using Origin 2018. Correlation analysis was conducted between MDS, DI, SI_{DI}, SI_{MDS} and meteorological factors; The relationships between MDS, DI, SI_{DI}, SI_{MDS} and SWP, RWC as well as between MDS, DI, SI_{DI}, SI_{MDS} and soil water content were analyzed through regression analyses. In all cases, the coefficient of determination (R^2) was used to assess the goodness of fit of the associations among variables.

3. Results

3.1. The Relative Variation of Stem Diameter under Different Stages

The relative variation (RV) curve of stem diameter under different stages showed a 24 h up and down cycle, and different irrigation amounts had different influences on the stem diameter of grapevine (Figure 1). The total increase in stem diameter was 0.128 mm under W1, while those of W2 and W3 fluctuated and decreased. The stem diameter of W3 began to decrease sharply after 7 April, and the total increase in stem diameter under W2 and W3 was −0.143 and −0.570 mm, respectively (Figure 1a). There was a certain difference in the RV curve between the vegetative and flowering stages, and the RV curves of stem diameter under three treatments showed an up and down growth trend (Figure 1b). In terms of the total increase in stem diameter, W1 was the largest (0.555 mm), and W2 and W3 showed values 69.55% and 32.79% of that under W1, respectively.

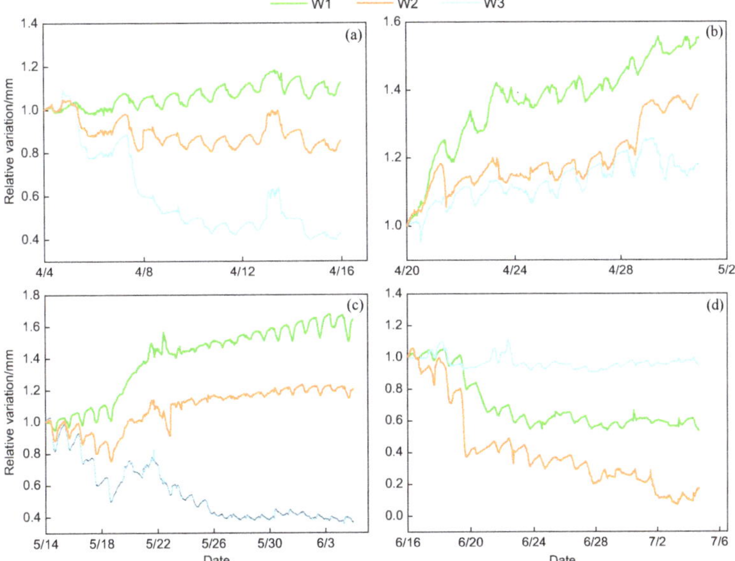

Figure 1. Relative variation curve of stem diameter during different stages. (**a**) Represents the vegetative stage, (**b**) represents the flowering stage, (**c**) represents the fruit expansion stage, and (**d**) represents the coloring mature stage.

There were significant differences in the RV curve of stem diameter among different treatments during the fruit expansion stage (Figure 1c). The RV curve of W1 showed an upward trend, and the RV curve of W2 fluctuated by approximately 1 mm. Before irrigation on 20 May, the RV curve of the W3 treatment showed a decreasing fluctuating trend, the diameter of the stem recovered after irrigation, and the recovery effect gradually weakened when the soil moisture content gradually decreased. The stem diameter recovered values under W2 and W3 were 86.28% and 72.79% of that under W1, respectively. The stem diameter RV under W3 remained stable at approximately 0.4 mm after 25 May. The RV curve of W1 and W2 showed a decreasing trend during the mature stage (Figure 1d), the RV curve of W3 fluctuated at 1 mm, but the three treatments still had significant shrinkage. The contractions under W1 and W2 were more pronounced than those under W3. The total increase in stem diameter among the three treatments was negative.

The daily change of stem diameter was the same under three treatments (Figure 2), it showed a trend of first increasing, then decreasing, and then increasing over 24 h. The MDS of the stem diameter showed significant differences under three treatments. The MDS of

W3 was the largest at 0.138 mm, and W1 was the lowest at 0.051 mm. The stem diameter of W1 could recovered to the maximum of the previous day and continued to grow. However, the stem diameter under W2, W3 could not recover to the maximum of the previous day owing to moisture stress, and growth of both W1 and W2 appeared negative. The MXSD and MNSD of three treatments appeared at the same time on rainy days, which included that rainy weather had no significant effect on the occurrence time of MXSD and MXSD under different treatments. The variation in R_a, T_a and RH on rainy days was smaller than that on sunny days, and the degree of stem diameter contraction was also lower on rainy days than that on sunny days, which showed meteorological factors may be the main force affecting the variation in stem diameter. We can conclude that MDS decreased with increasing irrigation amount, the influence of meteorological factors and soil moisture on the variation in stem diameter was interactive.

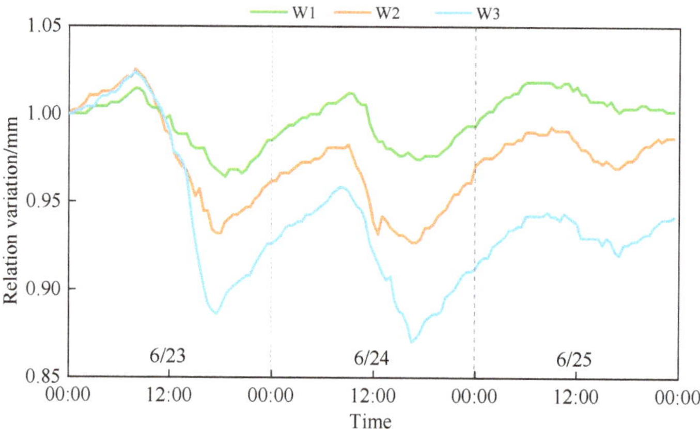

Figure 2. The daily change in stem diameter of grapevine under different weathers. Sunny days: 6/23, 6/24; Rainy day: 6/25.

3.2. Evaluation of Applicability as a Moisture Diagnosis Indicator

3.2.1. The Correlation of MDS and DI with Meteorological Factors

The MDS and DI are the main components of stem diameter variation and are affected by soil moisture and meteorological factors [22]. Correlation analysis was carried out between MDS (Figure 3) and DI (Figure 4) in stem diameter and meteorological factors. The change in MDS was similar to those in meteorological factors (R_a, T_a and VPD). The MDS increased with increasing R_a, T_a and VPD and decreased with increasing RH. The correlation of DI with meteorological factors was opposite to that of MDS. The correlations between the MDS and DI of the three treatments and meteorological factors were significant ($p < 0.05$). The correlation between MDS and meteorological factors decreased with increasing irrigation, and the correlation between the DI and meteorological factors increased with increasing irrigation. It can be seen from Table 4 that the correlation coefficients of MDS, DI and T_a were the highest, at 0.601–0.692 and 0.683–0.723, respectively.

3.2.2. The Correlation of MDS, DI with Stem Water Potential and RWC of Leaves

Stem water potential (SWP) and relative water content (RWC) of leaves are important indicators to characterize plant water status and exhibit the most direct response to drought during crop growth [37,38]. The models of MDS, DI and SWP and RWC were established. The coefficient of determination (R^2) is shown in Table 5. The regression equations of MDS and DI with SWP and RWC generally meet the significance test, which shows that both MDS and DI have obvious correlation with SWP and RWC, but R^2 values are not high. We found that, compared with W1 and W2, the R^2 values of MDS with SWP and RWC in W3 treatment were below 0.20, and the significance was weak. Therefore, MDS and DI were easily disturbed by meteorological factors and could not be directly used for the diagnosis of grapevine water status.

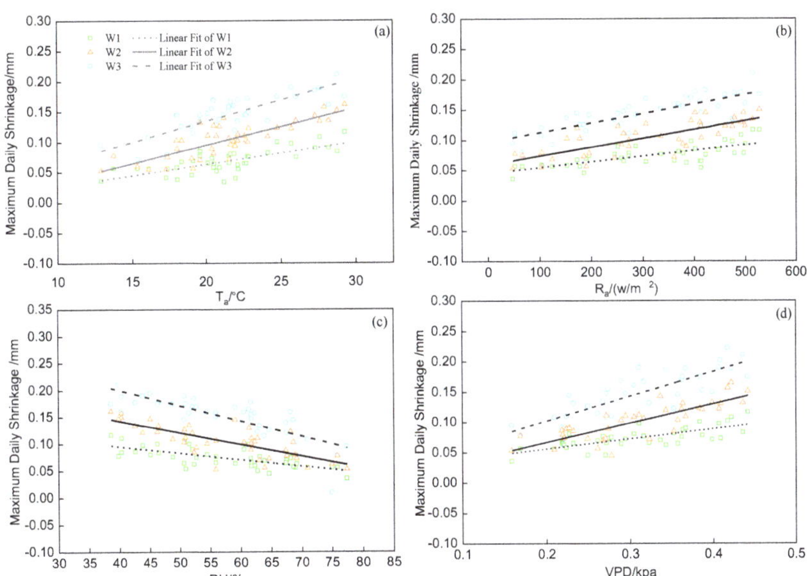

Figure 3. The relationship of MDS with T_a (**a**), R_a (**b**), RH (**c**) and VPD (**d**).

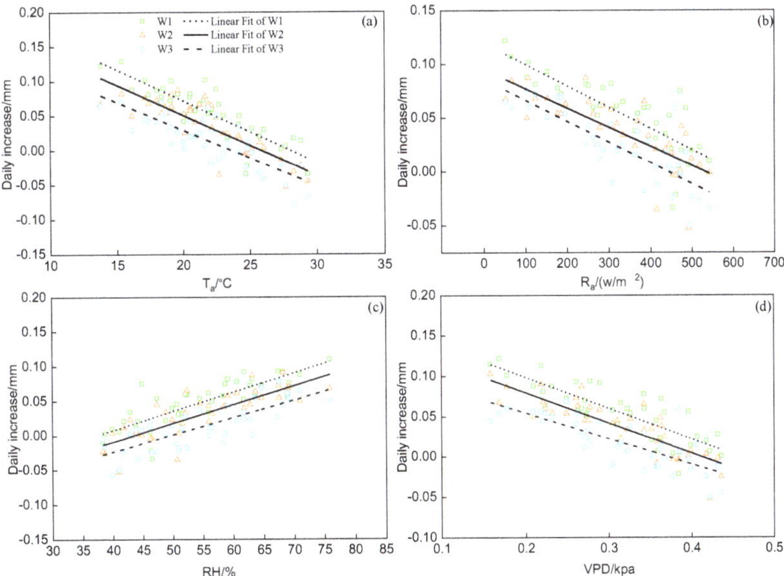

Figure 4. The relationship of DI with T_a (**a**), R_a (**b**), RH (**c**) and VPD (**d**).

Table 4. Correlation analysis of MDS and DI in different stages under different treatments.

Treatment	Air Temperature		Solar Radiation		Relative Humidity		Vapor Pressure Difference	
	MDS	**DI**	**MDS**	**DI**	**MDS**	**DI**	**MDS**	**DI**
W1	0.601 **	0.723 **	0.534 **	0.671 **	0.498 **	0.672 **	0.547 **	0.705 **
W2	0.674 **	0.710 **	0.594 **	0.606 **	0.585 **	0.663 **	0.673 **	0.721 **
W3	0.692 **	0.683 **	0.642 **	0.664 **	0.635 **	0.623 **	0.686 **	0.631 **

Note: The correlation between MDS and meteorological factors. ** Correlation is significant at the 0.01 level.

Table 5. Stem water potential and leaf relative water content model with MDS and DI.

Index	Stem Water Potential			Relative Water Content of Leaves		
	Model	**R^2**	***p***	**Model**	**R^2**	***p***
MDS	$SWP_{W1} = -0.162MDS + 0.027$	0.338 *	0.023	$RWC_{W1} = -0.004MDS + 0.438$	0.315 *	0.030
	$SWP_{W2} = -0.302MDS - 0.024$	0.498 **	0.003	$RWC_{W2} = -0.014MDS + 1.204$	0.385 *	0.014
	$SWP_{W3} = -0.064MDS + 0.017$	0.144	0.163	$RWC_{W3} = -0.010MDS + 0.837$	0.167	0.130
DI	$SWP_{W1} = 0.215DI + 0.105$	0.391 *	0.013	$RWC_{W1} = 0.005DI - 0.382$	0.296 *	0.036
	$SWP_{W2} = 0.126DI + 0.081$	0.356 *	0.019	$RWC_{W2} = 0.006DI - 0.484$	0.338 *	0.023
	$SWP_{W3} = 0.023DI + 0.046$	0.299 *	0.035	$RWC_{W3} = 0.005DI - 0.308$	0.410 *	0.010

* indicates a significance level of $p = 0.05$; ** indicates a significance level of $p = 0.01$.

3.2.3. The Correlation of MDS and DI with Soil Moisture under Different Stages

Soil moisture has been widely used as an indirect index of crop water deficit. It can be used to diagnose crop moisture status if stem diameter variation is sensitive to soil moisture. Regression analysis was carried out between MDS and DI in stem diameter and soil moisture (Figure 5). It can be concluded that the response of DI to soil moisture was more sensitive during the flowering stage, while that of MDS was more sensitive during the fruit expansion stage. In general, the R^2 values of the MDS and DI models were low under different treatments. The reason for this result is that the MDS and DI in grapevine stem diameter were easily affected by meteorological factors. Therefore, combined with the above conclusion, it is necessary to eliminate the interference of meteorological factors on the stem diameter variation.

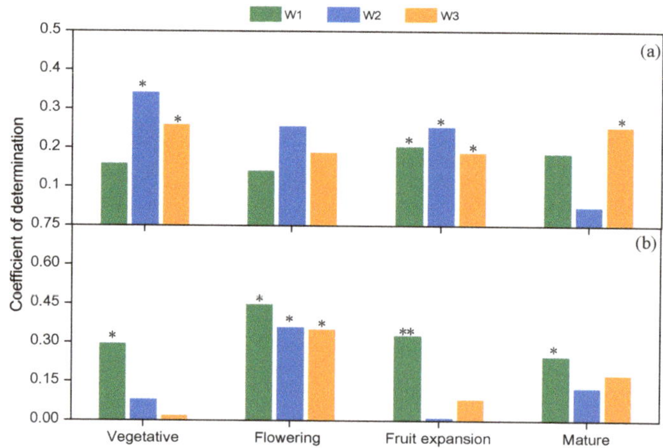

Figure 5. The correlation of MDS and DI with soil moisture content under different stages. (**a**) Represents the correlation of MDS with soil moisture content; (**b**) represents the correlation of DI with soil moisture content. * indicates a significance level of $p = 0.05$; ** indicates a significance level of $p = 0.01$.

3.2.4. Signal Intensity of Stem Diameter Indicator

The signal intensities of MDS (SI_{MDS}) and DI (SI_{DI}) under different treatments at different stages were significantly different (Figure 6). The SI_{MDS} fluctuated by approximately 1 mm from the vegetative stage to fruit expansion stage under W1, while it mostly fluctuated below 1 mm at the mature stage. The SI_{DI} fluctuated approximately 1 under W1 during the whole growth period. The degree of change in SI_{MDS} and SI_{DI} under W2 and W3 increased sequentially, and the signal values of the same stage were larger than those of the W1. It can be seen that with the increase in irrigation, the SI_{DI} and SI_{MDS} values of stem diameter tended to become stable. The SI_{MDS} of the three treatments peaked during the fruit expansion stage (Figure 6e), followed by the flowering stage (Figure 6c). The SI_{DI} of the W3 dropped sharply to below 0 on 26 May and was restored after rehydration on 27 May. Compared with those under the other growth stages, the SI_{MDS} and SI_{DI} of each treatment decreased to different degrees at the mature stage (Figure 6g,h), and that of W3 treatment decreased most significantly. The SI_{MDS} of the W3 decreased rapidly to below the W1 level within one week after irrigation stopped and then remained stable. The SI_{DI} was difficult to use to distinguish the moisture status of grapevines at the mature stage due to the large fluctuation and instability of the signal values. In conclusion, SI_{MDS} was preliminarily determined to be an appropriate indicator of plant moisture status, and SI_{DI} can be used as an indicator of moisture status at stages other than the mature stage.

3.2.5. The Correlation of SI_{MDS} and SI_{DI} with Meteorological Factors, Stem Water Potential and RWC

The correlation coefficients of the SI_{MDS} and SI_{DI} with meteorological factors during different growth stages are shown in Figure 7. In addition, while the individual correlation coefficient reached a significant level, the correlation of the SI_{DI} and SI_{MDS} of three treatments with meteorological factors during the whole growth period was not significant (Figure 7), which indicated that the influence of meteorological factors on SI_{DI} and SI_{MDS} had been excluded.

It can be seen from Table 5 that the correlations between MDS and DI and SWP and RWC are relatively low. Therefore, after eliminating the interference of meteorological factors, the fitting diagrams and equations of SI_{MDS} and SI_{DI} with SWP and RWC are established (Figure 8 and Table 6). With the increase in SI_{MDS}, SWP under W3 decreased most significantly, RWC under W1 decreased most significantly; with the increase in SI_{DI}, SWP and RWC under W1 had the most significant increase. On the whole, the R^2 of each equation was high, and the correlation reached a very significant level (Table 6). We concluded that the ability of SI_{MDS} and SI_{DI} to represent the deficit status of plants is greatly improved after eliminating the interference of meteorological factors.

Table 6. Stem water potential and leaf relative water content model with SI_{MDS} and SI_{DI}.

Index	Stem Water Potential			Relative Water Content of Leaves		
	Model	R^2	p	Model	R^2	p
SI_{MDS}	$SWP_{W1} = -0.221SI_{MDS} - 0.025$	0.723 ***	<0.001	$RWC_{W1} = -0.100SI_{DI} + 0.951$	0.710 ***	<0.001
	$SWP_{W2} = -0.362SI_{MDS} + 0.017$	0.762 ***	<0.001	$RWC_{W2} = -0.045SI_{DI} + 0.861$	0.695 ***	<0.001
	$SWP_{W3} = -0.484SI_{MDS} - 0.711$	0.719 ***	<0.001	$RWC_{W3} = -0.028SI_{DI} + 0.757$	0.599 **	0.001
SI_{DI}	$SWP_{W1} = 0.205SI_{MDS} - 0.386$	0.717 ***	<0.001	$RWC_{W1} = 0.073SI_{DI} + 0.803$	0.705 ***	<0.001
	$SWP_{W2} = 0.101SI_{MDS} - 0.519$	0.691 ***	<0.001	$RWC_{W2} = 0.024SI_{DI} + 0.776$	0.685 ***	<0.001
	$SWP_{W3} = 0.172SI_{MDS} - 1.633$	0.640 ***	<0.001	$RWC_{W3} = 0.015SI_{DI} + 0.700$	0.660 ***	<0.001

Note: ** indicates a significance level of $p = 0.01$; *** indicates a significance level of $p < 0.001$.

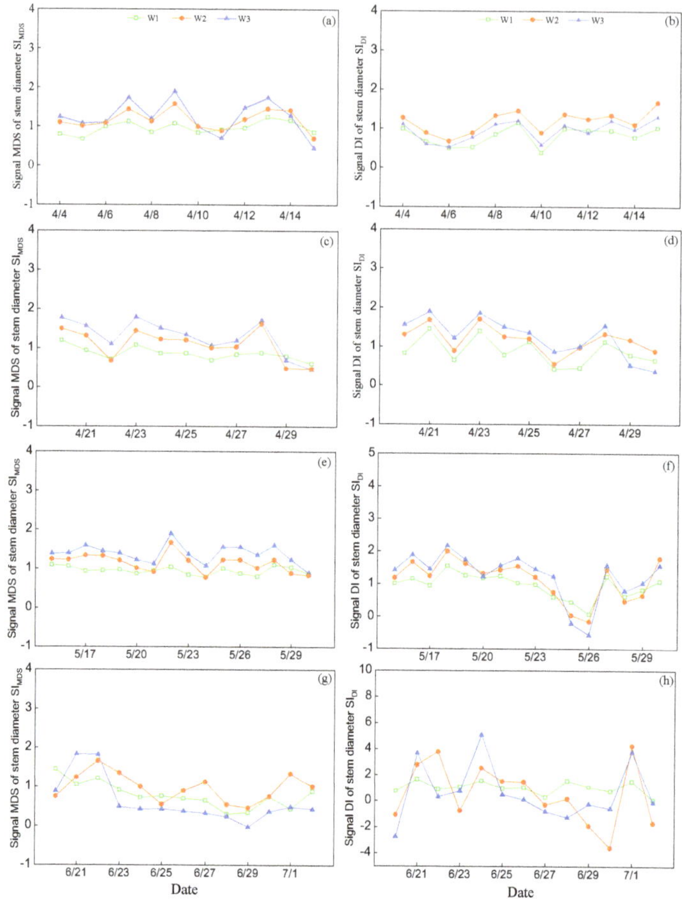

Figure 6. The change in SI_{MDS} and SI_{DI} under different stages of grapevine. SI_{MDS} represents signal MDS of stem diameter, and SI_{DI} represents signal DI of stem diameter. (**a,c,e,g**) Represents SI_{MDS} at the vegetative, flowering, fruit expansion and mature stage, respectively. (**b,d,f,h**) Represents SI_{DI} at the vegetative, flowering, fruit expansion and mature stage, respectively.

3.2.6. Adaptable Evaluation of Signal Intensity under Different Stages

In addition to using signal intensity of stem diameter to characterize the water status of plants, the signal intensity of stem diameter was also explored to monitor soil water content and diagnose whether grapevines were under water stress in real time. Therefore, the regression model between the SI_{MDS} and SI_{DI} of stem diameter and soil moisture was established (Table 7). It can be seen from Table 7 that the determination coefficient (R^2) of the SI_{MDS} model under three treatments were high during the growth stage, which indicated that SI_{MDS} showed a good diagnostic effect on soil moisture status. The R^2 of the SI_{DI} model was higher at the vegetative and flowering stages, and R^2 decreased to 0.022–0.232 ($p > 0.05$) at the mature stage. We can conclude that SI_{MDS} and SI_{DI} were more suitable for diagnosing soil water status than MDS and DI. However, the diagnostic effect of the two indicators was quite different under different stages, so it is necessary to further consider the moisture sensitivity under different stages to select the optimum indicator.

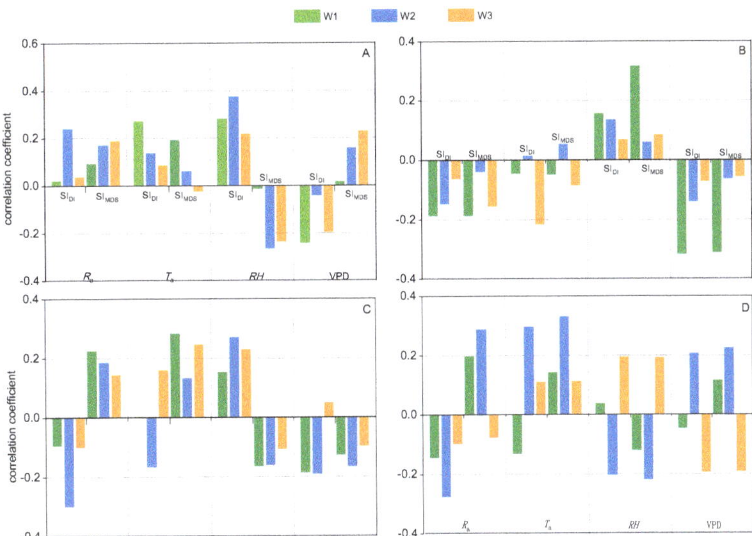

Figure 7. The correlation coefficients between SI_{MDS}, SI_{DI} and meteorological factors under different treatments at different stages.

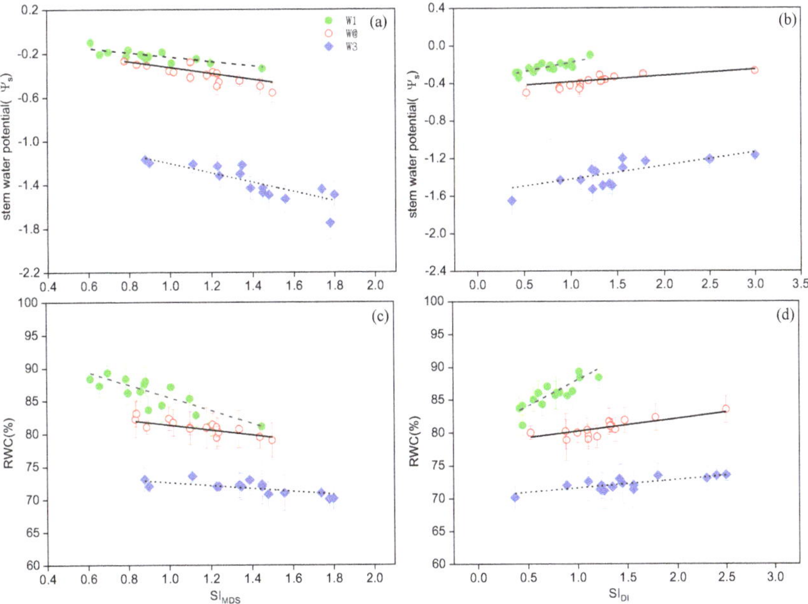

Figure 8. Relationships between SI_{MDS} and SI_{DI} and SWP and RWC under different treatments. (**a**) Represents the relationship between SWP and SI_{MDS}, (**b**) represents the relationship between SWP and SI_{DI}, (**c**) represents the relationship between RWC and SI_{MDS}, and (**d**) represents the relationship between RWC and SI_{DI}.

Table 7. The model of SI_{DI} and SI_{MDS} with soil water content at different growth stages.

Stage	Treatment	SI_{MDS}	R^2	p	SI_{DI}	R^2	p
Vegetative stage	W1	$SI_{MDS} = -39.485\theta + 6.304$	0.688 ***	<0.001	$SI_{DI} = 39.961\theta - 4.286$	0.739 ***	<0.001
	W2	$SI_{MDS} = -36.246\theta + 5.848$	0.566 ***	<0.001	$SI_{DI} = 187.96\theta - 22.077$	0.448 *	0.023
	W3	$SI_{MDS} = -53.413\theta + 7.639$	0.358	0.067	$SI_{DI} = 162.48\theta - 19.697$	0.451 *	0.019
Flowering stage	W1	$SI_{MDS} = -45.420\theta + 7.095$	0.514 ***	<0.001	$SI_{DI} = 83.604\theta - 9.626$	0.631 ***	<0.001
	W2	$SI_{MDS} = -37.140\theta + 5.382$	0.652 ***	<0.001	$SI_{DI} = 72.731\theta - 7.0531$	0.762 ***	<0.001
	W3	$SI_{MDS} = -18.224\theta + 3.042$	0.565 ***	<0.001	$SI_{DI} = 137.85\theta - 14.041$	0.676 ***	<0.001
Fruit expansion stage	W1	$SI_{MDS} = -37.788\theta + 5.795$	0.589 ***	<0.001	$SI_{DI} = 37.116\theta - 3.687$	0.389 *	0.038
	W2	$SI_{MDS} = -34.848\theta + 4.983$	0.560 ***	<0.001	$SI_{DI} = 27.010\theta - 3.140$	0.313	0.087
	W3	$SI_{MDS} = -53.532\theta - 8.870$	0.575 ***	<0.001	$SI_{DI} = 74.046\theta - 8.738$	0.495 *	0.016
Mature stage	W1	$SI_{MDS} = -13.717\theta + 2.259$	0.409 *	0.027	$SI_{DI} = 80.714\theta - 7.368$	0.022	0.433
	W2	$SI_{MDS} = -94.494\theta + 12.967$	0.621 ***	<0.001	$SI_{DI} = 143.48\theta - 15.637$	0.148	0.121
	W3	$SI_{MDS} = -18.736\theta + 2.387$	0.646 ***	<0.001	$SI_{DI} = 146.91\theta - 16.031$	0.232	0.885

Note: θ represents the soil moisture content. * indicates a significance level of $P_{0.05}$; *** indicates a significance level of $p < 0.001$.

The signal intensity of MDS and DI (SI_{MDS}, SI_{DI}) and the coefficient of variation of signal intensity under different stages are shown in Figure 9. The signal-to-noise ratio of SI_{MDS} and SI_{DI} under different treatments at different stages can be seen in Figure 10. The SI_{MDS} and SI_{DI} increased first and then decreased over the whole growth stage (Figure 9). We found that the average values of SI_{MDS} and SI_{DI} under different treatments were similar at the vegetative stage, but the variability of SI_{MDS} was greater than that of SI_{DI} (Figure 10), so SI_{DI} was a more suitable diagnostic indicator of grapevine water status and soil water status during the vegetative stage. The average signal intensity, sensitivity and signal-to-noise ratio were similar at the flowering stage, but SI_{DI} had a better correlation with soil moisture at the flowering stage (Table 7); thus, SI_{DI} should be selected as the diagnostic indicator of grapevine water status and soil water status at the flowering stage. The signal-to-noise ratio of SI_{MDS} during the fruit expansion and the mature stages was higher than those of SI_{DI} (Figure 10), and the sensitivity of SI_{MDS} to soil moisture was better than that of SI_{DI}. Therefore, SI_{MDS} was selected as the most suitable indicator of grapevine water status and soil water status during the fruit expansion and the mature stages.

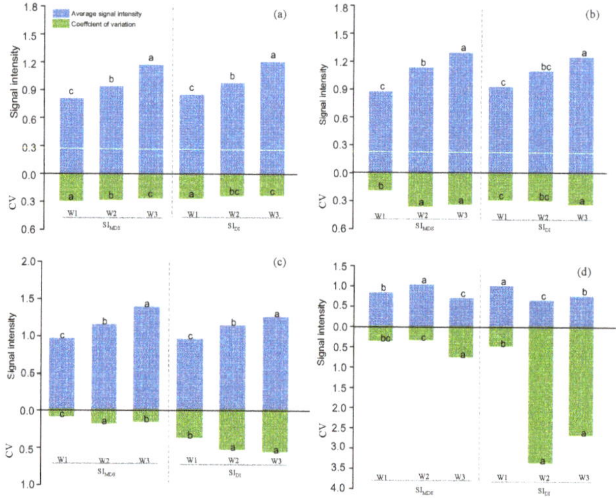

Figure 9. The average value of SI_{MDS} and SI_{DI} and coefficient of variation of signal strength under different growth stages. (**a**) Represents the vegetative stage, (**b**) represents the flowering stage, (**c**) represents the fruit expansion stage, and (**d**) represents the mature stage. Different lowercase letters indicate that there is a statistical difference at $P_{0.05}$ under different treatments.

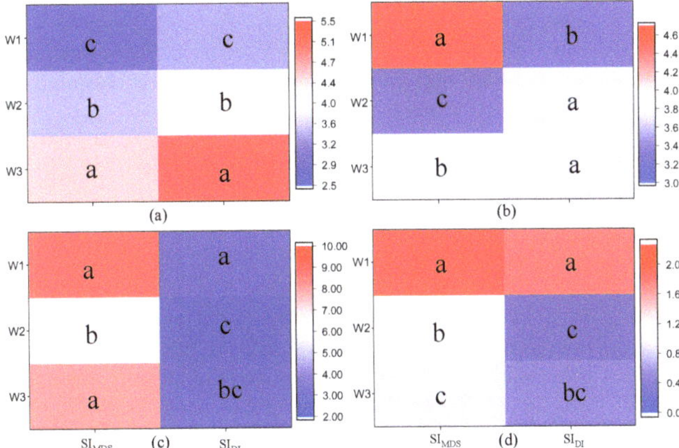

Figure 10. The signal-to-noise ratio of SIMDS and SIDI under different treatments at different stages. (**a**) Represents the vegetative stage, (**b**) represents the flowering stage, (**c**) represents the fruit expansion stage, and (**d**) represents the mature stage. Different lowercase letters indicate that there is a statistical difference of signal-to-noise ratio at $P_{0.05}$ in the same column.

4. Discussion

4.1. Relative Variation in Grapevine Stem Diameter

To achieve sustainable water use and efficient cultivation of crops, the moisture condition of crops is an important factor. Under both high and low soil moisture conditions, the grapevine stems shrunk in the daytime and recovered or expanded at night, and the microchange in stem diameter was closely related to the water status of the plant [39,40]. The present study showed that the stem diameter under the W3 treatment began to decrease sharply after 7 April, and the total increase in stem diameter was −0.570 mm. Because the growth of new shoots mainly depends on the absorption of soil moisture and nutrients by the root system and transport of these nutrients to the new shoots, the soil moisture under the W3 treatment was low. When transpiration stopped at night, the moisture absorbed by the root system was not sufficient to make up for the transpiration loss during the day, so the increase in stem diameter stopped or stems even exhibited negative growth, similar to the results of Xiong [41].

The difference in the MDS among the three treatments was not significant during the vegetative and flowering stages. The reason for this phenomenon is that the rapid growth rate conceals the short-term variation in stem diameter caused by water deficit, which indicates that MDS is not a suitable indicator of moisture status during early grapevine development [42]. During the mature stage, the growth of the stems slowed with the seasonal process. At the fruit growth stage, the stems also ceased growth or shrunk without water stress [43,44]. The results showed that the relative variations in stem diameter among the three treatments decreased gradually, similar to the results of Intrigliolo and Castel [44] and Girón et al. [44]. With the increase in water stress, the MDS under the W3 treatment was the lowest at the flowering and fruit expansion stages. This effect may be the result of the combination of the degree of water stress on the plant and the ability of the tissue to hold water against the water potential gradient [12]. Further research is needed to more accurately explain these findings.

4.2. The Correlations of MDS and DI with Meteorological Factors

Numerous studies have shown that MDS is sensitive to soil and plant moisture status, and this measure has been applied in production as a key indicator to guide fruit tree irrigation [31,45–47]. The DI reflects the growth rate of the stem, which is affected by the water supply in the root zone and the intensity of transpiration. The DI is also sensitive to the plant moisture status during the rapid growth stage of crops. Therefore, it is particularly important to analyze stem diameter microchanges for diagnosing crop water deficit. Previous studies have shown that the key meteorological factors affecting stem diameter of fruit trees under outdoor conditions are daily mean water pressure deficit (VPD) or daily maximum temperature (T_{max}) [16,23,48]. In the current study, we found that T_{max} and VPD are the key meteorological factors affecting MDS and DI, which is consistent with previous studies on outdoor fruit trees. In addition, the plastic film on the top of the greenhouse has good light transmittance, which can well transmit sunlight into the grapevine of the greenhouse, which also explains the reason why the indoor and outdoor results are similar.

Our results showed (Figures 3 and 4) that the positive correlation between MDS and VPD was extremely significant ($p < 0.01$), and the negative correlation between DI and VPD was significant ($p < 0.05$). Goldhamer et al. [16] indicated that VPD was the main factor affecting the stem diameter variation of almond trees. Moriana et al. [49] showed that T_a had the best correlation with MDS, followed by VPD, which was different from the results of this study, and these discrepancies may have been caused by the differences in plant type and test sites. Therefore, when the crop growth environment is changed, the relationship between the indicator of stem diameter and meteorological factors may change, so the reference equation obtained under a specific condition cannot be used to calculate the reference value, which needs to be further referenced with local meteorological data.

In addition, under the same experimental conditions, due to the different responses of different grapevine varieties to water stress, there may be differences in the stomatal resistance, transpiration rate, and photosynthetic rate under different varieties following water deficit, which may cause changes in sap flow in the stem, resulting in differences in the stem diameter indicators of different grapevine varieties [50,51]. Therefore, the reference equation obtained on a certain grapevine variety and the SI_{MDS} and SI_{DI} irrigation threshold values may no longer be applicable. When calculating the crop reference value, it is better to use the same crop variety under the same growth conditions as the reference crop.

4.3. Signal Value and Signal-to-Noise Ratio of the Stem Diameter

The stem diameter indicator is greatly affected by external meteorological factors. The model of the correlations between MDS and DI and stem water potential, RWC and soil moisture cannot exclude the interference of external factors [14]. According to the experimental results, the SI_{MDS} and SI_{DI} had better sensitivity, signal intensity and reliability in diagnosing the grapevine water content. In addition, the diagnostic applicability of SI_{MDS} and SI_{DI} was different in different growth stages of grapevine. A possible explanation might be the great coefficient of variation among grapevine plants growth rates masked the differences created by water stress on SI_{MDS} at early and middle growth stage. Thus, the SI_{DI} is a more appropriate indicator of the grapevine water content than SI_{MDS} during the vegetative and flowering stages, but as important plant growth indexes, the variation of SI_{MDS} and SI_{DI} should be taken into full consideration in practice as well. The signal value of the stem diameter variation indicator can eliminate the interference of meteorological factors, so an accurate reference value is critical for the use of signal intensities for guiding crop irrigation [52,53]. The SI_{MDS} and SI_{DI} values of the W1 treatment fluctuated up and down by approximately 1 (Figure 6). The SI_{MDS} and SI_{DI} maximum of the W2 and W3 treatments under water stress was approximately 2.0, and the water deficit led to a significant increase in SI_{MDS} and SI_{DI} compared with the full irrigation treatment. The reason for these findings may be related to the water absorption by the root system and the ability of the stem to transport water following water stress, but the details of these mechanisms should be researched in the future. The SI_{DI} at the mature stage showed an irregular curve with large fluctuations, which may have been caused by the large fluctuation of meteorological factors at the mature stage. However, the variation curve of SI_{MDS} at the mature stage was relatively stable, which may have been related to the stem growth characteristics of grapevine at the later growth stage; the relevant internal physiological mechanism needs to be further explored.

Although stem water potential and RWC measurements require frequent trips to the field and a considerable input of labor, these parameters are reliable plant-based water status indicators and have been used for irrigation scheduling in fruit trees [43,54–56]. The present results showed that there were weak correlations between MDS, DI and stem water potential and RWC. These results were due to the unstable meteorological factors during this period, which affected the short-term grapevine stem growth to a certain extent, resulting in the unstable changes in stem diameter. However, when the interference of meteorological factors on MDS and DI were eliminated, SI_{MDS} and SI_{DI} were not sensitive to meteorological factors. According to the fitting equations of grape stem water potential and RWC with SI_{MDS} and SI_{DI}, the fitting effect of each equation is good (Figure 8). The coefficient of determination (R^2) is approximately 0.7, which indicates that it is feasible to use the signal intensity of stem diameter variation to characterize crop water status. This approach can not only substantially reduce the labor required but also can be used to continuously and nondestructively carry out index observation, which is consistent with the research results of Badal et al. [17]. Furthermore, when using this index to diagnose the plant water content, the best time is on a fine day, as the stem diameter microchange indexes were small and not significantly different ($p > 0.05$) between stress and full irrigation treatments under poor weather conditions in this experiment [6].

It can be seen from Figures 9 and 10 that the larger coefficient of variation (CV) of SI_{MDS} during the vegetative and flowering stages, which resulted in lower signal-to-noise ratio. While SI_{DI} has a greater variability in the late growth stage, the larger CV of SI_{MDS} and SI_{DI} increased the uncertainty of judging water stress in grapevine. In order to further explore the applicability of SI_{MDS} and SI_{DI} at different growth stages, more sensors should be installed to acquire the real water content message in practice. Previous studies have shown that some factors could affect the plant CV of MDS such as the crop load, location of sensor installation, and so on [43,48,49,57]. The relative researches should be studied further in future experiments to facilitate practical operation of this technic.

5. Conclusions

This work shows that there were significant differences in stem diameter variation under different irrigation levels. Water shortage resulted in larger maximum daily shrinkage and smaller daily increase. The stem water potential and leaf relative water content of stress plants (W2 and W3 treatment) were significantly lower than that of the W1 treatment. Regression analysis between MDS, DI and meteorological factors revealed that the MDS of stem diameter was positively correlated with R_a, T_a and VPD and negatively correlated with RH, and the DI among three treatments decreased with the increase in R_a, T_a and VPD and increased with the increase in RH. The key meteorological factors influencing grapevine stem diameter variation in a greenhouse were VPD and T_a. The MDS and DI had a weak correlation with stem water potential and RWC, thus these measures cannot be directly applied as indicators of the moisture status of grapevine and soil. SI_{MDS} and SI_{DI} can distinguish the differences in the grapevine stem diameter indicators under different soil moisture conditions, eliminate the interference of meteorological factors, and were highly correlated with stem water potential, RWC and soil moisture content. At the vegetative and flowering stages, SI_{DI} has less variability and greater reliability than SI_{MDS}, it is more suitable for the diagnosis of grapevine water status in these two periods. At the fruit expansion and the mature stages, the signal-to-noise ratio of SI_{MDS} is significantly higher than that of SI_{MDS}, so it is more suitable to be used as a diagnostic index of water status in late growth stage of grapevine. In sum, compared with other plant water diagnosing indexes, the SIMDS and SIDI indexes had the advantages of sensitivity, signal intensity and reliability and were good indicators of the grapevine water content.

Author Contributions: C.R., X.H., W.W. and H.R. designed the experiments, C.R., Y.G. and T.S. performed research and date analysis, C.R. wrote the manuscript with contributions from all authors. All authors have read and agreed to the published version of the manuscript.

Funding: This work was supported by the National Key Research and Development Program of China (2017YFD0201508), the National Natural Science Foundation of China (51179163), Shaanxi Key Science and Technology Innovation Team Project (2016KTZDNY-01–05).

Institutional Review Board Statement: Not applicable.

Informed Consent Statement: Not applicable.

Data Availability Statement: Data is available upon request to the corresponding author.

Acknowledgments: The authors would like to acknowledge all the team members of key Laboratory of Agricultural Soil and Water Engineering in Arid and Semiarid Areas, Ministry of Education.

Conflicts of Interest: The authors declare that they have no conflict of interest.

References

1. Goodwin, I.; Boland, A.M. Scheduling deficit irrigation of fruit trees for optimizing water use efficiency. In *Deficit Irrigation Practices, Water Reports*; FAO: Quebec City, QC, Canada, 2002; Volume 22, p. 109.
2. Naor, A.; Gal, Y.; Peres, M. The inherent variability of water stress indicators in apple, nectarine and pear orchards, and the validity of a leaf-selection procedure for water potential measurements. *Irrig. Sci.* **2005**, *24*, 129–135. [CrossRef]
3. Jones, H.G. Irrigation scheduling: Advantages and pitfalls of plant-based methods. *J. Exp. Bot.* **2004**, *407*, 2427. [CrossRef]
4. Fernández, J.E.; Cuevas, M.V. Irrigation scheduling from stem diameter variations: A review. *Agric. For. Meteorol.* **2010**, *150*, 135–151. [CrossRef]
5. Argyrokastritis, I.G.; Papastylianou, P.T.; Alexandris, S. Leaf Water Potential and Crop Water Stress Index variation for full and deficit irrigated cotton in Mediterranean conditions. *Agric. Agric. Sci. Procedia* **2015**, *4*, 463–470. [CrossRef]
6. Wang, X.; Meng, Z.; Chang, X.; Deng, Z.; Li, Y.; Lv, M. Determination of a suitable indicator of tomato water content based on stem diameter variation. *Sci. Hortic.* **2017**, *215*, 142–148. [CrossRef]
7. Ojeda, H.; Deloire, A.; Carbonneau, A. Influence of water deficits on grape berry growth. *Vitis* **2001**, *40*, 141–145. [CrossRef]
8. González-Fernández, A.B.; Rodríguez-Pérez, J.R.; Marcelo, V.; Valenciano, J.B. Using field spectrometry and a plant probe accessory to determine leaf water content in commercial vineyards. *Agric. Water Manag.* **2015**, *156*, 43–50. [CrossRef]
9. Gao, Y.; Qiu, J.W.; Miao, Y.L.; Qiu, R.C.; Li, H.; Zhang, M. Prediction of Leaf Water Content in Maize Seedlings Based on Hyperspectral Information. *IFAC-Pap.* **2019**, *52*, 263–269. [CrossRef]
10. Li, B.; Zhao, X.T.; Zhang, Y.; Zhang, S.J.; Luo, B. Prediction and monitoring of leaf water content in soybean plants using terahertz time-domain spectroscopy. *Comput. Electron. Agric.* **2020**, *170*, 105239. [CrossRef]
11. Jiménez, M.N.; Navarro, F.B.; Sánchez-Miranda, A.; Ripoll, M.A. Using stem diameter variations to detect and quantify growth and relationships with climatic variables on a gradient of thinned Aleppo pines. *For. Ecol. Manag.* **2019**, *442*, 53–62. [CrossRef]
12. Ortuño, M.F.; García-Orellana, Y.; Conejero, W.; Ruiz-Sanchez, M.C.; Mounzer, O.; Alarcon, J.J.; Torrecillas, A. Relationships Between Climatic Variables and Sap Flow, Stem Water Potential and Maximum Daily Trunk Shrinkage in Lemon Trees. *Plant Soil* **2006**, *279*, 229–242. [CrossRef]
13. Leperen, W.V.; Volkov, V.S.; Meeteren, U.V. Distribution of xylem hydraulic resistance in fruiting truss of tomato influenced by water stress. *J. Exp. Bot.* **2003**, *54*, 317–324. [CrossRef]
14. Ortuño, M.F.; Conejero, W.; Moreno, F.; Moriana, A.; Intrigliolo, D.S.; Biel, C.; Mellisho, C.D.; Pérez-Pastor, A.; Domingo, R.; Ruiz-Sánchez, M.C.; et al. Could trunk diameter sensors be used in woody crops for irrigation scheduling? A review of current knowledge and future perspectives. *Agric. Water Manag.* **2010**, *97*, 1–11. [CrossRef]
15. Schepper, V.D.; Dusschoten, D.V.; Copini, P.; Jahnke, S.; Steppe, K. MRI links stem water content to stem diameter variations in transpiring trees. *J. Exp. Bot.* **2012**, *63*, 2645–2653. [CrossRef]
16. Goldhamer, D.A.; Fereres, E. Irrigation scheduling of almond trees with trunk diameter sensors. *Irrig. Sci.* **2004**, *23*, 11–19. [CrossRef]
17. Badal, E.; Buesa, I.; Guerra, D.; Bonet, L.; Ferrer, P.; Intrigliolo, D.S. Maximum diurnal trunk shrinkage is a sensitive indicator of plant water, stress in *Diospyros kaki* (Persimmon) trees. *Agric. Water Manag.* **2010**, *98*, 143–147. [CrossRef]
18. Doltra, J.; Oncins, J.A.; Bonany, J.; Cohen, M. Evaluation of plant-based water status indicators in mature apple trees under field conditions. *Irrig. Sci.* **2007**, *25*, 351–359. [CrossRef]
19. Velez, J.E.; Intrigliolo, D.S.; Castel, J.R. Scheduling deficit irrigation of citrus trees with maximum daily trunk shrinkage. *Agric. Water Manag.* **2007**, *90*, 197–204. [CrossRef]
20. Moriana, A.; Fereres, E. Plant indicators for scheduling irrigation of young olive trees. *Irrig. Sci.* **2002**, *21*, 83–90. [CrossRef]
21. Fereres, E.; Goldhamer, D.A. Suitability of stem diameter vriations and water potential as indicators for irrigation scheduling of almond trees. *J. Hortic. Sci. Biotechnol.* **2003**, *78*, 139–144. [CrossRef]

22. Intrigliolo, D.S.; Castel, J.R. Continuous measurement of plant and soil water status for irrigation scheduling in plum. *Irrig. Sci.* **2004**, *23*, 93–102. [CrossRef]

23. Ortuño, M.F.; García-Orellana, Y.; Conejero, W.; Pérez-Sarmiento, F.; Torrecillas, A. Assessment of maximum daily trunk shrinkage signal intensity threshold values for deficit irrigation in lemon trees. *Agric. Water Manag.* **2008**, *96*, 80–86. [CrossRef]

24. Katerji, N.; Itier, B.; Ferreira, I. Etude de quelques criteres indicateurs de l'etat hydrique d'une culture de tomate en region semi-aride. *Agronomie* **1988**, *8*, 425–433. [CrossRef]

25. Rötzer, T.; Biber, P.; Moser, A.; Schäfer, C.; Pretzsch, H. Stem and root diameter growth of European beech and Norway spruce under extreme drought. *For. Ecol. Manag.* **2017**, *406*, 184–195. [CrossRef]

26. Matimati, I.; Musil, C.F.; Raitt, L.; February, E.C. Diurnal stem diameter variations show CAM and C3 photosynthetic modes and CAM-C3 switches in arid South African succulent shrubs. *Agric. For. Meteorol.* **2012**, *161*, 72–79. [CrossRef]

27. Tognetti, R.; Giovannelli, A.; Lavini, A.; Morelli, G.; Fragnito, F.; d'Andria, R. Assessing environmental controls over conductances through the soil-plant-atmosphere continuum in an experimental olive tree plantation of southern Italy. *Agric. For. Meteorol.* **2009**, *149*, 1229–1243. [CrossRef]

28. Devine, W.D.; Harrington, C.A. Factors affecting diurnal stem contraction in young Douglas-fir. *Agric. For. Meteorol.* **2011**, *151*, 414–419. [CrossRef]

29. Du, S.Q.; Tong, L.; Zhang, X.T.; Kang, S.Z.; Du, T.S.; Li, S.; Ding, R.S. Signal intensity based on maximum daily stem shrinkage can reflect the water status of apple trees under alternate partial root-zone irrigation. *Agric. Water Manag.* **2017**, *190*, 21–30. [CrossRef]

30. Tuccio, L.; Piccolo, E.L.; Battelli, R.; Matteoli, S.; Remorini, D. Physiological indicators to assess water status in potted grapevine (*Vitis vinifera* L.). *Sci. Hortic.* **2019**, *255*, 8–13. [CrossRef]

31. Conesa, M.R.; Torres, R.; Domingo, R.; Navarro, H.; Soto, F.; Pérez-Pastor, A. Maximum daily trunk shrinkage and stem water potential reference equations for irrigation scheduling in table grapes. *Agric. Water Manag.* **2016**, *172*, 51–61. [CrossRef]

32. Gallardo, M.; Thompson, R.B.; Valdez, L.C.; Fernandez, M.D. Use of stem diameter variations to detect plant water stress in tomato. *Irrig. Sci.* **2006**, *24*, 241–255. [CrossRef]

33. Goldhamer, D.A.; Fereres, E. Irrigation scheduling protocols using continuously recorded trunk diameter measurements. *Irrig. Sci.* **2001**, *20*, 115–125. [CrossRef]

34. Miralles, J.; Franco, J.A.; Sánchez-Blanco, M.J.; Bañón, S. Effects of pot-in-pot production system on water consumption 0034, stem diameter variations and photochemical efficiency of spindle tree irrigated with saline water. *Agric. Water Manag.* **2016**, *170*, 167–175. [CrossRef]

35. Allen, R.G. Using the FAO-56 dual crop coefficient method over an irrigated region as part of an evapotranspiration intercomparison study. *J. Hydrol.* **2000**, *229*, 27–41. [CrossRef]

36. Abdelfatah, A.; Aranda, X.; Savé, R.; Herralde, F.D.; Biel, C. Evaluation of the response of maximum daily shrinkage in young cherry trees submitted to water stress cycles in a greenhouse. *Agric. Water Manag.* **2013**, *118*, 150–158. [CrossRef]

37. Kramer, P.J. *Water Relations of Plants*; Academic Press: New York, NY, USA, 1983.

38. Turner, N.C. Techniques and experimental approaches for the measurement of plant water status. *Plant Soil* **1981**, *58*, 339–366. [CrossRef]

39. Zhang, J.G.; He, Q.Y.; Shi, W.Y.; Otsuki, K.; Yamanaka, N.; Du, S. Radial variations in xylem sap flow and their effect on whole-tree water use estimates. *Hydrol. Process.* **2015**, *29*, 4993–5002. [CrossRef]

40. Irvine, J.; Grace, J. Continuous measurements of water tensions in the xylem of trees based on the elastic properties of wood. *Planta* **1997**, *202*, 455–461. [CrossRef]

41. Xiong, W.; Wang, Y.H.; Yu, P.T.; Liu, H.L.; Shi, Z.J.; Guan, W. Growth in stem diameter of *Larix principis-rupprechtii* and its response to meteorological factors in the south of Liupan Mountain, China. *Acta Ecol. Sin.* **2007**, *27*, 432–440. [CrossRef]

42. Zhang, J.Y.; Duan, A.W.; Meng, Z.J.; Liu, Z.G. Stem diameter variations of cotton under different water conditions. *Trans. Chin. Soc. Agric. Eng.* **2005**, *21*, 7–11.

43. Intrigliolo, D.S.; Castel, J.R. Evaluation of grapevine water status from trunk diameter variations. *Irrig. Sci.* **2007**, *26*, 49–59. [CrossRef]

44. Girón, I.F.; Corell, M.; Martín-Palomo, M.J.; Galindo, A.; Torrecillas, A.; Moreno, F.; Moriana, A. Feasibility of trunk diameter fluctuations in the scheduling of regulated deficit irrigation for table olive trees without reference trees. *Agric. Water Manag.* **2015**, *161*, 114–126. [CrossRef]

45. De la Rosa, J.M.; Conesa, M.R.; Domingo, R.; Torres, R.; Pérez-Pastor, A. Feasibility of using trunk diameter fluctuation and stem water potential reference lines for irrigation scheduling of early nectarine trees. *Agric. Water Manag.* **2013**, *126*, 133–141. [CrossRef]

46. Kanai, S.; Adu-Gymfi, J.; Lei, K.; Lto, J.; Ohkura, K.; Moghaieb, R.E.A.; El-shemy, H.; Mohapatra, R.; Mohapatra, P.K.; Saneoka, H.; et al. N-deficiency damps out circadian rhythmic changes of stem diameter dynamics in tomato plant. *Plant Sci.* **2008**, *174*, 183–191. [CrossRef]

47. Swaef, T.D.; Steppe, K.; Lemeur, R. Raoul Lemeur. Determining reference values for stem water potential and maximum daily trunk shrinkage in young apple trees based on plant responses to water deficit. *Agric. Water Manag.* **2009**, *96*, 541–550. [CrossRef]

48. Conejero, W.; Mellisho, C.D.; Ortuo, M.F.; Moriana, A.; Moreno, F.; Torrecillas, A. Using trunk diameter sensors for regulated deficit irrigation scheduling in early maturing peach trees-ScienceDirect. *Environ. Exp. Bot.* **2011**, *71*, 409–415. [CrossRef]

49. Moriana, A.; Moreno, F.; Girón, I.F.; Conejero, W.; Ortuño, M.F.; Morales, D.; Corell, M.; Torrecillas, A. Seasonal changes of maximum daily shrinkage reference equations for irrigation scheduling in olive trees: Influence of fruit load. *Agric. Water Manag.* **2011**, *99*, 121–127. [CrossRef]
50. King, G.; Fonti, P.; Nievergelt, D.; Büntgen, U.; Frank, D. Climatic drivers of hourly to yearly tree radius variations along a 6 °C natural warming gradient. *Agric. For. Meteorol.* **2013**, *168*, 36–46. [CrossRef]
51. Janbek, B.M.; Stockie, J.M. Asymptotic and numerical analysis of a porous medium model for transpiration-driven sap flow in trees. *Siam J. Appl. Math.* **2017**, *78*, 2028–2056. [CrossRef]
52. De la Rosa, J.M.; Conesa, M.R.; Domingo, R.; Pérez-Pastor, A. A new approach to ascertain the sensitivity to water stress of different plant water indicators in extra-early nectarine trees. *Sci. Hortic.* **2014**, *169*, 147–153. [CrossRef]
53. De la Rosa, J.M.; Domingo, R.; Gómez-Montiel, J.; Pérez-Pastor, A. Implementing deficit irrigation scheduling through plant water stress indicators in early nectarine trees. *Agric. Water Manag.* **2015**, *152*, 207–216. [CrossRef]
54. Shackel, K.A.; Ahmadi, H.; Biasi, W.; Buchner, R.; Goldhamer, D.; Gurusinghe, S.; Hasey, J.; Kester, D.; Krueger, B.; Lampinen, B.; et al. Plant water status as an index of irrigation need in deciduous fruit trees. *HortTechnology* **1997**, *7*, 23–29. [CrossRef]
55. Naor, A. Midday stem water potential as a plant water stress indicator for irrigation scheduling in fruit trees. *Acta Hortic.* **2000**, *537*, 447–454. [CrossRef]
56. Choné, X.; Cornelis, V.L.; Denis, D.; Gaudillère, J.P. Stem Water Potential is a Sensitive Indicator of Grapevine Water Status. *Ann. Bot.* **2001**, *4*, 477–483. [CrossRef]
57. Cuevas, M.V.; Torres-Ruiz, J.M.; Álvarez, R.; Jiménez, M.D.; Cuerva, J.; Fernández, J.E. Assessment of trunk diameter variation derived indices as water stress indicators in mature olive trees. *Agric. Water Manag.* **2010**, *97*, 1293–1302. [CrossRef]

MDPI
St. Alban-Anlage 66
4052 Basel
Switzerland
Tel. +41 61 683 77 34
Fax +41 61 302 89 18
www.mdpi.com

Horticulturae Editorial Office
E-mail: horticulturae@mdpi.com
www.mdpi.com/journal/horticulturae